Green by Design

Harnessing the power of bio-based polymers at interfaces

Online at: https://doi.org/10.1088/978-0-7503-6184-2

Green by Design

Harnessing the power of bio-based polymers at interfaces

Edited by
Kai Zhang
Department of Wood Technology and Wood Chemistry, Georg-August-Universität Göttingen, Göttingen, Germany

Philip Biehl
Department of Wood Technology and Wood Chemistry, Georg-August-Universität Göttingen, Göttingen, Germany

IOP Publishing, Bristol, UK

ISBN 978-0-7503-6184-2 (ebook)
ISBN 978-0-7503-6182-8 (print)
ISBN 978-0-7503-6185-9 (myPrint)
ISBN 978-0-7503-6183-5 (mobi)

DOI 10.1088/978-0-7503-6184-2

Version: 20241201

IOP ebooks

British Library Cataloguing-in-Publication Data: A catalogue record for this book is available from the British Library.

Published by IOP Publishing, wholly owned by The Institute of Physics, London

IOP Publishing, No.2 The Distillery, Glassfields, Avon Street, Bristol, BS2 0GR, UK

US Office: IOP Publishing, Inc., 190 North Independence Mall West, Suite 601, Philadelphia, PA 19106, USA

Contents

Preface

The urgency of addressing environmental challenges such as climate change, plastic pollution, ocean littering, and excessive energy consumption has made it critical to explore alternative materials. These alternatives must not only maintain our quality of life but also minimize their impact on society and the environment. Since the invention of Bakelite by Leo Baekeland in 1909 [1], fossil-based polymers have played a significant role in shaping modern life. However, the prevalence of plastic litter is now so widespread that researchers discuss the possibility of a new geological epoch based on the presence of litter in soil—the Anthropocene [2].

In response to these challenges, bio-based materials have emerged as a promising solution to reduce our dependence on fossil-based polymers. Unlike synthetic polymers, which often require complex and energy-intensive processes to achieve similar functions, bio-based polymers are naturally occurring and come with built-in advantages. Their molecular structures, such as those found in proteins and DNA, offer biocompatibility, biodegradability, and the presence of functional groups that enable diverse applications. These materials not only bring technical benefits but also align seamlessly with the broader goals of sustainability, supporting the necessary political and economic shifts required to address climate change, pollution, and environmental degradation.

As we transition toward more sustainable practices, bio-based materials resonate with modern design philosophies such as cradle-to-cradle, which promote economic systems that are not only environmentally sustainable but also economically viable and socially responsible. By supporting a circular economy, bio-based polymers naturally embody these principles. Furthermore, their utility can be expanded through chemical modifications, such as grafting and cross-linking, allowing for broader applications in fields such as biomedicine, packaging, and sustainable manufacturing. However, while these innovations are crucial, scientists often face limitations in directly influencing the political and economic decisions that shape environmental policy. Their strength lies in laying the base for future solutions by advancing the science of new materials. Although the direction of policy and market forces may be beyond their control, the innovations they create today can serve as the foundation for the sustainable applications of tomorrow. This book aims to contribute to this effort by providing a foundation for ongoing research, exploring practical applications of bio-based materials, and helping to pave the way for these materials to eventually replace fossil-based alternatives.

Maximizing the potential of bio-based materials requires a deep understanding of interfacial science. The interactions at the interface between materials—whether it is attraction, repulsion, adhesion, or cohesion—are crucial for determining how well polymeric-based materials perform in various applications. Understanding and manipulating these interactions is key to developing the next generation of materials with specific functionalities and improved performance. This book highlights interfacial aspects, offering insights into how interfacial science can be applied to optimize bio-based materials for a wide range of uses, from coatings and adhesives to biomedical devices and composite materials.

This book serves as a guide to bio-based polymeric materials, revealing the potential of bio-based polymers to replace fossil-based ones while exploring their broader applications. It aims to drive forward developments in polymer science, bio-based materials, and sustainability. By focusing on the use of renewable materials, this book seeks to support the development of a greener and more sustainable world.

Structured into chapters covering a broad range of topics, the book starts with an overview of bio-based polymeric materials in the context of interfacial science. Subsequent chapters explore specific applications, such as bio-based surface coatings, gas barrier materials, hydrogels, porous materials for EMI-shielding, and hemicellulose-based materials. Each chapter, written by experts in their field, provides valuable insights, guiding readers to connect foundational concepts with the latest innovations. This structure helps readers build a comprehensive understanding of the subject, from the basics to the latest innovations.

This book will appeal to researchers, professionals, and students in surface science, materials science, polymers, biomaterials, and environmental science, particularly those interested in developing and applying sustainable, bio-based materials. While the use of biomass is an important step toward sustainability, effectively addressing global warming and other environmental challenges will require a broader range of solutions. The next generation of scientists will need to innovate and develop interdisciplinary approaches to tackle these complex issues. This book aims to provide the knowledge and tools to support those efforts.

Bibliography

[1] Baekeland L H 1909 Method of making insoluble products of phenol and formaldehyde *Patent* USA US942699A

[2] Praet E Plastic pollution: archaeological perspective on an Anthropocene climate emergency *World Archaeol.* 1–19

Editor biographies

Kai Zhang

Kai Zhang is a Professor and Head of the Department for Wood Technology and Wood based Composites at Georg-August-University of Göttingen, Germany. He studied at Hefei University of Technology (BEng), China, and Dresden University of Technology (MS and PhD, 2011), Germany. After completing research at Dresden University of Technology, the Pennsylvania State University, and Darmstadt University of Technology, he joined the University of Göttingen in 2015. His group's research focuses on using bio-based renewable compounds to create functional materials for sustainable development.

Philip Biehl

Philip Biehl is a postdoctoral researcher at the Department for Wood Technology and Wood-based Composites at the Georg-August-University of Göttingen, Germany. He studied at Martin-Luther-University Halle-Wittenberg (BS), Germany, and Friedrich-Schiller-University Jena (MS and PhD, 2020), Germany. He joined the University of Göttingen in 2021. His main research area is the preparation of functional amphiphilic bio-based compounds for the sustainable development of surface coatings.

Book description

This book offers a comprehensive exploration of bio-polymeric materials, with a special focus on their interfacial properties. It covers the latest advancements in biopolymer applications at interfaces, examining their chemical structures, biological foundations, modifications, and diverse applications. The discussion highlights these materials as potential replacements for fossil-based products, while also demonstrating their broader capabilities that extend beyond simple substitution of petrochemical materials. The focus is on understanding the interfacial science of these materials, particularly the key functional groups, physical phenomena, and interactions that occur in specific environments. The book features contributions from experts in the respective fields, making it a reference for scientists, academicians, practitioners, and students who are involved in the development and application of sustainable materials.

List of contributors

Philip Biehl
Department of Wood Technology and Wood Chemistry, Georg-August-Universität Göttingen, Göttingen, Germany

Xinyan Fan
Key Laboratory of Bio-based Material Science and Technology, College of Materials Science and Engineering, Northeast Forestry University, Harbin, China

Yves Grohens
University Bretagne Sud, UMR CNRS 6027, IRDL, Lorient, France

Siyu Jia
Beijing Key Laboratory of Lignocellulosic Chemistry, MOE Engineering Research Center of Forestry Biomass Materials and Bioenergy, Beijing Forestry University, Beijing, People's Republic of China

Dongmei Liu
School of Life Sciences, Anhui University, Hefei, People's Republic of China

Ziwen Lv
Beijing Key Laboratory of Lignocellulosic Chemistry, MOE Engineering Research Center of Forestry Biomass Materials and Bioenergy, Beijing Forestry University, Beijing, People's Republic of China

Gopika G Nair
CMS College (Autonomous) Kottayam, Kerala, India

Feng Peng
Beijing Key Laboratory of Lignocellulosic Chemistry, MOE Engineering Research Center of Forestry Biomass Materials and Bioenergy, Beijing Forestry University, Beijing, People's Republic of China

Jun Rao
Beijing Key Laboratory of Lignocellulosic Chemistry, MOE Engineering Research Center of Forestry Biomass Materials and Bioenergy, Beijing Forestry University, Beijing, People's Republic of China

Xiangyu Tang
Key Laboratory of Bio-based Material Science and Technology, College of Materials Science and Engineering, Northeast Forestry University, Harbin, China

Sabu Thomas
International and Inter-University Centre for Nanoscience and Nanotechnology (IIUCNN) Mahatma Gandhi University, Kottayam, Kerala, India

Jiaxiu Wang
School of Life Sciences, Anhui University, Hefei, People's Republic of China

Yonggui Wang
Key Laboratory of Bio-based Material Science and Technology, College of Materials Science and Engineering, Northeast Forestry University, Harbin, China

Fanjun Yu
Key Laboratory of Bio-based Material Science and Technology, College of Materials Science and Engineering, Northeast Forestry University, Harbin, China

Suji Mary Zachariah
International and Inter-University Centre for Nanoscience and Nanotechnology (IIUCNN) Mahatma Gandhi University, Kottayam, Kerala, India

Haodong Zhang
Department of Chemistry, Wuhan University, Wuhan, People's Republic of China

Kai Zhang
Department of Wood Technology and Wood Chemistry, Georg-August-Universität Göttingen, Göttingen, Germany

Jinping Zhou
Department of Chemistry, Wuhan University, Wuhan, People's Republic of China

Chapter 1

Introduction to advances in bio-based polymers: chemical structures and functional properties at the interface

Philip Biehl and Kai Zhang

1.1 Introduction

Since the inception of fossil-based synthetic polymer production, these materials have permeated nearly every industrial sector and have become integral components of everyday products. Their exceptional and versatile performance has had a profound impact on our society and such materials are indispensable in the modern world. However, this class of materials has witnessed a decline in reputation in recent years, primarily due to pressing concerns such as white pollution [1], complex waste management issues [2], the widespread presence of microplastics [3], greenhouse gas emissions [4], the emission of environmentally active chemicals (such as plasticizers and volatile organic compounds (VOCs)) [5], and the broader social and political conversation about sustainability.

Even though plastic materials are highly valued for their lightweight nature, thermal stability, flexibility, durability, reusability, hygiene, moldability, resistance to corrosion, and high physical strength [6], many aspects of their production and lifecycle pose significant environmental challenges. Each year, global plastic production exceeds 380 million tones, with an annual growth rate of 4%. Since 1950, this has resulted in the generation of over 6300 million tons of plastic waste [7]. Growing awareness of the environmental consequences of plastic waste and its contribution to greenhouse gas emissions has led to a push for transitioning to a circular plastic economy. In this model, the emphasis is on minimizing the use of non-renewable resources and reducing waste production, while promoting reuse and recycling throughout the material lifecycle. The reliance on fossil resources for artificial polymers, the energy-intensive production process, and the non-biodegradable nature of the final products, which leads to significant waste problems, all raise questions about their overall beneficial impact.

doi:10.1088/978-0-7503-6184-2ch1

Following from this, during the last decade the importance of green and regenerative alternatives to conventional plastic has grown, which can be seen on product labels which certify the regenerative quality of plastics and in the increasing political measures to overcome these current issues. Plastic pollution, in particular the dramatic littering of the oceans and the public interest in it, have forcefully brought the issue of plastic to the forefront of the European political agenda in recent years. In early 2018, the EU introduced a plastic strategy that primarily focuses on recycling, reduction of plastic pollution in the environment, and reduction of carbon emissions resulting from plastic. This non-binding plastic strategy includes the goal of making plastic packaging 100% recyclable by 2030 [7]. Plastics span a broad range of application fields such as (food) packaging, coatings, construction, adhesives, electronics, biomedical devices, textiles, and more. In this regard researchers worldwide are trying to find replacements for fossil-based polymeric materials in the form of bio-based materials, which perform similarly to today's conventional systems.

Currently, the market for bio-based plastic materials is experiencing constant growth, though it still accounts for only 1% of global plastics production. In 2023, the worldwide production capacities of bio-plastics reached approximately 2.18 million tons, with forecasts predicting continued growth to 7.4 million metric tons by 2028. Within this capacity, 43%, or 973 400 tons, were allocated to the packaging market, the largest segment within the bioplastics industry [8]. Globally, Asia leads the production of bioplastics by mass, reflecting the region's dominance in overall plastic production [9]. The leading bio-based and biodegradable plastic by production mass is polylactic acid (PLA), which accounts for 31% of the global production capacity, followed by starch-containing polymer compounds at 6.4% [10]. Together, these two types comprise more than 50% of the globally produced bio-based and biodegradable plastics.

Bio-based polymers are the most promising materials that are either partially or completely biodegradable. They are synthesized from plants or microorganisms through metabolic or biochemical engineering processes [11]. Unlike conventional polymers derived from finite petrochemicals, bio-based materials are produced from sustainable and renewable biomass. The International Union of Pure and Applied Chemistry define bio-plastics as a: 'bio-based polymer derived from the biomass or issued from monomers derived from the biomass and which, at some stage in its processing into finished products, can be shaped by flow'[12].

However, bio-based materials should not be confused with biomaterials. While bio-based materials, which are the focus of this book, are derived from renewable resources and emphasize sustainability, in contrast biomaterials are specifically designed to be compatible with living systems. Biomaterials are commonly used in medical applications such as implants, drug delivery systems, and tissue engineering.

Bio-based materials are derived from living or once-living organisms. These materials can be classified in various ways, often based on their source, degradability, or field of application. In this discussion, we differentiate these materials based on their chemical structure into the following categories (figure 1.1):

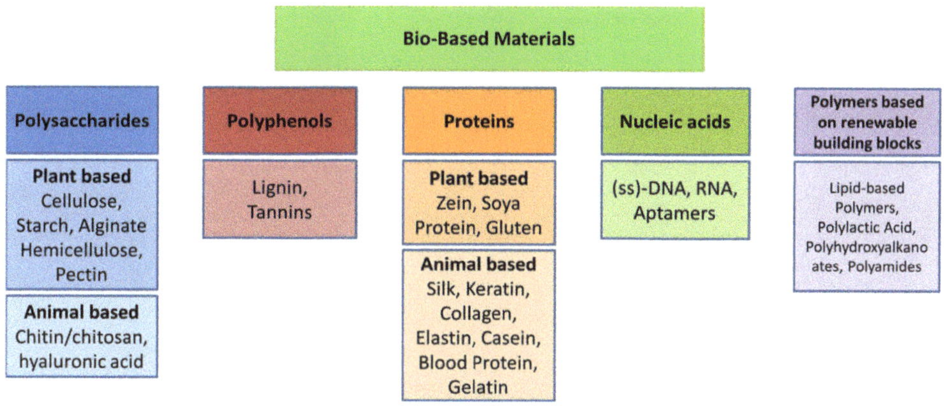

Figure 1.1. Classification of bio-based materials.

1. Polysaccharide-based biomaterials.
2. Phenolic compounds.
3. Protein-based biomaterials.
4. Nucleic-acid-based materials.
5. Polymers based on renewable building blocks.

As is evident from these categories, most materials, except the fifth category, are polymeric and naturally occurring in biological environments. This chapter will focus on polymers directly sourced from nature and their modifications. Monomers derived from biological resources that are subsequently converted into polymers will not be covered within the scope of this chapter.

In the context of eco-friendly solutions for today's problems, it is important to recognize that bio-based materials represent a class of materials derived from renewable resources. However, it is crucial to consider their entire lifecycle, including production and application, which often require significant energy and can release organic pollutants. Therefore, the label 'bio-based' alone is inadequate for judging the sustainability of a material. Instead, 'sustainability' should be emphasized, encompassing the environmental impact and costs throughout the material's entire lifecycle [13].

What unites all these polymeric materials is their foundation in renewable resources, with their synthesis involving living organisms within a relatively short timescale of a few years. The significance of time is crucial in defining renewability. Unlike the lengthy conversion process of carbon in fossil-based materials, which spans hundreds of thousands of years, the rapid carbon conversion in renewable materials facilitates a net-zero impact on our current biosphere.

Bio-based materials offer a broad variety of chemical groups and functionalities due to their unique and diverse molecular structures rooted in their biological origin. These materials can typically be assigned to one of the fundamental biological building blocks of life science, such as proteins, carbohydrates, lipids, or nucleic acids. This inherent relationship to fundamental functional building blocks allows

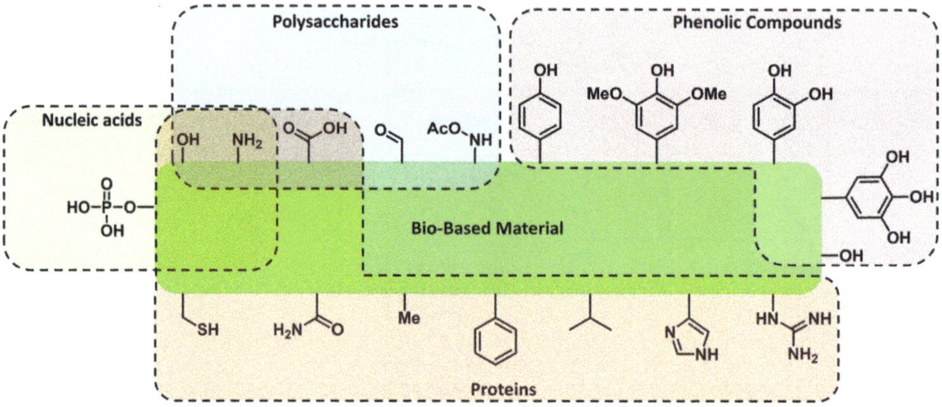

Figure 1.2. Surface functional groups in bio-based materials categorized by material group: polysaccharides, phenolic compounds, nucleic acids, and proteins.

for the determination of the potential biodegradability and chemical behavior of the respective materials, which will be discussed in detail for each material later in this chapter.

Figure 1.2 highlights the essential surface functional groups categorized by material groups: polysaccharides, phenolic compounds, nucleic acids, and proteins. These functional groups significantly influence the interactions, stability, and performance of bio-based materials in diverse applications and will be discussed in this chapter.

However, understanding the properties of bio-based materials requires more than just knowledge of their chemical structure. Bio-based materials exhibit properties based on a multiscale structure, including supramolecular arrangements, nano- and microscale structures, and an observable macrostructure.

The next chapter will further focus on the surface and interfacial properties of each individual class of bio-based materials, covering aspects such as interactions with the surrounding environment, hydrophilic and hydrophobic surface properties, adhesion, wettability, and the influence of these properties on material performance and applications. Additionally, it will explore potential methods for modifying and treating these surfaces to tailor their characteristics, optimizing the materials for specific applications.

The inherent properties of bio-based materials are highly advantageous, allowing them to replace most conventional polymers derived from fossil resources. Moreover, bio-based materials possess unique characteristics that can surpass those of conventional plastics, offering new possibilities for recycling, mechanical and optical properties, responsive behavior, and various application fields. These materials exhibit specific crystallinity, nanostructures, unique interactions with their environment, and supramolecular properties, making them a valuable resource for developing innovative polymeric materials. Despite the significant interest in bio-based materials as replacements for fossil-based counterparts, their inherent properties offer possibilities far beyond simple substitution.

1.2 Polysaccharide based biomaterials

Polysaccharides, as a subgroup of carbohydrates, are organic compounds made up of carbon, hydrogen, and oxygen in a ratio of 1:2:1, respectively [14]. As the most abundant organic compounds on Earth, they are one of the four main classes of biomolecules, alongside lipids, proteins, and nucleic acids. Carbohydrates in general are the principal source of energy for all cells. Polysaccharides represent a group of polymeric carbohydrates in which monosaccharides are linked by glycosidic bonds, extending up to thousands of monosaccharide units [15]. Polysaccharides are fundamental structural components of plants, where they occur as cellulose, and of animal exoskeletons, where they occur as chitin. They also play a crucial role in energy storage, as seen with starch in plants and glycogen in animals. Polysaccharides can be categorized into structural and storage types. Structural polysaccharides, such as cellulose and chitin, provide support and protection, contributing to the structural integrity of biological systems with their rigidity and strength. Storage polysaccharides, such as starch and glycogen, serve as energy reserves, breaking down to release glucose for metabolic processes when needed [15]. These natural roles result in interesting properties for bio-based polymeric materials. Structural polysaccharides such as cellulose exhibit high tensile strength and durability, making them suitable for applications requiring robust materials. The strong hydrogen bonding between cellulose chains contributes to its rigidity and reduced reactivity. Storage polysaccharides, such as starch, are excellent film-forming and coating materials due to their rapid biodegradability and solubility, which also makes them highly versatile for chemical modification.

In the following sections, we will discuss significant polysaccharides for applications in bio-based materials, highlighting their unique properties and potential uses with a focus on their properties at interfaces. The discussion will be limited to cellulose, starch, chitin, alginates, and hemicellulose, as these polysaccharides have the most well-documented and versatile applications in the scope of bio-based materials. Other polysaccharides, such as pectin and glycosaminoglycans, will not be covered due to their more specialized and less widespread use in bio-based materials, despite their exciting properties and potential in this field.

1.2.1 Cellulose

Cellulose, the most abundant naturally occurring biopolymer, offers immense potential for innovation and diverse applications due to its versatility and eco-friendly nature [16]. It serves as a valuable raw material for conversion into smaller building blocks for synthetic polymers, as well as a structural polymer with inherent properties that enable exciting possibilities for interfacial interactions. This chapter will first explore cellulose as a macromolecular material and then focus on nanoscale applications of the same.

Regardless of its source, cellulose consists of D-glucose monomeric units repeatedly linked together by beta-1,4-glycosidic bonds, resulting in a highly functionalized, linear stiff-chain homopolymer with distinct properties such as hydrophilicity, chirality, biodegradability, extensive chemical modification capability, and the ability

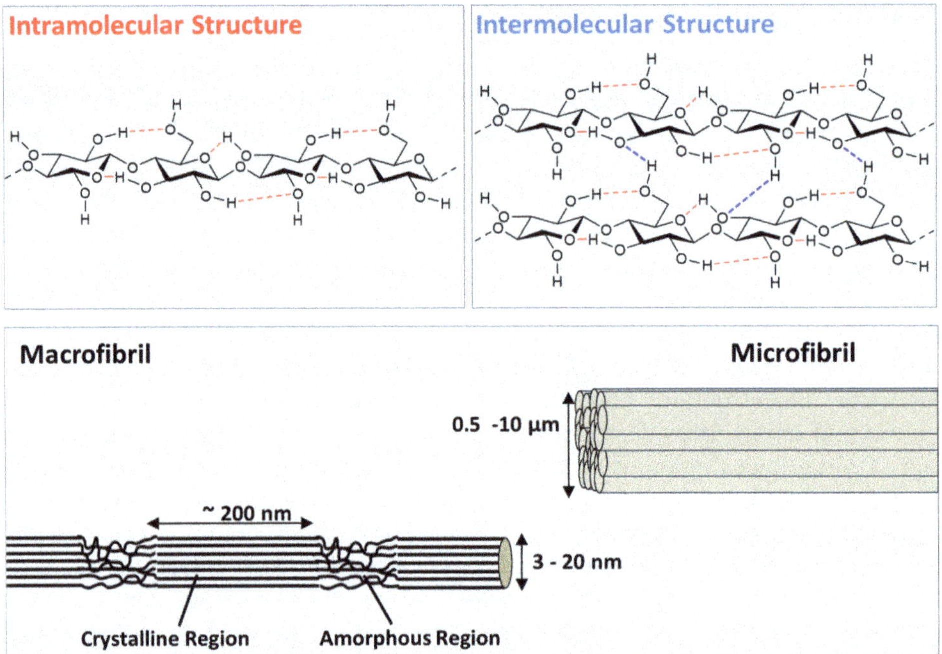

Figure 1.3. Structure and assembly of cellulose illustrating the hierarchical structure of cellulose and its interactions leading to the formation of macrofibrils and microfibrils.

to form diverse semicrystalline fiber structures [16]. The pristine surface of cellulose is determined by hydroxyl groups. The inter- and intra-molecular interactions of the single cellulose chains are mainly determined by hydrogen bond networks which join the individual polymers into macrofibrils and further microfibrils in a micrometer range (figure 1.3).

Further, these interactions lead to the crystalline regions within cellulose which are considered to be responsible for the insolubility of the material in a number of solvents including water [17]. The abundant hydroxyl groups in cellulose facilitate interactions with polar molecules such as water and contribute to the formation of hydrogen bond networks between the individual cellulose molecules. Despite the hydroxyl groups along its molecular structure making cellulose a hydrophilic material, aspects of its amphiphilic character remain controversial [18]. Even though generally considered hydrophilic, there are several aspects hinting towards a hydrophobic character of cellulose as well. In both native and regenerated cellulose, strong hydrophobic interactions between cellulose molecules cause them to associate, while hydrogen bonds form within this nonpolar environment but do not drive the association. Notably, the dissolution of cellulose in water is energetically unfavorable because the entropy loss from water–cellulose interactions is not compensated by the entropy gain from the increased flexibility of cellulose chains when they dissolve [19, 20]. The associated cellulose chains can thus further form partially crystalline domains, while the amorphous regions exhibit different structural characteristics. The insolubility of cellulose is often attributed to the extensive hydrogen bonding present in the crystalline regions of the material. However, this is not the exclusive reason for its insolubility; the so-called 'amorphous' regions may also contain hydrogen bonding, albeit presumably not to the same extent as in the crystalline regions. Despite this, amorphous cellulose is also insoluble in water, indicating that factors beyond hydrogen bonding contribute to the overall insolubility of cellulose [21]. Even though the discussion about the amphiphilic properties of cellulose remains unresolved, the overall surface properties of the bulk material will further be considered hydrophilic.

The use of cellulose in bio-based materials spans a wide range of fields. Given its abundance, cellulose is regarded as a highly promising candidate for replacing conventional fossil-based plastics and has already found applications for a long time in commercial products. Common commercially available bio-plastic materials derived from cellulose include regenerated celluloses (such as Cellophane, Rayon, and Lyocell), cellulose acetate, and cellulose ethers such as methylcellulose and hydroxypropyl cellulose [22]. A field of cellulose based materials where surface properties are particularly important is in cellulose based dialysis membranes. The development of cellulose-based dialysis membranes has evolved significantly since the introduction of Cuprophan® in Wuppertal, Germany, which was initially made from cotton and effective for small solute removal but had poor biocompatibility compared to synthetic polymer membranes [23]. To enhance biocompatibility, cellulose membranes were chemically modified by acylation of hydroxyl groups with acetate, resulting in cellulose acetate, cellulose diacetate (CDA), and cellulose triacetate (CTA). These modifications reduced free hydroxyl groups, minimizing binding to complement receptor C3b and subsequent complement activation, with CTA being the most biocompatible [24]. Further advancements included the incorporation of aromatic benzyl groups to create hydrophobic domains, and the development of Hemophan®, which incorporated functional tertiary amines and large diethylaminoethyl groups to improve biocompatibility by preventing reactions with blood cells [25].

The modification of the interfacial properties of cellulose is crucial for its application as a bio-based material. For example, its sensitivity to humidity presents a significant challenge for its use in various applications. Tailoring the surface properties by utilizing the individual hydroxyl groups on the cellulose surface enables a wide range of chemical modifications. Cellulose acetate, the most prominent example of chemically modified cellulose, is a thermoplastic polymer derived from cellulose. It is used in a wide range of applications, including film and sheet products, textiles, eyewear frames, and cigarette filters, as a bio-based alternative to conventional plastics. Several studies have demonstrated the degradability of cellulose acetate, establishing it as a biodegradable polymer. Cellulose acetate-based materials can undergo biodegradation and photodegradation through well-defined mechanisms. Various strategies, such as enzymatic treatment and the use of photocatalysts, can enhance these processes, thereby mitigating their environmental impact [26].

Further chemical modifications of cellulose include esterification, etherification, grafting, and surface modification. Esterification and acylation processes involve the use of acyl chlorides, anhydrides, and vinyl esters in the presence of catalysts or ionic liquids, such as acetyl chloride and acetic anhydride. Esters typically make cellulose more hydrophobic and improve thermal stability. The interfacial properties are thereby altered, improving the material's resistance to moisture. Esterbonds are generally more prone to hydrolytic degradation compared to ethers and make the materials better degradable. Etherification involves reacting alkaline cellulose with etherifying agents such as epoxides and alkyl halides. This modification leads to interfacial properties of the cellulose derivatives which improve their solubility in water and increase viscosity. As a result, etherified cellulose derivatives, such as carboxymethyl cellulose (CMC) and hydroxyethyl cellulose (HEC), are widely used in pharmaceuticals, food products, and personal care items for their thickening and stabilizing properties. Ethers generally increase the hydrophilicity of the material, making it more soluble in water. They are less prone to degradation than esters, providing better stability [27].

Grafting is a method that involves covalently bonding side chains to the cellulose backbone using techniques such as ring-opening polymerization (ROP) [28] and radical polymerization technologies [29]. In general, the initiation and anchoring processes in radical polymerization for cellulose involve the generation of free radicals that create reactive sites on the cellulose backbone. These sites serve as initiation points for the polymerization of monomers, resulting in the grafting of polymer chains onto cellulose [29]. This modification imparts new functionalities to cellulose without altering its inherent properties, such as improved thermal resistance and chemical stability. Grafted cellulose finds applications in the biomedical field, including drug delivery systems, and in the production of high-strength, lightweight materials [30]. Since other review articles have extensively covered the chemical modification of cellulose, we encourage readers to refer to those sources for a more detailed understanding [27, 29].

However, recent efforts in the chemical modification of cellulose for commercial applications have shifted focus from the resulting functional groups to the modification process itself. The emphasis is now on utilizing environmentally friendly solvents and catalysts to achieve more sustainable and efficient modifications. For example, the use of ionic liquids as reaction media and the development of green oxidation processes that combine ultraviolet light, hydrogen peroxide, and ozone are gaining importance. These methods not only enhance the efficiency of cellulose modification but also significantly reduce the environmental impact [31, 32].

In the packaging industry, cellulose-based materials are employed both directly as regenerated cellulose films, such as Cellophane, and as fibers and nanocellulose additives to enhance the gas barrier properties and strength in composite materials. Cellophane, a prominent example, offers excellent barrier properties against gas, oil, and liquid water, but it has high permeability to water vapor [33]. Despite its benefits, Cellophane's production process is not environmentally friendly and is relatively expensive. Nanocellulose, in the form of cellulose nanofibers (CNFs) and cellulose nanocrystals (CNCs), enhances the gas barrier properties in films but suffers from poor water vapor barrier properties due to moisture absorption of the hydrophilic cellulose surface [34]. Due to its nanoscopic structure, nanocellulose in reinforced composites can act as an effective gas blocking material. It hinders gas permeation by increasing the material's free volume, thus affecting gas diffusivity and solubility [35]. As a effective coating on paperboard, CNFs improved the mechanical strength as well as the gas impermeability, making them suitable for multilayer packaging [36]. The combination of several nanocomposites in one material can further improve certain interfacial properties for high-performance packaging, as in an example of CNCs which are combined with Ag nanoparticles in a hybrid material exhibiting excellent barrier properties, antibacterial activity, and low migration in food applications [37]. Additionally, nanocellulose–montmorillonite (MMT) composites offer reduced water vapor permeability and enhanced mechanical properties [38], while metallic hybrid nanomaterials formed with add regenerated cellulose (RC) and microcrystalline cellulose (MCC) show promising antibacterial properties for food packaging [39].

1.2.1.1 Nanostructured cellulose

Nanostructured cellulose has been gaining interest over the last few decades and has become one of the most prominent green materials of modern science. Nanostructured cellulose can be categorized into bacterial cellulose (BC), cellulose nanofibers (CNFs), and CNCs and is determined by having at least one dimension at the nanoscale [40]. These materials benefit from a high surface to volume ratio, making them a promising building block for innovative materials. While CNCs enable the dispersion of cellulose in the form of nanoparticular suspensions, the suspension behavior of CNFs and BC is determined by their fibrous properties and the resulting entanglements of these fibers [41]. The surface properties of nanostructured celluloses are significantly influenced by their preparation methods, as will be discussed in the following sections.

BC shares the same molecular formula as plant-derived cellulose but exhibits superior properties, such as higher purity and greater water-retaining capacity. It consists of microfibrils of indefinite length that are randomly interwoven, giving it a ribbon-like appearance. BC is primarily produced by bacteria such as *Acetobacter xylinum* [42]. Since the production of BC does not involve chemical modifications, its surface properties in the pristine state are determined by hydroxyl groups (–OH), similar to cellulose from natural sources [43]. The surface of BC can be tailored post-production through several methods: (i) substituting the hydroxyl groups with other functional groups, (ii) crosslinking with other polymer materials, (iii) forming composites with metal or metal chalcogenide nanomaterials, and (iv) integrating carbon-based nanocomposites [44].

CNFs are produced through mechanical processes, often combined with chemical or enzymatic treatments, to liberate the fibrous structure of cellulose from plant sources. The primary functional groups present on the surface of CNFs are again hydroxyl groups (–OH). Chemical modifications can further tailor the surface properties of CNFs. For instance, introducing carboxyl groups (–COOH) through processes such as TEMPO-mediated oxidation enhances the anionic nature of

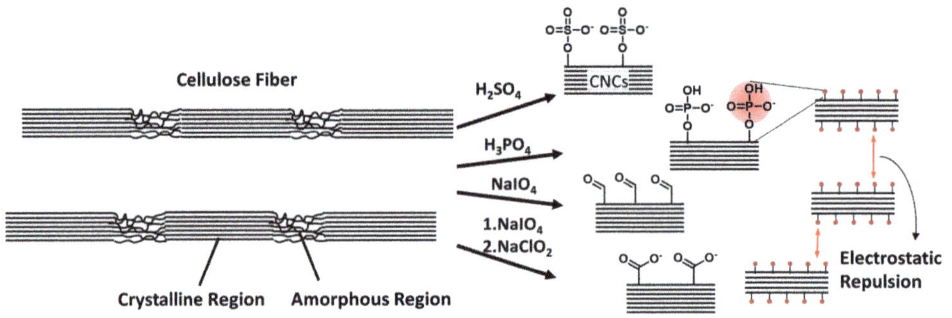

Figure 1.4. Illustration of the synthesis process and resulting surface modifications of CNCs. The figure highlights the introduction of functional groups such as sulfate, phosphate, aldehyde, and carboxylate.

CNFs, improving their dispersibility and stability in water. Similarly, sulfonation can introduce sulfonate groups ($–SO_3H$), providing negative charges that enhance colloidal stability and interaction with positively charged species. Enzymatic treatments can also modify the surface of CNFs by selectively removing certain components or altering the cellulose structure without harsh chemicals [45].

CNCs are derived from natural cellulose sources through processes such as acid hydrolysis, which removes amorphous regions and leaves behind crystalline regions. The surface of CNCs typically contains hydroxyl groups (–OH), which are responsible for the high surface energy and hydrophilicity of CNCs. Furthermore, the surface properties of CNCs are determined by charged groups such as carboxyl [46], sulfate [47], or phosphate groups [48] depending on the hydrolysis process which is applied to access CNCs (figure 1.4). For example, sulfuric acid hydrolysis commonly used in CNC production introduces sulfate half-ester groups (–OSO3–) on the surface [47]. On the other hand, TEMPO-mediated oxidation primarily introduces carboxyl groups, enhancing the anionic character of the CNCs [46]. These charged groups enhance the dispersibility of CNCs in polar solvents, improve compatibility with other materials, and add new functionalities for specific applications.

In addition to the surface groups introduced during the production of nanocellulose materials, chemical modifications can be applied to further tailor these surfaces. Similar to cellulose, these modifications primarily target the abundant hydroxyl groups on the surface, allowing the introduction of various functional groups. Common chemical modifications include anionization (e.g. TEMPO-mediated oxidation) [49], cationization (e.g. modification with glycidyltrimethylammonium chloride [50]), introduction of aldehyde groups by ring-opening of the cellulose molecules [51], and hydrophobization (e.g. esterification with fatty acids [52]).

Owing to the exceptional strength of nanostructured cellulose, these materials are often employed as reinforcing phases in composites to enhance material strength and flexibility [53]. Given their hydrophilic nature, the surface modification of cellulose nanomaterials is crucial for their effective integration in hydrophobic matrices. Trinh *et al* modified CNCs by grafting lauroyl chloride, a medium-chain fatty acid chloride, onto the cellulose surface. They showed that slightly modified CNCs with a degree of substitution (DS) of 0.2 significantly reinforce epoxy resins, enhancing the tensile strength and modulus by 77% and 44%, respectively, at 5% loading. The modification improves dispersion and interfacial adhesion between CNCs and the epoxy matrix, which is crucial for effective load transfer and stress distribution. Additionally, highly modified CNCs (DS 2.4) reduce water uptake, increasing the composite's durability in moisture-prone environments [53].

It should be noted that while surface modification of nanocellulose offers numerous advantages, there are scenarios where it might be better to use unmodified nanocellulose. For applications where the natural properties of cellulose, such as biodegradability and non-toxicity, are paramount, introducing chemical groups may compromise these attributes [54]. Additionally, surface modification processes can be costly and complex, making them less suitable for large-scale or cost-sensitive applications. In some cases, chemical modifications can lead to a decrease in mechanical strength; for instance, Cañas-Gutiérrez *et al* demonstrated that TEMPO oxidation beyond 60 min significantly disrupted the nanofiber network of bacterial nanofibrillar cellulose (BNC) scaffolds, resulting in notable alterations to both the macro- and microstructure, making it less suitable for high-strength composite materials [55]. Physically driven coatings in the form of adsorbed polymers, clays and nanoparticles can be useful alternatives to chemical modifications, since they overcome many of the mentioned drawbacks and are considered suitable for large-scale manufacturing [56–59].

A recent method utilizes the inherent interactions between cellulose, tannic acid, and alkyl cellulose derivatives to modify the surface properties of CNCs, enhancing their hydrophobicity. This approach involves noncovalent interactions to coat CNCs, enabling the creation of nanomaterials with adjustable wetting and emulsification characteristics. By using tannic acid as a primer and combining it with methyl cellulose or ethyl cellulose, this method improves the dispersion and interfacial adhesion of CNCs in hydrophobic matrices, significantly enhancing their mechanical properties and stability in non-aqueous environments. This process is also advantageous due to its simplicity, environmental friendliness, and effectiveness in maintaining the sustainability of the modified CNCs [59].

The high aspect ratio of CNCs allows their application as emulsifiers in Pickering emulsions. In Pickering emulsions the stabilization mechanism is attributed to the accumulation of particles at the oil–water interface to form a densely packed layer which is closely related to the long-term stability of emulsions [57]. The adsorption behavior of nanocellulose at the interface results in the formation of interfacial films, contributing to the long-term stability of emulsions. Ni *et al* showed that the application of nanocellulose of different lengths can significantly influence the stability of emulsions. They point out that one of the major factors in nanocellulose based Pickering emulsions is the formation of network structures of CNC interfacial layers with fibril entanglements at the interface of the droplets. Here cellulose nanoparticles with shorter lengths caused a greater reduction of interfacial tension due to an enhanced hydrophobicity of the particles which accompanies the reduction in particle size [60].

Overall cellulose, with its exceptional biocompatibility and versatile interfacial behavior, stands out as a highly promising biomaterial. Its natural abundance, biodegradability, and ability to form strong hydrogen bonds make it suitable for a wide range of biomedical applications. Nanocellulose, in particular, offers enhanced properties such as increased surface area, improved mechanical strength, and the potential for surface modifications. These attributes enable its use in advanced applications, including biomedical applications, reinforced composites, and emulsion stabilization. The unique interfacial behavior of nanocellulose, including its ability to form stable interfacial films, further underscores its potential in creating stable emulsions and other complex formulations. As research and development continue, the functional versatility and environmental benefits of cellulose-based materials position them as key components in the future of biomedical innovation.

1.2.2 Starch

Starch represents another interesting candidate in the scope of polysaccharide-based materials due to its biodegradability, annual renewability, and abundance. Its economic advantages and compatibility with conventional plastic processing equipment have sparked significant interest in starch-based materials for various applications [61–63]. As a versatile biopolymer, starch is already widely used across diverse industrial sectors, with approximately 50% of commercially used bioplastics being prepared from starch [64]. The production of starch-based bioplastics is straightforward, making them popular for packaging applications [65]. Further, in paper manufacturing, starch enhances paper strength and printability as a sizing agent [66]. Similarly, in the textile industry, it improves yarn strength and weaving efficiency [67]. In pharmaceuticals [68], starch functions as a binder, disintegrant, and coating material [69]. Additionally, starch-based adhesives are used in various applications, providing a renewable and biodegradable alternative to traditional fossil-based materials [70].

Many properties of starch as a biomaterial derive from its molecular structure. Unlike the repetitive cellobiose units in cellulose, the monomeric units in starch are connected by alpha glycosidic bonds, which significantly alter its properties in biological environments. This difference in connectivity results in starch being more readily digestible and metabolized by organisms, in contrast to the more resistant nature of cellulose [71]. Starch as a polysaccharide consists of the two polymers amylopectin and amylose. While the chemical linear structure of amylose leads to a helical formation of the polymer, amylopectin, on the other hand, is a branched polymer composed of multiple shorter chains of glucose connected by α-1,4-glycosidic bonds, with these short chains being further interconnected by α-1,6-glycosidic bonds, resulting in a highly branched molecule. The interactions between amylopectin branches and amylose generally lead to a more amorphous structure of starch (the crystallinity of native starch granules varies from 15% to 45%) [72, 73]. Unlike cellulose, starch represents a highly hydrophilic polymer that dissolves well in water and exhibits poor thermo-mechanical properties [74]. Starch's excellent solubility in water facilitates easy chemical modification of this biomaterial but makes it vulnerable to humidity at the same time. Furthermore, its relatively weak thermal and mechanical stability restricts its direct application in various fields. To overcome these obstacles, material modifications are necessary to enhance starch's applicability across a broader range of applications.

The main focus of starch in the substitution of petroleum-based materials lies in the area of adhesives, plastic substitution and film-forming materials. The huge impact of starch in this field can be put in numbers, as already today about 50% of commercially produced bio-plastics are prepared from starch [65, 75]. The field of starch-based adhesives sets a focus on the interfacial properties between the substrate and adhesive. Native starch, comprising predominantly hydroxyl groups, can bind by hydrogen bonds to all sorts of substrates which offer polar surfaces. This includes mineral surfaces, such as those involved in wallpaper paste or bottle labeling [76], as well as the adhesion to cellulose based surfaces such as paper and wood [77, 78]. Here the adhesives are typically applied as an aqueous solution as cold-press adhesives and develop their strength by loss of water [76]. In contrast, chemically modified starch demonstrates tailored properties such as enhanced water resistance, or specific binding sites. Esterified starches, which incorporate hydrophobic substituents, are readily manufactured and effectively utilized to formulate bio-based adhesives that exhibit superior water resistance [69]. The most common esterified starch is acetylated starch [79, 80]. It is synthesized with acetic acid and acetic anhydride as esterification reagents and can be divided based on the DS into three types of acetylated starches: low, medium and high DS starch [81]. Of these, low DS starch is typically applied for adhesives (DS0.01–0.2).

Similar to acetylated starch, fatty acids can be introduced to starch through esterification. For example, sodium octenyl succinate plays an important role as a modifier for starch that enhances its emulsifying and encapsulating properties [82]. Similarly, Sun *et al* demonstrated that dodecenyl succinic anhydride (DDSA) can be effectively used as a modifier for starch, successfully applying it in starch-based wood adhesives [83]. This modification improves the adhesive properties, making it a viable option for various industrial applications. The long-chain olefin structures of DDSA enhance the water resistance of the adhesive. Additionally, the incorporation of fatty acids as substituents significantly influences the overall hydrogen bonding in starch, thereby altering its adhesive properties. Fatty acids break the intra- and intermolecular hydrogen bonds of the polymers, opening the starch network. This process increases the free volume of starch, thereby lowering the glass transition temperature (T_g) [84]. In addition to imparting thermoplastic properties to starch, fatty acids are known to have antibacterial properties [85]. Besides esterification, another common chemical substitution uses etherification to result in carboxymethyl- [86], hydroxyethyl- [87], and hydroxypropyl-starch [87]. In the case of hydroxethyl-cellulose the retrogradation of starch is slowed down in addition to enhanced water stability of the adhesive [88].

In addition to the introduction of hydrophobic groups, starch can also be modified to create more hydrophilic structures, such as cationic or oxidized starch. Cationic starch is typically used to enhance ionic interactions with substrates such as glass or paper, significantly improving the sheet strength of paper [88]. These ionic interactions influence the binding strength and internal properties of starch-based materials to substrates. Oxidized starch is employed in adhesives to introduce carbonyl and carboxyl groups. This oxidation process decreases the molecular weight of starch, increasing its solubility and reducing its viscosity, which enhances the adhesive's spreadability and penetration into substrates [89]. The interfacial properties are significantly improved, as oxidized starch forms stronger bonds with various materials due to the increased availability of reactive sites. A special case is the oxidation of starch to dialdehyde starch using sodium periodate. This oxidation equips the starch with reactive aldehyde groups, which can form covalent hemiacetal bonds with hydroxyl group-bearing substrates such as wood, introducing reactive functional groups into the starch [90].

Films made from pristine starch typically exhibit very brittle and difficult-to-handle properties. This issue can be addressed by using amylose-only starches, which have been shown to be a more efficient raw material for bio-plastic and can be obtained through effective methods in plants via transgenic techniques [91]. The molecular modification of common starch presents an opportunity to enhance the processability and gas barrier properties of starch-based films. Recent research demonstrated that adipic acid cross-linking effectively modifies starch, resulting in films with superior water vapor barrier properties. The water vapor permeability of these cross-linked films was significantly lower than that of native or oxidized starch films ($P < 0.05$) [92]. This cross-linking alters the interaction of gas molecules with the bulk material, thereby hindering the permeation of hydrophilic molecules.

An even more common way to overcome mechanical issues and water vulnerability is the addition of plasticizers such as citric acid, sorbitol, carboxymethyl cellulose, or glycerol [93, 94]. This modified starch is often referred to as thermoplastic starch (TPS). TPS is procured through the process of plasticization of starch with water and plasticizers, followed in some cases by blending with tougher polymers or biopolymers.

TPS stands out as one of the most important processes for film and plastic preparation due to its ability to be processed under heat and pressure into a moldable plastic film. If the compounded TPS is subjected to high temperatures (90 °C–180 °C) and shear it undergoes a transformation process by readily melting and flowing, similar to typical synthetic thermoplastic polymers. This enables its utilization in extrusion, injection molding, or blowing processes, mirroring the versatility of conventional synthetic thermoplastics [95]. The plasticizers interfere with the strong inter and intra-molecular H-bonding formed by the amylose and amylopectin of starch, partially depolymerizing the starch backbone and further leading to a decrease in the melt temperature of starch below degradation temperature (figure 1.5) [95]. However, on plasticization with glycerol, starch can undergo retrogradation over time and increase crystallinity, which affects the functional properties of TPS. It is important for an efficient plasticizer to have the ability to suppress retrogradation during aging, and it should enhance the flexibility. Therefore the use of multiple plasticizers can be beneficial and enhance the plasticization efficiency, preventing the retrogradation process of starch effectively [96, 97]. The effectiveness of TPS is governed by its T_g. Although the exact T_g of

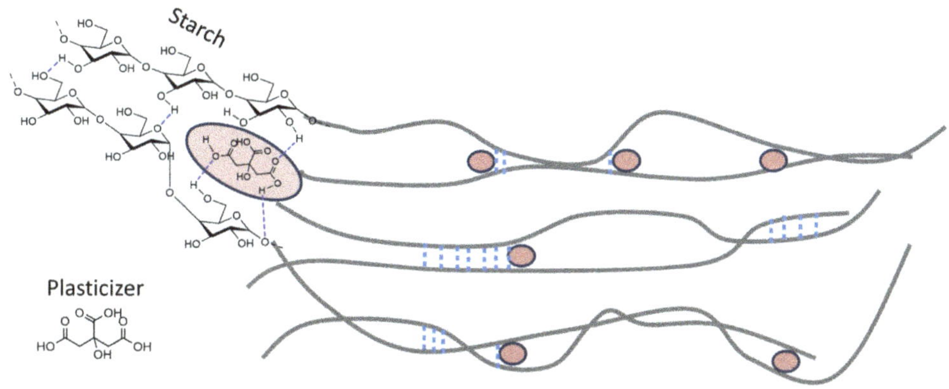

Figure 1.5. Molecular interactions in starch films and the effects of plasticizer (citric acid) application on hydrogen bonds.

starch is not definitively established, it is generally believed to fall within a range of approximately −75 °C–10 °C, depending on the water content conducive to gelatinization and subsequent processing [98].

Some of the latest research concerning TPS makes use of 3D-printing technology. The general difficulty in using hot extrusion 3D printing technologies is that the raw printing material requires appropriate rheological properties to be extruded and maintain its shape. Native starch gel has poor shape retention and lacks good self-support and strength, causing collapse and loss of the original shape when printing objects with multiple layers. To address these challenges, recent studies have focused on modifying the rheological properties of TPS to improve its printability. For instance, the research by Chen *et al* demonstrated that starch suspensions, particularly those of potato, rice, and corn starches, show shear-thinning behavior and possess sufficient mechanical integrity when heated to specific temperatures. These properties are crucial for achieving high-resolution prints and preventing deformation during the printing process [99]. Additionally, Haryńska *et al* explored the use of PLA-potato thermoplastic starch filaments, highlighting their potential as a sustainable alternative to conventional PLA filaments [100]. The modified TPS exhibited enhanced stability and strength, making it suitable for various 3D printing applications [99]. Moreover, Ju *et al* investigated the use of TPS in 3D printing, focusing on optimizing the blend composition to enhance the mechanical properties and printability of the material. Their study found that incorporating plasticizers and other additives significantly improved the flexibility and durability of TPS, making it more adaptable for complex printing tasks [101].

Further starch-based coatings and films play a significant role in sustainable packaging [102]. The integration of fatty acids into starch films can enhance the barrier properties of these films, as shown by Jiménez *et al* [103]. Additionally, the modification of starch films with nanomaterials can enhance their properties in multiple ways, such as antimicrobial and mechanical properties. Therefore various nanomaterials such as nanoclay [104, 105] and cellulose nanofibers [106] have been employed. For example, incorporating montmorillonite nanoclay improves the mechanical properties and thermal stability of starch-based nanocomposites [104]. The integration of chitosan or nano-silver can impart antibacterial properties resulting in active packaging materials [105, 107]. Cellulose nanofibers significantly increase the tensile strength and Young's modulus of the films; however, excessive amounts can lead to agglomeration, diminishing these enhancements [107].

Starch, as an abundant material, holds great promise for further application as a bio-based material. Its processability and low cost make it one of the most outstanding materials for broad commercial applications. Today, many starch-based products are already on the market for everyday use. Future research in this area will likely focus on further improving mechanical properties, resistance to humidity, and reducing costs. This research field is not yet sufficiently explored and offers significant potential for innovation and development in the area of sustainable materials.

1.2.3 Chitin/chitosan

Chitin its derivative chitosan represent a special group of amino-polysaccharides and are a common building block in nature. This linear homopolymer of N-Acetyl-d-glucosamine and its deacetylated form (chitosan) have various biological and physiochemical properties such as biodegradability, biocompatibility, and bioactivity [108].

Chitin and chitosan are polysaccharides composed of N-acetylglucosamine (GlcNAc) and glucosamine (GlcN) residues linked together through ($\beta 1 \rightarrow 4$) bonds, similar to the linkage found in cellulose. These molecules exhibit complex interactions involving acetyl, amino, and hydroxyl groups, which form both inter- and intramolecular hydrogen bonds [109]. In the case of chitin these properties lead to a highly aggregated and crystalline material [110]. The acetyl groups enable the formation of stable H-bonds along the inter-sheet direction that are hard to break [109]. Similar to cellulose, chitin exhibits hydrophilic surface properties due to the presence of numerous hydroxyl and amino groups in its structure, while maintaining insolubility in water due to entropy-driven self-assembly of the chitin chains through hydrogen bonds and hydrophobic interactions [111]. Its insolubility in most common solvents poses challenges for processing, limiting its industrial applications. Recent research has identified ionic liquids [112, 113] and deep eutectic solvents [114–116], as well as basic aqueous urea solvent systems [111, 117, 118] as promising solvents for chitin. Although these solvents show potential, they are currently expensive and associated with potential safety risks. Despite these challenges chitin can be transformed into a variety of materials, including chitin microspheres [117], hydrogels [117, 119], nanomaterials (see below), and films [120]. These materials exhibit properties such as enhanced mechanical strength, chemical resistance, and the ability to form biodegradable and biocompatible structures [121]. For instance, chitin dissolved in green solvent systems can produce functional chitin nanocrystals (ChNCs) and chitin nanofibers (ChNFs), which are used to create films and hydrogels suitable for drug delivery, food packaging, and environmental applications [122]. Additionally, chemical modifications of chitin, such as deacylation [118], and carboxylation [123], expand its functional potential. Modified chitins can be employed as adsorbents, drug carriers, and components in functional foods.

Their commercial application is already happening across industries, including medicine and pharmaceuticals. Products such as HemCon Bandages and ChitoFlex Dressings utilize chitosan for wound care due to its hemostatic and antimicrobial properties [124]. Chitosan capsules are popular dietary supplements for weight loss, binding fats in the digestive system [125]. In agriculture, chitosan-coated seeds enhance germination and resist fungal infections, while chitosan-based sprays promote plant growth. Commercial flocculants such as LaChiPur utilize chitosan in water treatment for its ability to aggregate and settle suspended particles and heavy metals from solution [126]. In cosmetics, chitosan appears in anti-aging creams and hydrating lotions for its moisturizing and antimicrobial qualities [127].

When chitin undergoes deacetylation to approximately 50%, it becomes soluble in aqueous acidic media and is referred to as chitosan. Due to a reduced number of acetyl groups it exhibits a less ordered structure [128]. Its potential to be dissolved in dilute aqueous acids makes it more suitable for chemical modifications and application in various fields [129]. Both, chitin and chitosan exhibit many biological activities such as antitumoral, antimicrobial, antioxidant, and anti-inflammatory activities [130].

The overall surface properties of chitosan are linked to its linear polycationic structure with a high charge density, reactive hydroxyl and amino groups as well as extensive hydrogen bonding. The pH dependent solubility relies on the pK_a value of the amines (approximately 6.5 [131]) along the polymeric backbone which become protonated upon dissolution at pH < 6 to form cationic amine groups ($-NH_3^+$), making it a pH responsive polymer. The resulting intermolecular electric repulsion results in the solubility of this polyelectrolyte [130].

The cationic groups along the partially deacetylated backbone of chitosan confer inherent antimicrobial and hemostatic properties against different microorganisms, including bacteria, filamentous fungi, and yeast [132]. The positively charged amino groups interact with the globally negatively charged cell membranes of cells and bacteria, which are determined by phosphate groups (figure 1.6). This interaction

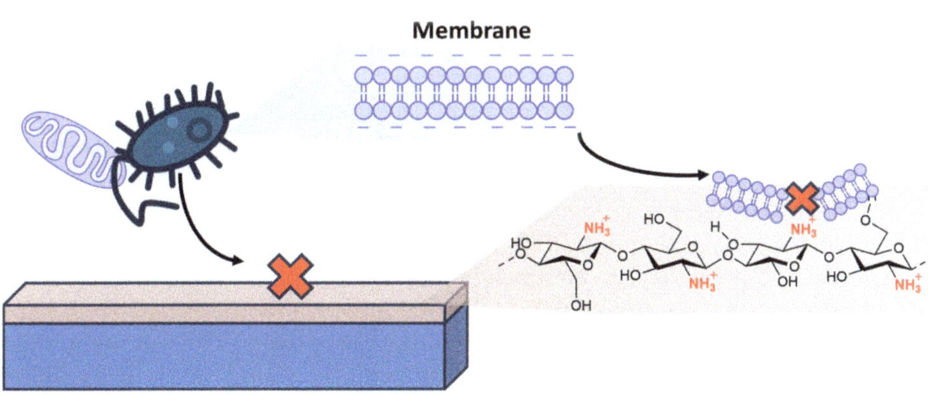

Figure 1.6. Relationship between the antimicrobial properties of chitin and chitosan and their surface charges.

disrupts their membranes, resulting in antimicrobial effects [130]. The same mechanism equips chitin with its antibacterial properties. However, the higher amount of acetylated amines in chitin consequently also lowers its antibacterial properties compared to chitosan [133].

Chitosan is not only considered a cationic polymer but also an amphiphilic material due to its non-polar acetyl groups associating with nonpolar phases, while its polar/charged hydroxyl and amine groups interact with a polar environment. Desbrières *et al* demonstrated that the hydrophilic–hydrophobic balance of chitosan can be adjusted through factors such as pH and the DS of ionic and alkyl groups. Using two distinct strategies, they applied electrostatically bound surfactants (surfactant-polyelectrolyte complexes (SPECs)) or N-alkylation to alter the amphiphilic character of chitosan. They highlighted that SPECs can be used as highly effective emulsifiers with better properties than alkylated derivatives of chitosan (polysoaps), even at low SPEC concentrations [134]. However, recent research suggests that the limited emulsifying and surface activity of commercially available chitosans may be attributed to their low degree of acetylation, which predominantly influences their electrostatic rather than their amphiphilic characteristics [135].

Other studies emphasize that the molecular weight of chitosan plays a critical role in determining its surface active properties. Payet *et al* investigated the interfacial behavior of chitosan with a degree of acetylation of 0.2, finding that it showed negligible surface activity at low concentrations. At higher concentrations, however, they observed a gradual decrease in surface tension, indicating slow adsorption kinetics at the oil–water interface and the progressive arrangement of hydrophilic and hydrophobic segments within the chitosan chains at the interface. Despite this, the reduction in surface tension was modest compared to typical surfactants [136]. Conversely, Mai-Ngam reported substantial surface activity and typical surfactant behavior in low molecular weight chitosan hydrochloride (Mw 5000 g mol^{-1}, degree of acetylation of 0.035), and suggested the formation of a chitosan monolayer at the air–water interface [137].

Chitosan's antimicrobial and antifungal properties make it ideal for use in active packaging materials. Often utilized as an active compound, chitosan exhibits excellent film-forming capabilities, which are further enhanced by the addition of phenolic compounds, improving both its mechanical and barrier properties [138]. Intelligent chitosan-based films act as pH indicators, changing color in response to pH variations caused by microbial growth and oxidative degradation of food. These materials primarily serve as indicators of food freshness and pH levels, with some also capable of monitoring CO_2 production by correlating pH changes with CO_2 concentration. The color change in these films signals alterations in pH, indicating potential spoilage or changes in the food's condition [139]. The overall oxygen permeability of these films is low (similar values to those of commercial polyvinylidene chloride) [140]. However, it should be noted that chitosan films exhibited relatively low water barrier characteristics due to their high hydrophilic nature. To overcome some of the shortcomings of chitosan films, it is common to produce chitosan composites. Due to its inherent hydrophilic properties chitosan is often combined with polar macromolecules. When combined with polysaccharides such as cellulose, starch, or xanthan gum, chitosan's properties are modified depending on the polysaccharide's structure and interactions [140]. For instance, blending with cellulose [141] or starch [142] can enhance the tensile strength. Similarly, blending with proteins such as gelatin [143], quinoa protein [144], or whey protein [145] alters the mechanical properties, typically reducing the tensile strength but increasing elongation at break due to specific interactions such as hydrogen bonding. These tailored composites enable diverse applications in food packaging, biomedical materials, and environmental sectors. Further plasticizers such as lipids, polyols (such as glycerol and sorbitol), sugars (such as glucose and sucrose) are frequently used to enhance characteristics such as elongation at break and water vapor permeability. However, it should be noted that plasticizers often improve one property while worsening another, as can be seen in the example of glycerol [94], which increases flexibility and elongation of chitosan films but can also elevate water vapor and oxygen permeability over time.

The application of chitin in bio-based materials spans various fields. It is frequently utilized as a cationic counterpart to materials with negatively charged surfaces in film-forming applications. In the biomedical sector, chitin's antimicrobial properties and its ability to serve as a counter ion for drug and nucleic acid delivery are particularly noteworthy [146].

Additionally, chitin is valuable in agriculture, enhancing soil fertility and plant growth through mechanisms such as nutrient retention and elicitation of plant defense responses against pathogens [147].

1.2.3.1 Nano-chitin and -chitosan

Similarly to cellulose, in the case of chitin and chitosan we also find a broad scientific field investigating biomass-based nanomaterials derived from these materials. The nanomaterials can be divided into nanofiber (NF) based materials, nanocrystals (NCs), and nanoparticular systems (NPs) consisting of the respective material. While the generation of chitin based nanomaterials can either be performed by top down or bottom up approaches, chitosan based nanomaterials are always generated by bottom up approaches [133].

The surface properties of nanochitin are determined by its unique cationic character, which arises from hydrolyzed acetyl groups on its surface. Below pH 6, these ammonium ions carry a positive charge in water, facilitating long-range electrostatic interactions over short-range hydrogen bonding. Above pH 6, hydroxyl, acetyl, and amine groups contribute to an intricate network of hydrogen bonds within and between chitin nanocrystals [148]. Additionally, less polar acetyl groups likely confer amphiphilic properties. Compared to nanostructured cellulose, nanochitin's amine groups allow straightforward modification using diverse functionalization methods. These groups, alongside primary hydroxyls, offer dual functionalities for orthogonal chemistry, leveraging higher reactivity [149]. Chitin modification examples include adding hydrophobic groups to reduce hydrophilicity or enhance compatibility with other surfaces [150, 151]. Beyond altering the surface energy, amine groups facilitate coordination bonds with heavy metals, making chitin suitable for applications such as water purification and heavy metal removal [152].

One distinguishing property of these nanomaterials is the surface charge, measured as zeta potential in suspensions. Chitosan-based nanomaterials exhibit pH-dependent zeta potentials due to their amine functionality, reaching up to +60 mV in acidic conditions and around −10 to −20 mV at higher pH (>8). In contrast, chitin nanomaterials show less variation in zeta potential across pH ranges, reflecting fewer protonatable amines [153].

Similar to chitosan in the bulk form, nanoparticles (ChsNPs) exhibit enhanced antimicrobial and antioxidant properties due to their high surface charge density, which facilitate strong interactions with microbial cell membranes. Due to the high surface area and nanoscale of ChsNPs, these properties are particularly potent compared to bulk chitosan, which is reflected in a antimicrobial activity increase by factors of 1.25–4.00 [154]. The effect is caused by ChsNPs disrupting bacterial and fungal cells and leading to cytosolic leakage and cell lysis. Further ChsNPs also show significant antioxidant activity by scavenging free radicals [155, 156].

Chitosan and chitin, as versatile biopolymers, represent a highly valuable class of resources for bio-based materials. Their cationic surface properties enable their use in fields benefiting from antimicrobial characteristics, such as biomedical applications, water treatment, and food preservation. Although these materials already find application in commercial products, there is still significant potential for improvement. Overcoming challenges such as the high cost and resource intensity of extraction and purification processes, as well as addressing solubility limitations, will enhance their broader implementation and effectiveness [157].

1.2.4 Alginates

Besides chitin/chitosan-based materials, alginate represents another important polyelectrolyte in the field of polysaccharides. Unlike chitosan, alginate polymers exhibit negatively charged carboxylic acid groups along their polymeric backbone. Alginate is primarily derived from brown seaweeds such as *Laminaria hyperborea*, *Laminaria digitata*, and *Ascophyllum nodosum* [158], as well as from certain soil bacteria [159]. Chemically, alginate consists of (1,4) β-D-mannuronic acid (M) and α-L-guluronic acid (G) units arranged in homopolymeric (MMM or GGG) and heteropolymeric (MG) sequences [160]. This biopolymer is highly suited for gel formation through electrostatic interactions, particularly with divalent cations such as calcium, resulting in the well-known 'egg-box' structure (figure 1.7) [161]. The solubility of alginate depends on the type of salt it forms: sodium, potassium, and ammonium alginates are water-soluble, whereas magnesium and calcium alginates are not. Alginate possesses several noteworthy properties, including non-toxicity, biocompatibility, biodegradability, and ease of chemical modification, making it a versatile material for various applications. At pH levels below its pK_a (<3.4), the carboxylic acid groups are non-ionized, leading to an insoluble structure. Conversely, at pH levels above 4.4, the carboxylic groups become ionized, increasing electrostatic repulsion among the negative charges, causing polymer chain expansion and swelling of the hydrophilic matrix, with the highest swelling observed around pH 7.4 [162].

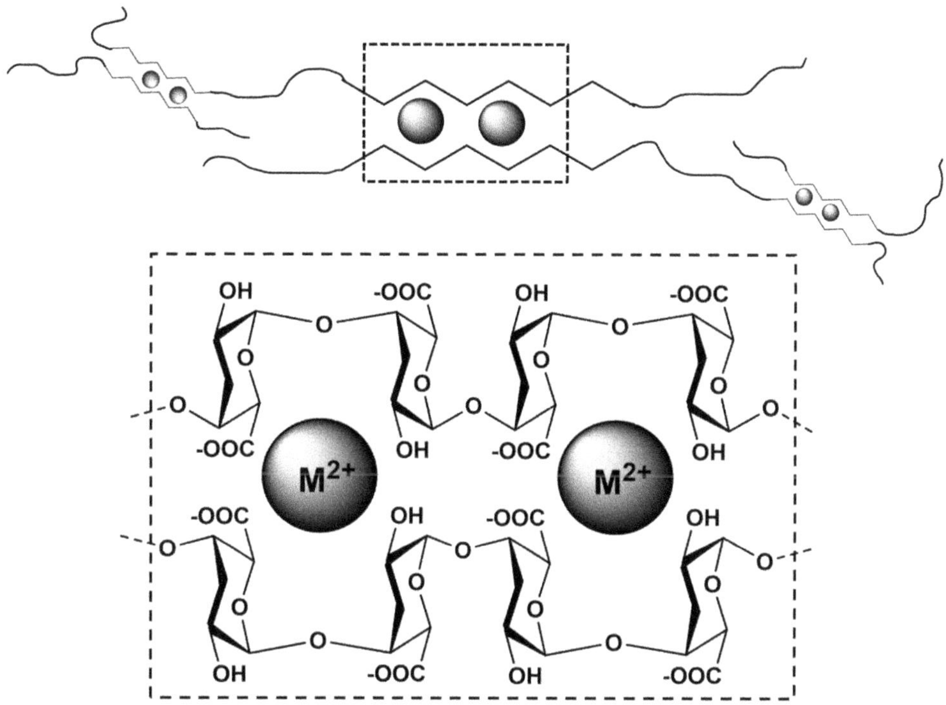

Figure 1.7. Schematic of the 'egg-box' structure in alginate hydrogel crosslinked with divalent metal ions (e.g. Ca^{2+}) illustrating the arrangement of alginate polymer chains, the positioning of Ca^{2+} ions, and the resulting crosslinking points that stabilize the hydrogel structure.

These properties make it highly versatile, with applications spanning various fields. In the food industry, alginate is used as a thickener and stabilizer [163]. In pharmaceuticals and medicine, it is utilized for drug delivery systems, wound dressings, tissue engineering, and as a component in prolonged-release drugs [164]. Alginate's film-forming ability also makes it valuable in food packaging and as a coating material for fruits and vegetables to delay spoilage and microbial damage [165]. Additionally, alginate is employed in water treatment due to its ability to adsorb heavy metals, leveraging the abundance of hydroxyl and carboxyl groups in its structure [166].

The interfacial properties of alginate-based materials are crucial for their applications, significantly influencing the performance of alginate films and coatings. Alginate's hydrophilicity, surface energy, and adhesion are determined by its chemical structure, which includes abundant carboxyl groups that impart charge. These functional groups allow alginate to form hydrogen bonds and ionic interactions, contributing to its high surface energy and strong adhesion. Alginate films typically exhibit water contact angles around 40°, demonstrating their strong hydrophilicity [167].

As one application field of alginate-based materials strongly connected to its surface properties and chemistry, alginate barrier films are of interest. Alginate-based films are notable for their blocking properties due to the dense network formed by intermolecular hydrogen bonds between the hydroxyl and carboxyl groups in alginate. This network significantly restricts oxygen diffusion, contributing to the low oxygen permeability of these films [168]. However, their high stiffness and low flexibility result in poor mechanical properties, while the hydrophilic nature of alginate results in high water vapor permeability, posing a challenge for certain applications.

To improve the mechanical properties and reduce water vapor permeability of alginate films, plasticizers are introduced into the film matrix. They are generally categorized into hydrophilic and hydrophobic types, each influencing the films' characteristics differently [168].

Hydrophilic plasticizers, such as glycerol, sorbitol, and polyethylene glycol (PEG), are frequently used to enhance the flexibility of alginate films. They achieve this by intercalating between the polymer chains, which reduces the strength of the intermolecular hydrogen bonds and allows the film to become more pliable. As in the case of starch-based films glycerol is also widely used as a hydrophilic plasticizer due to its ability to create strong hydrogen bonds with alginate, significantly increasing the film's flexibility. However, glycerol's small molecular size also makes the films more hygroscopic which leads to higher water vapor permeability (WVP). This can further lead to water incorporation in the films, whereby water itself can act as a plasticizer, altering the mechanical properties of the film [169, 170]. Sorbitol, another hydrophilic plasticizer, can also improve flexibility but tends to form more crystalline structures than glycerol, which can affect the transparency and mechanical properties of the films [170]. In a study by Pongjanyakul and Puttipipatkhachorn [171], the effects of glycerin and PEG400 on the physicochemical properties of sodium alginate-magnesium aluminum silicate (SA-MAS) micro-composite films were investigated. Both plasticizers improved the physicochemical properties of SA-MAS films, with FTIR spectroscopy suggesting interactions through hydrogen bonding. Considering the mechanical properties, glycerine was a better plasticizer than PEG400, providing more flexibility due to its smaller molecular size. PEG400, however, increased film rigidity due to its higher crystallinity. Both plasticizers reduced the WVP of alginate films at concentrations of 10%–30%, likely due to increased tortuosity of pore channels caused by plasticizer intercalation. The low WVP of films using PEG as plasticizer have been highlighted in further studies [172]. Hydrophobic plasticizers, such as triacetin [173, 174], triethyl citrate [175], and vegetable oils [176, 177], improve the barrier properties of alginate films by creating a more hydrophobic environment, which reduces water uptake and WVP. These plasticizers work by blocking voids and forming a tortuous path that hinders moisture diffusion. For example, the incorporation of essential oils such as cinnamon [178, 179] not only imparts antimicrobial properties but also enhances the hydrophobicity of the films, reducing WVP and improving shelf life for food packaging applications. However, in most research studies, glycerol or other low molecular weight plasticizers were the main plasticizers used, while oils were introduced to control hydrophobicity and barrier properties, as well as the bacterial and microbial activity of the films, and they also had a plasticizing effect, usually decreasing tensile strength and increasing elongation at break of the polymer.

The choice of plasticizer significantly affects the properties of alginate-based films. Hydrophilic plasticizers such as glycerol and PEG enhance flexibility but can increase WVP due to their hygroscopic nature. Hydrophobic plasticizers can be added to improve barrier properties by reducing water uptake and enhancing rigidity.

Further common possibilities to alter the properties of alginate films incorporate crosslinking by divalent cations, enhancing mechanical strength and decreasing WVP. Agents such as calcium chloride reduce swelling in solvents, improving structural integrity. The blending of alginate with bio-based polymers (pectin [169, 180], lignin [181], carrageenan [182], starch [183]) results in films with enhanced flexibility and mechanical properties. Often these macromolecules act as plasticizers, reducing alginate–alginate interactions and improving flexibility. Further bio-based materials for incorporation in these films encompass nanomaterials, such as cellulose nanocrystals (CNC) [172]. The reinforced films show improved mechanical strength and reduced WVP.

The negatively charged backbone of alginate can further be used for the preparation of composite films with chitosan. The blending of the two creates a stable polyelectrolyte complex due to the electrostatic attraction. This interaction leads to the formation of a dense network that enhances the film's structural integrity, making it less permeable to water vapor and gases [165]. These composite films exhibit superior mechanical properties, including high tensile strength and reduced solubility in water (compared to the individual materials), which are crucial for their application in food packaging and biomedical fields. The individual biopolymers can thereby be applied in a layer-by-layer assembly [184]. The primary challenge is the inherent brittleness and lack of flexibility in some formulations. Moreover, the dependence on the precise balance of electrostatic interactions means that any changes in the pH or ionic strength of the surrounding environment can significantly affect the film's properties and performance [165].

Interfacial properties in adhesives are mainly driven by the ionic interactions, hydrogen bonding, the possibility of metal-ion complexation, and in some cases covalent bonding. As discussed above, the carboxyl groups can interact with divalent cations, forming ionic crosslinks that create a stable three-dimensional bio adhesive, which has enhanced mechanical integrity and adhesion strength. Additionally, the hydroxyl groups in alginate can form hydrogen bonds with amino and hydroxyl groups present on the tissue surface, increasing adherence and facilitating a robust attachment. In some modified forms oxidized alginate bearing aldehyde groups can form covalent bonds with amino groups on the tissue surface, leading to strong, stable adhesion [185]. In this context alginate has gained considerable attention as a muccoadhesive material in the biomedical field [162, 186, 187]. Despite the electrostatic repulsion between the negative charges of alginate and mucin, adhesion still works because bio adhesion is achieved through intra- and intermolecular hydrogen bonds that form during the wetting, swelling, and interpenetration of the mucin with the polymer chains [162]. The presence of carboxyl and hydroxyl groups in alginate further allows for hydrogen bonding and electrostatic interactions with drug molecules, enhancing muccoadhesion and prolonging drug residence time in targeted areas such as the gastrointestinal tract [162].

While alginate is predominantly hydrophilic due to its charged polysaccharide backbone, the chains of alginate can exhibit hydrophobic interactions under certain conditions, particularly when associated with specific ions. Making use of this, oil-in-water (O/W) Pickering emulsions can be generated using alginate's amphiphilic nature. This allows it to migrate to the oil–water interface, where it can undergo *in situ* crosslinking upon the release of further calcium ions [188, 189]. This crosslinking process forms a polysaccharide shell around oil droplets, significantly enhancing the physical stability of the emulsions against environmental stresses such as centrifugation and freeze–thaw cycles. The ability of alginate to form strong hydrogen bonds and ionic interactions at the interface further contributes to the stability and functionality of these emulsions [189]. Moreover, the crosslinking of alginate at the interface of water-in-water (W/W) Pickering emulsions has been shown to enhance the structural and rheological properties of the emulsions. The addition of sodium alginate and calcium carbonate ($CaCO_3$) to these emulsions results in the formation of a dense, porous network upon crosslinking, improving the viscosity and reducing droplet size, which collectively contribute to the stability and efficacy of the emulsion [188].

Alginate stands out as a versatile polyelectrolyte in the field of polysaccharides due to its negatively charged carboxylic acid groups and unique structural properties. These carboxylic acid groups are most characteristic for this polysaccharide and its interactions with its surroundings. Its hydrophilic interfacial characteristics and multiple adhesion sites make it valuable for biomedical applications. The ability to modulate its hydrophilic properties through the complexation of multivalent metal ions is key to various approaches, including barrier films, emulsions, and the artificial generation of microcapsules. This versatility makes alginate promising for applications beyond its primary use in the food industry.

1.2.5 Hemicellulose

Hemicelluloses are a diverse group of polysaccharides found in the cell walls of plants, and represent a significant and renewable source of biopolymers. They are present in both primary and secondary cell walls, spanning various plant tissues and species [190]. Structurally categorized into four main groups—xylans, mannans, β-glucans with mixed linkages, and xyloglucans—hemicelluloses exhibit less crystallinity compared to cellulose due to their variable sugar monomer composition and branching patterns [191]. This lack of crystallinity makes hemicellulose much more water soluble and allows a more facile chemical modification of hemicellulose. In addition it represents the most hydrophilic and most unstable polymer of wood, leading also to the lowest thermal resistance [192]. As a highly functionalized polymer that contains many hydroxyl groups at the C-2, C-3, or/and C-6 positions, its surface properties are determined to be highly hydrophilic.

The inherent structure of hemicelluloses allows them to act as a connecting link between lignin and cellulose in lignocellulosic biomass, playing a crucial role in providing structural support and enhancing the flexibility of plant cell walls. They form a matrix that binds with cellulose fibers and lignin, contributing to the wall's integrity and strength. Hemicellulose interacts with cellulose through hydrogen bonds and van der Waals forces due to abundant hydroxyl groups along its backbone. The equatorial β-1,4 linkages in hemicellulose enable alignment with cellulose fibers. Additionally, hemicellulose interacts covalently with lignin, particularly through ferulate-mediated radical coupling, forming strong cross-links that enhance cell wall stability (figure 1.8) [193]. Hemicelluloses further affect the porosity and water retention properties of the cell wall, facilitating interactions among different cell wall components and helping plants adapt to environmental stress [190].

The polymeric, amorphous, and hydrophilic properties of hemicelluloses make them ideal for academic applications in materials such as biodegradable material in films, and hydrogels. However, commercially, hemicelluloses are primarily used as additives and ingredients rather than standalone products. Their primary applications in commercial production are as strengtheners in paper products, essential ingredients in bio-based goods, chemicals, and biofuel, functional additives in the food and beverage industry, and for fiber modification in textiles [194]. In addition to the mentioned applications, it has to be mentioned that in the conventional kraft pulping process, hemicelluloses are typically degraded into oligosaccharides or monosaccharides, dissolved along with lignin in black liquor, and subsequently burned as fuel, resulting in poor utilization of these valuable biopolymers at the moment [195].

As already mentioned, hemicelluloses find applications as strengtheners due to their ability to facilitate fiber bonding [196]. They significantly influence intrinsic fiber properties, such as swelling, fibrillation, bonding ability, and hornification tendency. It was shown that xylan can increase fiber tensile strength and flexibility by acting as a natural spacer that minimizes the interfibrillar cross-linking of

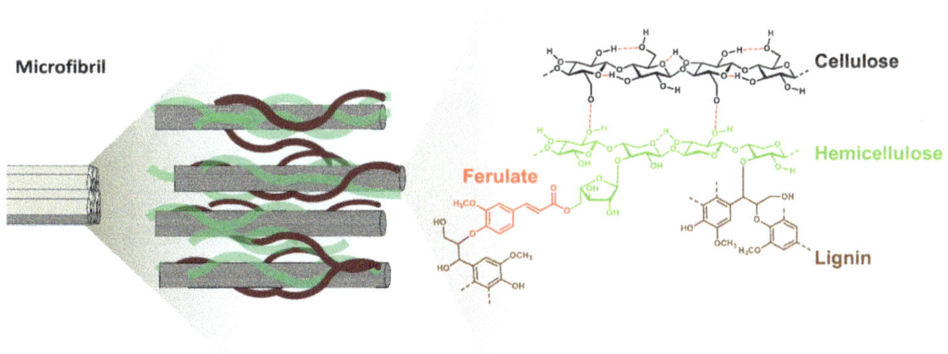

Figure 1.8. Function of hemicellulose connecting lignin and cellulose in lignocellulosic biomass.

cellulose, thereby maintaining the specific surface area and flexibility of fibers [197]. This already shows that hemicellulose exhibits an interesting interfacial behavior, especially with lignocellulose materials, where it interferes with the surface of the respective materials.

In terms of interfacial interactions, the side groups of hemicellulose can form various types of bonds with the free hydroxyl groups on the surface of cellulose fibrils. These interactions induce specific shear behaviors, which affect the overall mechanical properties and stability of the cell wall. Consequently, the nature and structure of hemicellulose play a critical role in determining these interfacial interactions, significantly impacting the robustness and functionality of lignocellulosic materials [196]. The interactions of hemicellulose are being used in the paper industry, where some remains of hemicellulose are left in the fibers to improve the strength of the paper products [197]. In a recent study by Pękala *et al*, the influence of hemicellulose molecular weight and sidechain composition on the adsorption to cellulose surfaces was investigated. In accordance with other studies, they found that polysaccharide adsorption is driven by entropy changes due to water displacement, rather than hydrogen bonding [198, 199]. Thus lower molecular weight polysaccharides adsorbed more effectively. Sidechains with fucose or galactose showed enhanced binding through better fit and hydrogen bonding, while neutral sugar sidechains increase interaction via hydrophobic and van der Waals forces. A high substitution can cause steric hindrance, whereas light substitution allows better adsorption (figure 1.9). In addition, lower concentrations caused chain rearrangement, while

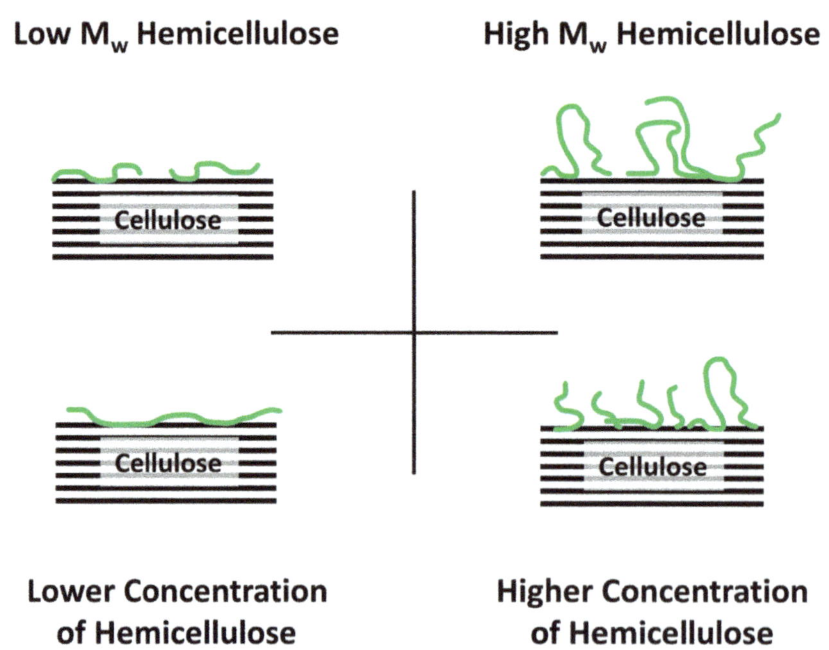

Figure 1.9. Schematic of hemicellulose adsorption onto cellulose, influenced by the molecular weight and concentration of hemicellulose in the solution. (Adapted from [199]. CC BY 4.0.)

higher concentrations led to loops and strings, indicating different adsorption mechanisms. Overall, hydrophobic forces primarily drive these interactions, with hydrogen bonds and van der Waals forces providing stabilization [199].

The interesting interfacial properties of hemicellulose endow it as a promising stabilizer for emulsions due to its stability across a wide pH range, ion strengths, and thermal conditions [200]. However, its limited natural amphiphilicity in contrast to cellulose, which on the basis of different crystal planes exhibits amphiphilic behavior, restricts its emulsifying properties. Enhancing the amphiphilicity of hemicellulose can be achieved through two main methods: the introduction of lignin by regulating the isolation method and molecular modification. Referring to the first approach it was found that that crude extracts of hemicellulose are more effective in emulsion stabilization than purified ones, challenging the necessity for intensive purification [201]. The heterogeneous composition of these extracts contributes to their enhanced functionality, combining the beneficial and complementary attributes of various compounds [201]. Hemicelluloses, such as galactoglucomannan (GGM) and glucuronoxylan (GX), provide robust steric stabilization of emulsions under diverse pH and ionic strength conditions. The fusion of lignin with hemicelluloses leads to the so-called lignin carbohydrate complex (LCC) [202]. These complexes form through covalent and non-covalent interactions, and exhibit naturally occurring amphiphilic macromolecules [203].

The chemical modification of hemicelluloses to form macromolecules with amphiphilic properties typically involves the introduction of various hydrophobic fatty acids by esterification [204]. For an application as emulsifier it was shown that the carboxymethylation modification of hemicelluloses can increase the surface charge, enhancing the emulsifying property [205]. Further it was shown in another publication that the emulsifying effect of modified hemicellulose can be steric repulsion assisted by Pickering-type stabilization [206]. Understanding these interfacial interactions is one key to optimizing hemicellulose for industrial applications [207].

The adsorption of hemicelluloses and their derivatives onto biopolymer surfaces, particularly cellulose, have been studied extensively [197, 208–210]. Key factors influencing this adsorption include hemicellulose molar mass, conformation, substituting units, and degree of substitution. Studies indicate that higher molar masses generally enhance adsorption due to increased aggregation tendencies, while substituting units impact solubility and interactions with cellulose. For example, arabinose substitutions in arabinoxylans facilitate adsorption by promoting aggregation, whereas acetyl groups on xylans hinder adsorption by causing steric hindrance [211, 212]. Fucose units in xyloglucans positively influence cellulose interaction, whereas galactose units in glucomannans significantly affect adsorption independently of molar mass [208–210, 213].

Overall, hemicelluloses, as hydrophilic and amorphous biopolymers, show great promise for interfacial applications such as emulsification and surface modification, particularly for lignocellulosic materials. Their high hydroxyl group content facilitates strong interactions at interfaces. Although naturally limited in amphiphilicity, hemicelluloses can be chemically modified to improve their emulsifying properties. Modifications, such as introducing hydrophobic groups or incorporating lignin, enable tailored functionalities that enhance performance in emulsifying systems and expand their application potential across various industries.

1.3 Phenolic based biomaterials

Compared to other bio macromolecules such as polysaccharides the general structure of phenolic biomaterials (polyphenols and phenolic polymers) is rather difficult to describe. Phenolic compounds are known for their antioxidant properties and are widely found in plants. The main phenolic compounds by mass in woody plants are lignins and tannins. While lignins are often considered waste materials in the pulping process used for paper fabrication, tannins are typically obtained as a desired product through the extraction of tannin-rich plant materials. This is also one of the main reasons why more than 95% of lignin is used as low-value fuel [214].

Lignins and tannins have long been used to replace phenol in thermosetting materials. Herein they are often used to substitute building-blocks in (poly)phenols as phenol-derivative [215]. Phenol represents an important petrochemical component used in phenol formaldehyde (PF) resins and is also a key precursor for bisphenol-A (BPA), which is essential in the production of polycarbonates and epoxy resins. Due to the health risks and environmental concerns associated with BPA, the interest in alternatives is particularly relevant for the development of PU-based systems within the industry [216, 217]. However, beyond the potential to replace phenolic building blocks in thermoplastics and thermosets, the inherent properties of tannins and lignins—such as their strong antioxidant, antimicrobial, and anti-inflammatory effects—make them valuable in sustainable materials science [214].

The interfacial interactions of phenolic compounds are primarily determined by their surface chemistry, characterized by phenolic structures bearing multiple hydroxyl (–OH) groups attached to aromatic rings. These large aromatic systems contribute to π–π interactions with surfaces possessing aromatic character, such as graphene or graphite. Additionally, their high reactivity enables the formation of hydrogen bonds with these surfaces.

1.3.1 Lignin

Lignin, a major by-product of the pulp industry, is a large-scale raw material derived from woody plants, constituting about one-third of their biomass [218]. In biological systems, lignin provides structural integrity and rigidity to cell walls, aids in mineral transportation, and offers protection against pathogens [219]. Unlike other biopolymers such as cellulose and starch, lignin has a complex, cross-linked polymeric structure that is difficult to define precisely, as isolating native lignin without altering its chemical structure is challenging [220].

This intricate structure is composed of phenolic polymers interconnected by various bonds, including ether, carbon–carbon, and ester bonds, with β-O-4′ ether linkages comprising more than 50% of the connections between monomers [221]. The primary building blocks of lignin are the *p*-hydroxyphenyl, guaiacyl, and syringyl units, linked by carbon–carbon and ether bonds (figure 1.10). The ratio

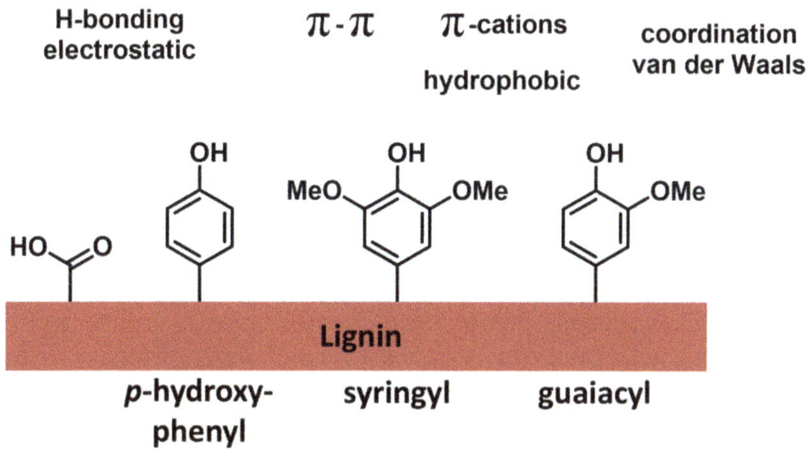

Figure 1.10. Overview of lignin's key surface groups and their corresponding interfacial interactions.

of these monomers varies among plant species, ages, and different parts of the plant, such as leaves and stems [219]. Lignin's molecular architecture features diverse functional groups, including methoxy, phenolic hydroxy, alcoholic hydroxy, and carbonyl groups, each in varying proportions, significantly influencing its properties [222].

Depending on the industrial isolation process, commercially available lignin is categorized into lignosulfonates, kraft lignin, organosolv lignin, and soda lignin, each with distinct properties. Soda lignin, produced with sodium hydroxide, is high in purity and free of sulfur contamination. Lignosulfonates, on the other hand, are water-soluble and contain sulfur in the form of sulfite ions. Kraft lignin, the most commercially dominant, involves breaking down ether and ester bonds, creating new functional groups and making the lignin soluble in basic conditions, though it typically contains sulfur and other impurities. Organosolv lignin, produced using organic solvents, is of high quality and low polydispersity but faces technological and economic challenges. These variations in lignin types and their respective surface chemistries influence their suitability for different applications and the extent of additional processing required [223].

The complex phenolic structure of lignin endows it with various beneficial properties, including UV protection, antioxidant activity, and antimicrobial effects. The phenolic hydroxyl groups enable free radical scavenging, enhancing its antioxidant capabilities and contributing to its antimicrobial properties. Additionally, the polyaromatic backbone of lignin allows for extended absorption of sub-visible wavelengths of light, underpinning its UV protective qualities [224]. However, it should be noted that this same ability makes lignin prone to UV degradation.

Lignin's three major functional groups—hydroxyl, phenolic, and carboxyl—enable its modification for various applications. Hydroxyl groups facilitate grafting reactions such as phosphorylation, sulfonation, esterification, and amination, which alter lignin's inherent properties. Phenolic groups are explored for replacing phenols in resins, coatings, and foams [225], allowing lignin to act as a macromonomer in the production of polyurethanes [226, 227], polyesters [228], epoxy resins [229], and phenolic resins [230, 231]. Other studies examine the direct processability of technical lignin without further chemical modification [232, 233]. Due to its inherent thermoplastic properties and a glass transition temperature between 120 °C and 190 °C [234], lignin can be processed into various materials, such as hydrophobic, polar, charged, and functionalized compounds [235].

The chemical structure of lignin fundamentally influences its surface properties. As a polyaromatic material, lignin is less hydrophilic than polysaccharide-based biopolymers [236]. However, due to the diverse functional groups and chemical structures within lignin, its polarity can vary significantly, allowing it to exhibit either hydrophobic or hydrophilic characteristics [237]. As already mentioned, the isolation process plays a crucial role in determining lignin's hydrophilic behavior. For instance, lignosulfonates, despite their high molecular weight, are water-soluble [238], whereas kraft lignin tends to be hydrophobic [239]. Lignin's functional groups, such as sulfonates, phenolates, and carboxylates, not only affect its water solubility but also enable various non-covalent interactions, including hydrogen bonding, π–π stacking, electrostatic interactions, and hydrophobic interactions (figure 1.10). The amphiphilic nature of lignin, due to aliphatic linkages between aromatic groups, allows it to interact with a wide range of matrices and surfaces, from amphiphilic carbonaceous materials to polar oxides. Additionally, lignin's ability to interact with multivalent metal ions and carbohydrates reflects its natural role in plant cell walls [240]. The surface properties of lignin, including wettability, have been extensively investigated through the preparation of thin films using various techniques such as spin-coating and Langmuir–Blodgett [220, 241–246]. These studies suggest, that lignin can be used to formed continuous and smooth films when prepared via spin-coating, exhibiting rather hydrophilic surface properties, with contact angles between 40° and 60° [220]. Another publication pointed out that even though lignin coated surfaces cannot be considered strictly hydrophobic, exhibiting contact angles below 90° and a surface energy comparable to that of cellulose, it can effectively reduce the wettability and water uptake of wood-based products, enhancing their water resistance [247]. This further supports the role of lignin in providing water-proofing in the wood cell wall. The surface-modifying properties of lignin make it a highly promising candidate for coating applications. Investigations into lignin's use as a surface coating have demonstrated its potential across various substrates, with wood being a predominant focus [248–253]. Research by Henn *et al* highlighted the application of lignin nanoparticles as wood coatings, offering exceptional resistance to UV irradiation, abrasion, solvents, and stains. Their study introduced amphiphilic lignin-based colloids to create water-based, solvent-free, multi-resistant surface coatings for wood and metal, achieving durability and breathability without a binding matrix [253].

One of the major challenges in utilizing lignin across various fields is activating its abundant functional groups. Commercially available lignin typically has high molecular weights and a limited number of vacant reactive sites due to substantial condensation during extraction from biomass [254, 255]. Efforts to enhance lignin's reactivity have included methods such as methylolation (hydroxymethylation) [256, 257], amination [258, 259], and phenolation [260, 261]. In the field of adhesives, hydroxymethylation and phenolation are widely used due to their cost-effectiveness and ability to increase the number of reactive sites, thereby improving the bonding strength and curing speed of lignin-based adhesives [262]. Additionally, demethylation, often achieved through enzymatic or chemical means, enhances lignin's reactivity by increasing the number of phenolic hydroxyl groups available for cross-linking with other active adhesive components [262, 263]. A recent study by Yang *et al* explored the potential of using lignin directly as an adhesive, leveraging its inherent surface properties in an uncondensed form. By separating uncondensed or slightly condensed lignins from biomass and applying a suspension of lignin and water as an adhesive during the hot pressing of wood veneers, the study demonstrated lignin's strong binding capacity without the need for further chemical modification [264].

The application of polyphenolic polymers as a resource for carbon-based materials such as carbon fibers, spheres, and foams represents a another promising way towards high-value utilization of phenolic biomass with interesting surface properties, particularly in fields such as energy storage and environmental applications [265]. Due to their phenolic structure, high carbon content (>60%), aromaticity, and abundant oxygen functional groups that offer good tunability in chemical structure, lignins and tannins are considered as a valuable resource for the next generation of renewable activated carbons (ACs) [265, 266]. These ACs with high surface area and hierarchical porosity can be synthesized from lignins and tannins using chemical activation methods such as treatment with KOH or $ZnCl_2$, or K_2CO_3 [267, 268]. The intrinsic phenolic structure of these materials enables thus the formation of robust and highly porous carbon networks with high surface area. By using processes such as melt-blowing and carbonization, lignin-derived carbon fibers (LCFs) can achieve specific surface areas as high as 923 $m^2\ g^{-1}$ [269]. Additionally, these materials demonstrate excellent conductivity, making them suitable for use in energy storage devices such as batteries [265, 267, 269]. The properties of these materials can further be tailored to be either hydrophilic or hydrophobic, depending on the functional groups present and the processing methods used. While higher carbonization temperatures tend to reduce the number of hydrophilic functional groups [265], leading to a more hydrophobic material, a higher hydrophilicity can be achieved through specific chemical activation. For instance, activating lignin with phosphoric acid can increase the presence of phosphate groups, enhancing hydrophilicity. Conversely, treatments that introduce more aromatic carbon structures tend to increase hydrophobicity [270].

1.3.2 Tannins

Tannins are explored for use in eco-friendly adhesives, coatings, and resins. Unlike lignin, tannins are primarily found in the tender tissues of plants, such as needles, leaves, or bark, from which they are extracted [271]. Conventionally, tannins are extracted from sources such as barks, wood chips, and leaves through mechanical processes such as milling and grinding. Common solvents for extraction include water or aqueous solutions of sodium salts, and various solvent mixtures such as acetone/water and methanol/water, while the latest state-of-the-art extraction methods include supercritical fluid extraction (SCE), pressurized water extraction (PLE), ultrasound-assisted extraction (UAE), and microwave-assisted extraction (MAE) [272]. Historically, tannins derived their name from their primary use in the tanning process, where animal skins are transformed into leather through the application of tannin-rich aqueous extracts from plants. However, their modern applications extend far beyond tanning [271]. Tannins can be structurally categorized into two groups: hydrolyzable and condensed tannins. Condensed refers to the polymerization or linking together of smaller phenolic units to form larger, more complex structures. This process typically involves the formation of multiple intermolecular bonds, such as ether or ester linkages, between the phenolic units. While condensed tannins, known for their higher molecular weight and greater structural stability, are less reactive but more suited for applications requiring durability, such as adhesives and coatings, hydrolyzable tannins are more reactive due to their simpler and more hydrolyzable structure and are thus often utilized in applications such as leather tanning and medical formulations [273]. Condensed tannins, comprising up to 90% of commercial tannin production, consist of oligomers and polymers of flavonoid units within the tannin family. The degree of polymerization varies among different tannin extracts; for instance, mimosa and quebracho tannins exhibit a high degree of polymerization, while pine and gambier tannins feature medium and low degrees of polymerization, respectively [274].

The interfacial interactions of tannins are guided by their chemical structure, consisting of multiple aliphatic hydroxyls and phenolic hydroxyl groups. These functionalities equip tannins with the ability to form complexes with proteins and polysaccharides as well as to interact with inorganic salts. They show a strong tendency to adsorb and reduce metal ions [275], while at the same time non-covalent π–π interactions can occur between the tannins' aromatic rings with aromatic systems such as graphene oxide [276]. The strong interaction between tannins and proteins/enzymes is related to their phenolic sites which crosslink with proteins by non-covalent and covalent bonding. While many factors influence the interaction of tannins with proteins, such as the isoelectric point of proteins, pH, and ionic strength, a key factor is the presence of proline-rich proteins (PRPs). These proteins react strongly with tannins, forming robust complexes [277].

The broad scope of interactions tannins can express can be used in multiple fields beyond protein crosslinking. In contrast to the previously discussed materials, tannins are particularly suitable for forming composite materials through their interactions with other materials, while they are less often used as homogeneous materials. They exhibit remarkable potential especially in enhancing the mechanical properties of various composites while imparting additional functionalities. One notable advantage is their ability to form physical bonds, leading to improved mechanical properties in materials such as elastomers. These enhancements, sometimes up to 190% in toughness and 200% in strength, are comparable to those achieved with commercial fillers but at significantly lower loadings [278]. Moreover, tannins can be utilized in the creation of flexible hydrogel networks with enzyme-mimetic catalytic crosslinking, offering versatility in applications such as nanocomposites with natural rubber.

In addition to mechanical enhancements, blending tannins into polymers can equip them with added properties. For instance, tannins' radical scavenging nature grants materials resistance to oxidative species and enhanced UV-resistance [279]. The incorporation of tannins also leads to the development of materials with antioxidant properties, making them suitable for various applications, such as multilayered films and solvent cast membranes [280]. Furthermore, tannins can induce morphological changes in polymers, influencing properties such as crystallization behavior and polymer morphology, due to their strong physical interactions with the polymers [281]. Tannins also offer opportunities for the fabrication of coatings and polyelectrolyte complexes with diverse functionalities, ranging from metal adsorption to drug delivery.

Due to their exceptional interfacial interactions, tannins are considered a promising candidate for next-generation sustainable adhesives, especially for wood-based materials. They can serve as a renewable substitute for petroleum-based phenol and resorcinol-containing wood adhesives [272]. Tannin-based adhesives utilize condensed tannins, which react with active compounds such as formaldehyde to form resins. These condensed tannins, rich in catechol and pyrogallol groups, exhibit high reactivity. Unlike lignin-based adhesives, tannin adhesives can undergo self-condensation and form stable, water-resistant bonds under alkaline conditions [282]. The free hydroxyl groups in tannins facilitate hydrogen bonding with wood substrates, enhancing adhesive strength. Additionally, tannins can be modified with compounds such as furfuryl alcohol, glyoxal, and polyethylenimine to improve their thermal stability, moisture resistance, and mechanical properties, thus offering a sustainable and effective solution for wood adhesives [272].

These versatile properties make tannins valuable candidates for creating advanced hybrid materials with a wide range of applications. From their role in catalytic applications to their use in stimuli-responsive materials and 3D/4D printing inks, tannins continue to demonstrate their importance in various fields of science and technology. Overall, the unique characteristics and broad applicability of tannins suggest a promising outlook for future research and development in emerging technological areas.

1.4 Protein based biomaterials

Proteins are one of the fundamental building blocks of nature, fulfilling various functions in living organisms. They are essential for the structure, function, and regulation of tissues and organs, with structural proteins playing a crucial role in maintaining tissue integrity [283]. In contrast to the previous polymeric materials, proteins represent a highly diverse class of bio-based materials due to the varying sequence and properties of their monomeric units, amino acids.

Based on their origin, proteins can be categorized into animal proteins, such as silk, keratin, collagen, elastin, resilin, and reflectin, and plant proteins, including corn zein, soy, and wheat gluten [284, 285]. Additionally, there are protein peptides derived from recombinant biotechnology, although this section will focus on naturally occurring animal and plant proteins. The early use of these proteins in bio-based materials is still evident today in products such as silk and wool textiles and leather goods [285]. The unique properties of these materials, often challenging to replicate with synthetic alternatives, contribute to their ongoing relevance in the modern market.

Proteins are unique in their amphiphilic nature, containing both hydrophilic and hydrophobic domains. This dual characteristic and the interaction with water of these domains is fundamental to their ability to self-assemble into complex structures and interact dynamically with their environments [286]. The hydrophilic regions typically contain amino acids with polar side chains, while the hydrophobic regions are composed of non-polar amino acids. This amphiphilicity is crucial for the formation of various protein structures and their subsequent properties and function in biological systems. The origin of these properties lies in the protein's amino acid sequence. Each protein's specific sequence dictates its folding pattern and the resulting three-dimensional structure. This sequence–structure relationship determines the protein's functionality, stability, and interaction with other molecules.

Following from their sequence and interactions to their surroundings, proteins exhibit a hierarchical structure. The hierarchical organization is mainly responsible for determining the final mechanical properties of protein-based materials [287]. For instance, the β-sheets in silk fibroin provide exceptional tensile strength and elasticity, making silk a highly sought-after material for various high-performance applications [288, 289]. Similarly, collagen, which is abundant in connective tissues, has a hierarchical structure where three α-helical chains form a triple helix, which then assembles into fibrils and fibers [290]. This organization imparts collagen with remarkable tensile strength and resistance to stretching, making it essential for applications in medical implants and tissue engineering.

The interactions and interfacial properties of proteins are fundamental to their functionality in biological systems. Most proteins share the common ability to interact with a wide range of molecules, including other proteins, nucleic acids, small molecules, peptides, carbohydrates, and fatty acid chains. These interactions are essential for their activities within the crowded cellular environment, ensuring specificity and efficiency [291]. Proteins can adsorb onto surfaces, mediate cell attachment, and facilitate molecular recognition processes [292]. They can form stable interfaces with a wide range of substrates, including metals, polymers, and biological tissues. This remarkable versatility arises from the diverse chemical functionalities of amino acid side chains, which enable the formation of various non-covalent interactions such as hydrogen bonds, ionic bonds, and hydrophobic interactions [291].

Proteins' ability to undergo conformational changes in response to environmental stimuli adds another layer of functionality. This capability allows proteins to become more hydrophilic or hydrophobic on demand, which is particularly useful in creating smart materials that adapt to changing conditions, such as pH-responsive drug delivery systems or temperature-sensitive materials [293]. However, it should be noted that these changes are often irreversible, resulting in a one-directional response.

The diverse sequences of amino acids and resulting structures of proteins enable a wide range of material properties. Silk fibroin's β-sheet structure provides high tensile strength and elasticity [288, 289], while proteins used in film formation create effective barrier systems for packaging and coatings. Collagen's structure offers remarkable tensile strength and extensibility, making it ideal for biomedical scaffolds [294]. Additionally, proteins' tunable hydrophilicity allows for responsive interactions with water, enhancing smart materials [293].

Deriving from this interesting pool of diverse biopolymers, protein-based materials hold special promise in that they can be tuned and altered in their chemical composition as well as their morphology. As polymeric materials, they hold great promise in replacing polymer-based fibers [295], packaging (films and coatings) [296], adhesives [297], air and water filtration [298], creating high-performance materials. The importance of interfacial properties, including adhesion to various surfaces and interaction with molecules, further underscores the versatility and potential of protein-based materials as sustainable alternatives to synthetic materials.

1.4.1 Fibers

Protein-based fibers exhibit a remarkable combination of tensile strength and extensibility. These polymeric materials have sometimes remarkable properties which exceed the limits of what can be reached by conventional polymers. For example, silk, known for its outstanding mechanical properties [299], has seen a surge in artificial production through biotechnology in the past decade [300]. Fibrous proteins such as collagen, keratin, and silk fibroin are valued in textiles for their strength, flexibility, and aesthetic qualities. Their mechanical properties stem from the structural compositions of multiple tandem repeats of short amino acid sequences with hydrophobic and hydrophilic domains, such as the polyalanine (contributing to high tensile strength) and glycine-proline regions (facilitate elasticity through hydrogen bonding between crystalline β-sheets) in spider silk [301].

The most important proteins for protein based fibers are collagen, keratin, and fibroin.

Keratin, found in wool, hair, and feathers, is known for its strength and resilience. The high sulfur content from cysteine residues forms disulfide bonds, enhancing durability and elasticity. These properties make keratin ideal for fiber production. Recently keratin-based nanofibers produced via electrospinning have received attention. Ramirez *et al* demonstrated that keratin or keratin composites can be spun from aqueous or non-aqueous solutions, resulting in fibers supporting cell growth and tissue engineering applications due to their similarity to the extracellular matrix [302].

Silk fibroin, present in silkworm and spider silk, offers high tensile strength and elasticity. Fibroin's strength, flexibility, and smooth texture make it valuable for high-performance textiles, surgical sutures, and tissue engineering scaffolds.

The fibers resulting from fibron have a substructure of oriented β-sheet-rich nanofibrils (90–170 nm in diameter) which are cross-linked by a disulfide bond [303]. It was recently shown that the combination of fibron with graphene oxide or carbon nanotubes significantly enhances the mechanical strength and electrical conductivity of the resulting fibers. These composites are particularly useful in creating flexible electronic devices and sensors [304, 305].

Collagen, the main protein in leather and connective tissues, provides flexibility and biocompatibility. Its triple-helix structure assembles into fibrils and fibers, offering high tensile strength and toughness. In addition to collagen, keratin, and fibroin, other prominent proteins used in fiber fabrication include elastin and resilin. Elastin is known for its exceptional elasticity and resilience, making it ideal for applications requiring flexible and stretchable fibers. Due to its origins near the extracellular matrix, elastin is often a choice for vascular grafts and skin substitutes and other materials coming into contact with blood [306, 307].

The fabrication of protein-based fibers can be categorized into two main approaches: regenerated protein fibers and recombinant protein-based fibers. Regenerated protein fibers are produced by dissolving proteins in solvents and spinning them into fibers using techniques such as electrospinning, wet-spinning, and dry-spinning [287]. This method replicates the hierarchical structures of natural fibers. The fabrication of recombinant protein-based fibers, on the other hand, involves genetic engineering to produce the desired proteins, which are then purified and spun into fibers [287].

Protein-based fibers have long been used by mankind and remain highly valuable products today. Their inherent properties, such as mechanical performance, biocompatibility, and biodegradability, make them highly interesting for the field of bio-based materials. One particularly promising application of protein fibers is in the field of filtration. Protein-based materials can be engineered to create highly efficient filtration systems for air and water purification. The following sections will explore how these natural materials are utilized in developing advanced filtration technologies.

1.4.2 Filtration

As an application field in which interfacial processes have specific importance, the filtration of air and water is an emerging area related to bio-based protein materials. The rapid development in recent years of various forms of protein-based membranes with nanometer-scale thickness and unique membrane structures already enable high-flux, efficient and selective separation [308]. Protein membranes, particularly those derived from abundant natural sources such as soy protein [309], zein [310], gelatin [311], and silk [312], have shown significant promise in both liquid and gas filtration due to their unique surface and interfacial properties [313].

In liquid filtration, proteins such as lysozyme [314], bovine serum albumin (BSA), and ferritin are utilized to create nanofilm coatings with controllable thickness and morphology, essential for achieving high water permeance and solute rejection. These protein-based membranes exhibit functional groups such as hydroxyl (–OH), carboxyl (–COOH), and amine (–NH_2), which interact strongly with various pollutants, enhancing the membrane's ability to capture particulate matter and toxic chemicals dissolved in water [298].

Amyloids, aggregates of proteins with characteristic fibrillar morphology, serve as templates or building blocks in separation membrane materials for wastewater treatment applications [315]. Functional amyloid fibers produced by *Escherichia coli* are integral to biofilms and exhibit strong adhesion properties, which enhance their stability and functionality [316]. These amyloid fibrils, typically 7–10 nm in diameter and about 1 μm in length, form robust structures through noncovalent interactions. Proteins such as β-lactoglobulin, can be converted into amyloid fibrils with significant mechanical strength and adhesiveness [317]. Leveraging these properties, hybrid membranes have been developed that efficiently adsorb heavy metals and radionuclides, demonstrating a versatile ion-binding capacity [317]. Additionally, attaching metal nanoparticles to β-lactoglobulin fibrils creates membranes that act as efficient catalytic materials in water purification [318]. The high mechanical strength and strong adhesion to various substrates make amyloid-based membranes particularly effective. However, the complexity of their production currently limits commercial implementation.

In gas filtration, the surface chemistry and functional groups of protein-based materials play a critical role, since the pollutants are typically too small to be captured by size exclusion. In this perspective instead of conventional physical or size-based filtration mechanisms, intermolecular interactions are used to adsorb pollutants from air [310]. Proteins such as soy protein and zein can form nanofibers that expose hydrophobic groups, which are effective in adsorbing non-polar substances such as oils and organic chemicals through hydrophobic interactions [319]. Typically denaturation of the proteins increases the exposure of hydrophobic groups such as alkyl (–CH_3) and benzene rings. Lin *et al* used this process to create a filter system based on denatured zein nanofibers for the effective filtration of oils and organic chemicals [319]. Polar and reactive pollutants on the other hand can be separated by (charged) functional groups present on proteins. An example can be found in soy protein, rich in amine groups, which interacts effectively with dissolved toxic chemicals such as formaldehyde, facilitating their capture and removal from air [320]. Further, cross-linked gelatin nanofabrics were shown to interact with airborne pollutants. Due to the presence of particular functional groups in the structure of gelatin molecules, specifically the positively charged amines, nanofabrics can disrupt the cell walls of bacteria and inactivate viruses. This provides antimicrobial properties essential for creating hygienic air filtration systems [321].

In addition to small molecules, the separation of particulate matter (PM) is also of interest in the field of air filtration. The ability of proteins to form nanofibers with large surface areas and specific functional groups enables the effective capture of airborne particulate pollutants [298, 319, 322]. The electrostatic interactions between charged functional groups on the protein fibers and airborne pollutants are crucial for trapping fine particulate matter.

1.4.3 Barrier materials

In addition to the field of filtration, other barrier materials are represented in the form of packaging materials, where proteins can fulfill an excellent job by being used for edible or non-edible coatings and films. Since proteins show good film forming properties they are applied in protective and packaging films for food. In contrast to fossil-based film materials, some protein based preservation films exhibit the benefit of being edible [323]. The respective films, derived from sources such as zein [324, 325], collagen [326, 327], soy [328], and whey [328–330], offer excellent barrier properties, including selective permeability to gases and moisture. In comparison to other bio-based film materials (such as polysaccharide based films) they generally exhibit higher gas barrier qualities and a better mechanical performance [331]. However, a major drawback of proteins is their susceptibility to moisture. To address this, crosslinking strengthens films by chemically binding proteins (by $CaCl_2$, EDC/NHS, or glutaraldehyde), while adding plasticizers or polymers enhances flexibility and elongation [332, 333]. It should be mentioned that protein-based barrier systems are often integrated into composite materials to enhance their performance. By combining proteins with polysaccharides and lipids, these composites improve strength, reduce water vapor transport, and ensure structural integrity, addressing the limitations of single-material applications.

Similar to the barrier systems discussed earlier, protein-based films can be further enhanced with additives such as antioxidants, antimicrobials, and probiotics to improve food quality and extend shelf life. These active packaging materials not only protect food but also interact with it or the environment to release beneficial compounds, preventing oxidation and microbial growth [334]. Often referred to as 'smart' or 'active' packaging, these materials perform additional functions beyond basic protection. Innovations in this field have led to the development of intelligent packaging systems equipped with indicators that monitor food freshness in real-time, providing consumers with valuable insights into the safety and quality of the packaged food. The chemical mechanisms behind these indicator systems are similar to those discussed in section 1.2.2.

Zein, a prominent protein derived from maize, serves as an efficient coating material due to its relatively low water vapor permeability, grease resistance, and biocompatibility. Due to abundant nonpolar amino acids such as leucine, proline, and alanine, zein exhibits a relatively hydrophobic nature [324]. Its strong interactions with starch make it a viable compound in starch composite films. The blending of starch and zein has garnered attention because they possess complementary properties—starch is hydrophilic, while the zein prolamin protein is hydrophobic—are both biodegradable, and are sourced sustainably. Research by Masanabo *et al* demonstrated that adding NaOH to melt-processed thermoplastic starch–zein composite films significantly improves the compatibility and material properties by reducing zein aggregate size, enhancing plasticization, and increasing elongation at break by about 28%. This enhancement in thermoplastic properties results from the improved interaction between starch and zein, leading to more efficient plasticization [325].

Collagen is valued for its good shrinkage and stretchability, making it an ideal candidate for coatings. However, its application has been limited by poor thermal stability, weak mechanical properties, and high hydrophilicity. Wang *et al* have addressed these limitations by successfully scaling up the production of collagen/sodium alginate blend films, which demonstrate significantly improved mechanical properties, enhanced thermal stability, and better water vapor barrier performance, making them suitable for industrial use [326]. Similarly, Peng *et al* have tackled collagen's inherent drawbacks by developing high-strength collagen-based composite films with regulated interfacial microstructures, resulting in enhanced mechanical strength and barrier properties. These advancements show that collagen-based materials can be effectively engineered to overcome their limitations, paving the way for their broader application in sustainable food packaging [327].

Whey protein is known for its excellent film-forming and oxygen barrier properties, but its use in packaging is limited by poor mechanical strength and water vapor resistance. Song *et al* recently tackled the limitations of whey protein by developing multi-layer films coated with whey protein isolate (WPI), which significantly enhances both the mechanical strength and barrier properties. Their research demonstrated that these WPI-coated films are highly effective in preserving the quality of frozen marinated meat during extended storage [335].

In addition to commonly used proteins, recent advances have been made with more exotic keratin-based films, as demonstrated by Shubha *et al.* They successfully produced and characterized bioplastic films derived from human hair keratin, incorporating novel plasticizers such as ethanediol, diethylene glycol, and triethylene glycol. These keratin-based films exhibit desirable properties, including smooth surface morphology, and strong structural integrity, making them promising candidates for sustainable packaging and other industrial applications as alternatives to conventional plastics [336].

1.4.4 Protein-based adhesives

Protein-based adhesives have emerged as promising alternatives to synthetic adhesives, with their interfacial behavior being crucial, particularly in challenging environments such as wet surfaces.

A prominent and extensively studied system in this field is derived from marine mussels. These mussels exhibit remarkable underwater adhesion through proteins known as *Mytilus* foot proteins (Mfps), which contain the amino acid DOPA [337]. This amino acid facilitates strong binding via catechol chemistry, allowing mussels to displace interfacial water and adhere to substrates through hydrogen bonding and coordination with metal ions [338]. Mfps have been commercialized as adhesives by companies such as ACROBiosystems, based on naturally extracted and recombinant proteins from *Mytilus edulis*. This unique adhesion mechanism is ideal for developing bioadhesives effective in wet conditions, such as surgical adhesives and wound dressings [339].

In addition to these outstanding adhesives, the resolubility of most protein-based adhesives is a drawback that needs to be addressed. There are mainly four ways to reduce the resolubility of protein-based adhesives and enhance their water resistance and bond strength: cross-linking, chemical modification, physical treatments, and additive reinforcement [340]. Cross-linking networks are created using synthetic agents such as polyamidoamine-epichlorohydrin [341] or bio-based agents such as tannic acid [342], which form stable, insoluble networks within the protein structure. Chemical modifications introduce hydrophobic properties through agents such as epoxidized oleic acid [343], decreasing the adhesive's water solubility. Physical treatments applied to the adhesives modify the protein structure, exposing hydrophobic groups and thereby reducing the protein's solubility. Thermal treatment of wheat protein, for example, has been confirmed as an effective method for unfolding the glutenin protein structure, enhancing aggregation and promoting intermolecular disulfide/sulfhydryl exchange reactions [344]. Additive reinforcement with fillers such as sodium montmorillonite [345], lime [346], organic pigments, or tannins enhances mechanical properties and water resistance by creating a robust composite material. Due to its high aspect ratio and unique layered nanoscale structure, montmorillonite is widely employed as a reinforcing and nano-filler material [347]. Additionally, removing hydrophilic components, such as carbohydrates, through alkali treatment or selective precipitation, decreases water absorption and improves durability [340]. These combined methods effectively enhance the water resistance and overall performance of protein-based adhesives, making them suitable for applications where moisture exposure is a concern.

Commonly used proteins in adhesives include casein from milk, collagen, soy protein, blood protein, keratin, wheat gluten, and zein. Casein adhesives are favored for their heat stability and calcium bridging properties, enhancing their mechanical strength [346]. Collagen, as the most abundant structural protein, known for its triple-helical structure [348], provides flexibility and strong interactions between chains, making it effective for bonding. The many polar groups along collagen provide good interaction between the chains and to polar surfaces. Tanning agents can be used to increase the water resistance of collagen adhesives to alter their properties and prevent them from redissolving under humid conditions. Soy protein is widely used due to its availability and adhesive properties. An current example is the application of soy protein isolate in combination with keratin to form strong adhesives for wood, utilizing disulfide bond interactions instead of petroleum-based crosslinkers, which significantly improves its wet shear strength [349]. Blood proteins represent a large waste product in industry, and can be used to produce blood-based adhesives for wood fiberboards [350]. These adhesives are known for their moisture resistance and were historically used to enhance the water resistance of other protein adhesives [351]. The fibrous protein keratin is an interesting material for adhesives. Its high content of hydrophobic amino acids gives it excellent moisture resistance, while its fibrous structure provides strong mechanical properties and durability. It has been used to develop formaldehyde-free wood adhesives that offer good water resistance and mechanical properties. It is also often used in combination with other proteins to enhance the water resistance of the final glue [349]. However, it should be noted that keratin is not typically solubilized by ordinary methods, but under low pH and with reducing/oxidizing agents, it becomes more water-soluble and chemically reactive due to its disulfide, amino, and carboxylic acid moieties [352].

A by-product of wheat starch processing, wheat gluten contains gliadins and glutenins, which contribute to its adhesive properties [353]. It is used for wood adhesives due to its cohesive and viscoelastic properties. Modification methods such as enzymatic hydrolysis and heat treatment can enhance its adhesive performance [353].

Similar to keratin, zein represents a hydrophobic protein due to its high non-polar amino acid content. Zein-based adhesives can be enhanced by chelation with metal cations, resulting in good adhesive strength and water resistance [354].

The binding mechanisms of protein-based adhesives typically involve multiple interactions which are critical for adhesion to various substrates. Hydrogen bonds form between protein adhesives and hydroxyl groups present in substrates such as wood and cellulose, providing substantial strength in dry conditions. Electrostatic interactions arise from the attraction between charged groups on the protein and oppositely charged groups on the substrate, enhancing adhesion through interactions such as those between positively charged amino groups in the protein and negatively charged surfaces [355]. Covalent crosslinking occurs when covalent bonds form between the protein and substrate, leading to strong and durable adhesion. This can be achieved through enzymatic crosslinking, such as using transglutaminase with silk fibroin [356], or chemical crosslinking agents that significantly improve the adhesive's strength and moisture resistance. Hydrophobic interactions, particularly with proteins such as keratin and zein that are rich in hydrophobic amino acids, interact with non-polar surfaces to enhance water resistance and durability [349, 354, 357]. These interactions help bind to substrates with low surface energy. Incorporating metal ions such as calcium or magnesium with proteins such as zein enhances adhesion to metallic and mineral substrates through the formation of coordination complexes with protein side chains [354, 358].

Additionally, mechanical interlocking contributes to the overall adhesion strength by allowing protein molecules to physically interlock within the microstructure of porous substrates such as wood or paper [355].

Proteins have long been used by mankind as effective and durable materials in commercial products. Current research suggests that this trend will continue due to the abundance and diverse properties of proteins. This group of extremely heterogeneous molecules holds great promise in various fields where interfacial interactions are of immense importance. By leveraging their interactions with other substances, such as metal ions or tannins, researchers can tailor the properties of protein-based materials to meet specific needs. The outstanding mechanical properties and sequence-guided functionalities endow these materials with optimistic research opportunities for the future.

1.5 Nucleic acid based biomaterials

Nucleic acids (RNA and DNA) are fundamental biological macromolecules that carry genetic information essential for the functioning of all living organisms.

By separating these negatively charged macromolecules from their initial biological purpose and using it as a bio based raw material, exciting new materials can be obtained through various chemical interactions. Since the late 1980s, inspired by the pioneering work of Nadrian Seeman, nucleic acids have been utilized as materials for constructing nanostructures, heralding the inception of DNA nanotechnology [359]. This innovative approach has broadened their applications beyond traditional biology, leading to a fruitful field of research and growing interest in the interactions between nucleic acids and various materials. The versatility and unique properties of nucleic acids are driving advancements in medicine, biotechnology, materials science, and environmental applications.

In contrast to the other bio-based materials discussed in this chapter, nucleic acid materials are not used to create substitutes for already existing fossil-based materials, but broaden the scope of artificial nanometer sized manufacturing. Furthermore, while nucleic acids can be considered abundant since DNA and RNA are present in all living organisms and are becoming increasingly common in sectors such as medicine, their overall commercial abundance and accessibility are not yet anywhere near the levels of the previously discussed materials. The production of nucleic acids is primarily driven by the biopharmaceutical and research sectors [360].

The surface properties of nucleic acid-based materials are determined by exact sequence programmability, structural controllability, and exceptional molecular recognition. Unlike other nanomaterials used in the biomedical field, nucleic acids have demonstrated exceptional biocompatibility. Cationic polymer surfaces are often cytotoxic, dendritic polymer nanocarriers can increase immunotoxicity *in vivo*, and inorganic nanoparticle residues are challenging for the body to catabolize. In contrast, nucleic acids offer low toxicity and high programmability [361]. The unique properties of nucleic acids facilitate various chemical interactions at interfaces with diverse materials. These interactions involve nearly all kinds of chemical interactions, encompassing non-covalent interactions such as π–π stacking, hydrogen bonding, van der Waals forces, electrostatic interactions, and hydrophobic interactions, as well as covalent interactions such as covalent binding and coordination linkages [362]. The following points briefly illustrate how these interactions occur:

1. **π–π stacking interaction**: Nucleic acids, composed of nitrogenous bases such as adenine and guanine, exhibit π-electron-rich structures akin to aromatic rings. These structures enable π–π stacking interactions with materials such as grapheme [363], enhancing stability and functionality for applications in sensing [364].
2. **Hydrogen-bonding interaction**: Hydrogen bonds form between nucleic acids and materials with high electronegative atoms (O, F, N). This interaction is significant in DNA nanostructures, contributing to the formation and stability of complexes such as melamine-mediated DNA assemblies (thymine-melamine-thymine triplet) [365].
3. **van der Waals interaction**: These weak forces arise from dipole moments and electron movement, facilitating interactions between nucleic acids and molecules with similar compositions. For instance, as shown by several studies, MoS_2 can adsorb single stranded DNA (ssDNA) through van der

Waals forces, aiding in the development of DNA sensing strategies [366–368].

4. **Electrostatic interactions**: The negative charge of nucleic acids allows them to adsorb positively charged molecules and materials. This interaction facilitates the assembly of nucleic acid-based nanostructures, as seen with cationic species such as L-arginine and L-lysine, enhancing the stability and functionality of DNA nanoarchitectures [369].
5. **Hydrophobic interactions:** DNA self-assembly and the formation of the double helix are significantly driven by hydrophobic interactions. The bases in DNA are hydrophobic and their interactions drive the self-assembly process, while hydrogen bonds provide specificity in base pairing. Hydrophobic interactions also influence DNA interactions with various additives and surfactants, affecting its stability and behavior at interfaces [18].
6. **Covalent binding**: Strong covalent bonds form between nucleic acids and other molecules through routes such as Au–S bonds, disulfide bridges [370], and click reactions [371]. These bonds enable the functionalization of nucleic acids with diverse groups, crucial for applications in drug delivery and photothermal therapies [372].
7. **Coordination linkage**: Coordination bonds, where one atom provides both electrons for the bond, occur between nucleic acids and metal ions such as Hg^{2+} and Ag^{+} [373]. These interactions form hybrid structures used in sensing and bioimaging, expanding the utility of nucleic acids in environmental monitoring and biomedicine [374].

These different interactions enable nucleic acids to integrate with novel materials, creating hybrid composites with specific structural and functional properties. Nucleic acid composites are typically nano-composites, often involving nanoparticles modified with DNA or RNA. Common materials combined with nucleic acids include inorganic nanoparticles such as gold, silver, iron oxide, silica, and quantum dots, as well as organic nanoparticles such as liposomes, polymers, and proteins, and carbon-based materials such as carbon nanotubes [375].

These nano-composites can be produced through various methods that utilize different interfacial interactions to combine DNA or RNA with their respective counterparts. In the synthesis of AuNP–DNA composites, the binding of DNA to AuNPs is facilitated by the electrostatic attraction, which is further stabilized by adding salts to shield repulsive forces among negatively charged components, also known as the salt-aging method [376]. Further covalent bonding can be used, particularly in the attachment of DNA to nanoparticle surfaces. DNA can be functionalized with thiol groups that form strong covalent bonds with gold surfaces [377]. Also hydrophobic interactions are particularly relevant when using liposomes or hydrophobic polymers as cores for DNA composites. For instance, the hydrophobic tails of DNA molecules can embed into liposome membranes, facilitating the formation of stable DNA–liposome composites [378]. Similarly, hydrophobic polymers can interact with DNA to form micelle-like structures that enhance the delivery and stability of the composite [379]. Another method which can be used for specific DNA-functionalized nanoparticles is rolling circle amplification (RCA). For instance, silica nanoparticles (SiNPs) and carbon nanotubes (CNTs) are functionalized with oligonucleotide primers, which are then extended through RCA to form complex DNA nanocomposites. In RCA, a short DNA primer is hybridized to a circular single-stranded DNA template attached to the nanoparticle surface. A DNA polymerase enzyme initiates DNA synthesis at the primer and continuously adds nucleotides complementary to the circular template [380].

The resulting composite materials exhibit enhanced properties, benefiting from synergistic effects that enhance their overall performance. For instance, integrating CNTs into SiNPs using RCA increases the mechanical stiffness and enhances surface properties. The interactions between the DNA, silica, and CNTs are crucial for the structural integrity and functionality of the final composite material, influencing its entanglement properties and overall performance [381].

Typical application fields of nucleic acid composites are found in diagnostics and therapeutics. They provide the necessary specificity for recognizing target molecules, forming the basis of highly sensitive biosensors for early disease detection [380, 382]. In therapeutics, DNA combined with biocompatible materials ensures precise drug delivery to disease sites [383], minimizing side effects. Additionally, nucleic acid-functionalized nanoparticles enable real-time bioimaging for detailed cellular visualization [384].

In addition to composites, there is a broad field of purely nucleic acid-based materials. This field is focused mainly on nucleic acid nanotechnology. Due to their highly diverse and yet controllable structures, which are precisely determined by their sequences, nucleic acids can be manufactured in a desired way through self-assembly. This concept dates back to Paul Rothemund's groundbreaking discovery of folding DNA to create nanoscale shapes, a technique that pioneered the field of DNA origami [385]. This research field has since expanded significantly, enabling the generation of various 1D, 2D and 3D nanoobjects through DNA folding. These structures are formed by designing specific sequences of short single-stranded DNA (ssDNA), known as staples, which bind to a long ssDNA scaffold, guiding it into the desired shape [386]. The most widely used ssDNA scaffold uses derives from viral genomes such as M13 bacteriophage [386]. 1D structures include linear nanowires and nanotubes, primarily formed through the alignment of DNA tiles or repeated units [387]. 2D structures, such as lattices and grids, are created by arranging DNA tiles in a planar fashion, allowing the formation of patterns such as arrays and sheets [386]. 3D structures, including polyhedra, hydrogels, and hierarchical assemblies, are achieved by folding the DNA scaffold into complex three-dimensional shapes using computer-aided design tools such as caDNAno and vHelix [388].

The surface properties of nucleic acid origami are highly programmable due to the sequence specificity of DNA. These properties allow for molecular recognition, biocompatibility, and further functionalization. As known from its biological origin, DNA structures exhibit high specificity in molecular interactions through Watson–Crick base pairing. However, Woo and Rothemund showed further that DNA origami can be used to create diverse bonds using geometric arrangements of blunt-end stacking interactions. This approach contrasts with the traditional Watson–Crick base pairing and expands the capability of DNA to form orthogonal, isoenergetic interactions [389]. The inherent biocompatibility of DNA ensures that DNA-based nanostructures interact well with biological systems, making them suitable for biomedical applications without eliciting significant immune responses. This biocompatibility can be leveraged in vaccine development, as demonstrated by Oktay *et al*, who created DNA origami structures presenting viral antigens and adjuvants on their surfaces. By chemically conjugating antigens, such as the SARS-CoV-2 receptor binding domain (RBD), and immune-stimulating adjuvants such as CpG molecules, these DNA nanostructures mimic viral components, enhancing antigen presentation to immune cells. This triggers a robust, targeted immune response, effectively stimulating the production of neutralizing antibodies against the virus [390]. As discussed for DNA composites, the surfaces of DNA origami can be functionalized with various chemical groups, nanoparticles, or biomolecules, enabling tailored interactions and enhancing functionality. For instance, thiol-modified DNA origami can form covalent bonds with gold nanoparticles, allowing precise and stable surface modification [391]. Additionally, click chemistry is used to attach various functional entities to DNA origami through azide–alkyne reactions, enabling precise and stable modifications for diverse applications [392].

Overall DNA origami is one key technology for the precise generation of nanoscale objects with programmable and predictable surfaces. This nanometer-resolution addressability enables the construction of intricate frameworks, where specific materials such as metals, silica, lipids, or polymers can be precisely positioned. This technology holds great promise for developing advanced structures and nanosystems in fields such as nanophotonics and nanoelectronics [388]. Notably, the connection to electronic computing is significant, with some researchers suggesting that the greatest commercial opportunities for nucleic acid nanotechnology lie in bridging biology and semiconductor technology, creating innovative interfaces between biological systems and electronic devices [393].

DNA hydrogels are soft materials composed of polymeric networks formed by cross-linked DNA chains, categorizing them as typical supramolecular hydrogels. These hydrogels are typically formed through the self-assembly of nucleic acid strands or by the enzymatic polymerization of oligonucleotides [394]. However, the primary starting materials for these hydrogels include ssDNA or double-stranded DNA (dsDNA) and RNA. Based on their components, DNA hydrogels can be classified by size into macroscopic bulk hydrogels and sub-microscopic nanogels [395]. These gels can either be formed purely by DNA or as hybrid gels. While the construction of hydrogels using purely DNA relies on hydrogen bonding, physical entanglement, or enzymatic reactions between the chains, hybrid gels use either physical and chemical crosslink to generate the three dimensional gel. Physical crosslinked hydrogels are formed through interactions like electrostatic interactions, coordination interactions, and π–π stacking, offering excellent responsiveness. For instance, DNA can form hydrogels through electrostatic interactions with cationic surfactants such as cetyltrimethylammonium bromide (CTAB) [396] or with metal complexes such as platinum(II) complexes [397]. Chemical crosslinked hydrogels, with covalent bonds, have stable structures and high strength. The covalent binding involves polymerization and amine-epoxidation reactions [398]. Further methods use bottom up approaches involving RCA, which extend DNA primers to create long DNA strands that entangle and form hydrogels [399].

The surface properties of nucleic acid hydrogels are defined by their high water content and biocompatibility. Surface modification can be used to generate specific properties for interactions. For example, DNA hydrogels can be modified with bioactive peptides, proteins, or other signaling molecules to promote cellular adhesion, proliferation, and differentiation [398]. Due to their biocompatibility and ability to encapsulate and release therapeutic agents in a controlled manner, these hydrogels are excellent candidates for drug delivery systems. The programmability of nucleic acids allow the respective hydrogels to act as smart drug delivery system. Encapsulate therapeutic agents can thereby be released in a controlled manner in response to specific stimuli, such as changes in pH, temperature, or the presence of certain enzymes [400]. The hydrogen bonds between complementary base pairs in DNA can be disrupted by changes in pH or temperature, causing the hydrogel network to break down or swell. A notable example of such hydrogels was presented by Lu *et al* who constructed a gel from *Y*-shaped DNA nanostructures that form triplex structures. At pH 5.0, C–G·C + bridges form, leading to gelation, while at pH 7.0, they disassemble, transitioning the hydrogel to a sol state. Similarly, T–A·T bridges form at pH 7.0 and disassemble at pH 10.0 [401].

Nucleic acid based materials have potential applications in environmental monitoring and green chemistry. Their ability to integrate with various materials through diverse interaction mechanisms results in new characteristics and applications, ranging from biosensing and drug delivery to advanced therapeutic techniques.

1.6 Summary and perspectives

Fossil-based synthetic polymers have significantly shaped society over the last 100 years; however, this development has also brought about serious environmental challenges, such as waste management and pollution, prompting increased interest in bio-based materials as a more sustainable alternative in recent decades. In this chapter, we have delved into the exciting potential of bio-based polymeric materials The diversity of bio-based polymers is vast, as shown in this chapter. But it is not just their variety that is impressive. We have focused on how their unique interfacial properties can help us shrink the environmental footprint of synthetic polymers and drive forward a new wave of eco-friendly innovation. Bio-based polymers, including polysaccharides such as cellulose and starch, protein-based materials, and phenolic compounds such as lignin, bring valuable characteristics such as biodegradability and a reduced environmental footprint. In this chapter, we have provided an overview of these materials, showcasing innovations in their modification—such as the introduction of functional groups, grafting, and the creation of composite materials—which significantly improve their mechanical strength, thermal stability, and biodegradability.

These materials have been categorized according to their chemical structures and the resulting properties, emphasizing the significance of their supramolecular structures in determining specific functionalities. The range of properties spans from highly charged polyelectrolytes, such as chitosan and alginate, to hydrophobic protein structures, such as fibroin and keratin. These materials can be tailored into practical products, with nucleic acids exemplifying biopolymers that offer complex structures beyond conventional macromolecular chemistry.

The interfacial properties play a crucial role for the implementation of these materials in applications such as adhesives, coatings, and films. These properties—such as hydrophilicity, adhesion, and surface energy—determine how materials interact with their environment and affect their functionality. For example, the surface hydroxyl groups in cellulose enable hydrogen bonding, enhancing its effectiveness as a coating or adhesive, while the amphiphilic nature of lignin and the cationic surface of chitosan make them valuable in creating stable interfaces in packaging and biomedical devices. Bio-based polymers are increasingly replacing petrochemical products, although there is still room for improvement in their production and modification. Future advancements in novel solvent systems, such as deep eutectic solvents and ionic liquids, are expected to optimize processing, leading to materials with even lower environmental impact and reduced energy consumption. This chapter also highlights the versatile roles that bio-based polymers can play, whether as the primary material, integrated as co-monomers in polymer systems, or functioning as nanoparticulate stabilizers across various specialized applications. Future research should focus on enhancing production efficiency, material properties, and scalability while reducing environmental impact to maintain the sustainability of these materials. In this context, the chapter underscores the potential of environmentally friendly solvents and green chemistry approaches to lessen the ecological footprint of modification processes. Furthermore, designing novel supramolecular structures and adopting advanced techniques such as 3D printing and molecular self-assembly are paving the way for the development of next-generation bio-based materials.

Bio-based polymeric materials offer substantial potential for a sustainable future. We are convinced that bio-based materials are indispensable in developing alternatives to fossil-based polymers, shaping the future of material science and driving the innovation of novel functional materials across society.

Bibliography

[1] Hameed M, Bhat R A, Singh D V and Mehmood M A 2020 White pollution: a hazard to environment and sustainable approach to its management *Practice, Progress, and Proficiency in Sustainability* ed R A Bhat, H Qadri, K A Wani, G H Dar and M A Mehmood (Hershey, PA: IGI Global) pp 52–81

[2] Chang N-B, Pires A and Martinho G 2011 Empowering systems analysis for solid waste management: challenges, trends, and perspectives *Crit. Rev. Environ. Sci. Technol.* **41** 1449–530

[3] Osman A I *et al* 2023 Microplastic sources, formation, toxicity and remediation: a review *Environ. Chem. Lett.* **21** 2129–69

[4] Shen M, Huang W, Chen M, Song B, Zeng G and Zhang Y 2020 (Micro)plastic crisis: un-ignorable contribution to global greenhouse gas emissions and climate change *J. Clean. Prod.* **254** 120138

[5] Hahladakis J N, Velis C A, Weber R, Iacovidou E and Purnell P 2018 An overview of chemical additives present in plastics: migration, release, fate and environmental impact during their use, disposal and recycling *J. Hazard. Mater.* **344** 179–99

[6] Othman A R, Hasan H A, Muhamad M H, Ismail N I and Abdullah S R S 2021 Microbial degradation of microplastics by enzymatic processes: a review *Environ. Chem. Lett.* **19** 3057–73

[7] Chinthapalli R, Skoczinski P, Carus M, Baltus W, De Guzman D, Käb H, Raschka A and Ravenstijn J 2019 Biobased building blocks and polymers—global capacities, production and trends, 2018–2023 *Ind. Biotechnol.* **15** 237–41

[8] Vert M, Doi Y, Hellwich K-H, Hess M, Hodge P, Kubisa P, Rinaudo M and Schué F 2012 Terminology for biorelated polymers and applications (IUPAC recommendations 2012) *Pure Appl. Chem.* **84** 377–410

[9] European Bioplastics 2024 Applications for bioplastics *European Bioplastics* https://european-bioplastics.org/market/applications-sectors/

[10] Jaganmohan M 2024 Global bioplastics industry—statistics and facts *Statista* www.statista.com/topics/8744/bioplastics-industry-worldwide/#topicOverview

[11] Cottet C, Ramirez-Tapias Y A, Delgado J F, De La Osa O, Salvay A G and Peltzer M A 2020 Biobased materials from microbial biomass and its derivatives *Materials* **13** 1263

[12] Mederake L, Hinzmann M and Langsdorf S 2020 Hintergrundpapier: Plastikpolitik in Deutsch-land und der EU *Report* Ecologic Institut, Berlin

[13] Packham D E 2009 Adhesive technology and sustainability *Int. J. Adhes. Adhes.* **29** 248–52

[14] Benalaya I, Alves G, Lopes J and Silva L R 2024 A review of natural polysaccharides: sources, characteristics, properties, food, and pharmaceutical applications *Int. J. Mol. Sci.* **25** 1322

[15] Mohammed A S A, Naveed M and Jost N 2021 Polysaccharides; classification, chemical properties, and future perspective applications in fields of pharmacology and biological medicine (a review of current applications and upcoming potentialities) *J. Polym. Environ.* **29** 2359–71

[16] Klemm D, Heublein B, Fink H and Bohn A 2005 Cellulose: fascinating biopolymer and sustainable raw material *Angew. Chem. Int. Ed.* **44** 3358–93

[17] Kamide K, Okajima K, Kowsaka K and Matsui T 1985 CP/MASS ^{13}C NMR spectra of cellulose solids: an explanation by the intramolecular hydrogen bond concept *Polym. J.* **17** 701–6

[18] Lindman B, Medronho B, Alves L, Norgren M and Nordenskiöld L 2021 Hydrophobic interactions control the self-assembly of DNA and cellulose *Quart. Rev. Biophys.* **54** e3

[19] Bao Y, Qian H, Lu Z and Cui S 2015 Revealing the hydrophobicity of natural cellulose by single-molecule experiments *Macromolecules* **48** 3685–90

[20] Bergenstråhle M, Wohlert J, Himmel M E and Brady J W 2010 Simulation studies of the insolubility of cellulose *Carbohydr. Res.* **345** 2060–6

[21] Etale A, Onyianta A J, Turner S R and Eichhorn S J 2023 Cellulose: a review of water interactions, applications in composites, and water treatment *Chem. Rev.* **123** 2016–48

[22] Wang J, Wang L, Gardner D J, Shaler S M and Cai Z 2021 Towards a cellulose-based society: opportunities and challenges *Cellulose* **28** 4511–43

[23] MacLeod A, Daly C, Khan I, Vale L, Campbell M, Wallace S, Cody J, Donaldson C and Grant AThe Cochrane Collaboration 2001 Cellulose, modified cellulose and synthetic membranes in the haemodialysis of patients with end-stage renal disease *Cochrane Database of Systematic Reviews* (Chichester: Wiley) p CD003234

[24] Clark W R, Hamburger R J and Lysaght M J 1999 Effect of membrane composition and structure on solute removal and biocompatibility in hemodialysis *Kidney Int.* **56** 2005–15

[25] Chen Y-A, Ou S-M and Lin C-C 2022 Influence of dialysis membranes on clinical outcomes: from history to innovation *Membranes* **12** 152

[26] Puls J, Wilson S A and Hölter D 2011 Degradation of cellulose acetate-based materials: a review *J. Polym. Environ.* **19** 152–65

[27] Aziz T *et al* 2022 A review on the modification of cellulose and its applications *Polymers* **14** 3206

[28] Carlmark A, Larsson E and Malmström E 2012 Grafting of cellulose by ring-opening polymerisation—a review *Eur. Polym. J.* **48** 1646–59

[29] Kumar M, Gehlot P S, Parihar D, Surolia P K and Prasad G 2021 Promising grafting strategies on cellulosic backbone through radical polymerization processes—a review *Eur. Polym. J.* **152** 110448

[30] Tosh B and Routray C 2014 Grafting of cellulose based materials: a review *Chem. Sci. Rev. Lett.* **3** 74–92

[31] Zhang J, Qi Y, Shen Y and Li H 2022 Research progress on chemical modification and application of cellulose: a review *Mater. Sci.* **28** 60–7

[32] Ge W, Shuai J, Wang Y, Zhou Y and Wang X 2022 Progress on chemical modification of cellulose in 'green' solvents *Polym. Chem.* **13** 359–72

[33] Ferrer A, Pal L and Hubbe M 2017 Nanocellulose in packaging: advances in barrier layer technologies *Ind. Crops Prod.* **95** 574–82

[34] Brinchi L, Cotana F, Fortunati E and Kenny J M 2013 Production of nanocrystalline cellulose from lignocellulosic biomass: technology and applications *Carbohydr. Polym.* **94** 154–69

[35] Wu Y, Liang Y, Mei C, Cai L, Nadda A, Le Q V, Peng Y, Lam S S, Sonne C and Xia C 2022 Advanced nanocellulose-based gas barrier materials: present status and prospects *Chemosphere* **286** 131891

[36] Bideau B, Loranger E and Daneault C 2018 Nanocellulose-polypyrrole-coated paperboard for food packaging application *Prog. Org. Coat.* **123** 128–33

[37] Zhang H, Yu H-Y, Wang C and Yao J 2017 Effect of silver contents in cellulose nanocrystal/silver nanohybrids on PHBV crystallization and property improvements *Carbohydr. Polym.* **173** 7–16

[38] Garusinghe U M, Varanasi S, Raghuwanshi V S, Garnier G and Batchelor W 2018 Nanocellulose-montmorillonite composites of low water vapour permeability *Colloids Surf. A* **540** 233–41

[39] Shankar S, Oun A A and Rhim J-W 2018 Preparation of antimicrobial hybrid nanomaterials using regenerated cellulose and metallic nanoparticles *Int. J. Biol. Macromol.* **107** 17–27

[40] Ullah M W, Rojas O J, McCarthy R R and Yang G 2021 Editorial: Nanocellulose: a multipurpose advanced functional material *Front. Bioeng. Biotechnol.* **9** 738779

[41] Qi Y, Guo Y, Liza A A, Yang G, Sipponen M H, Guo J and Li H 2023 Nanocellulose: a review on preparation routes and applications in functional materials *Cellulose* **30** 4115–47

[42] Lin D, Liu Z, Shen R, Chen S and Yang X 2020 Bacterial cellulose in food industry: current research and future prospects *Int. J. Biol. Macromol.* **158** 1007–19

[43] Aditya T, Allain J P, Jaramillo C and Restrepo A M 2022 Surface modification of bacterial cellulose for biomedical applications *Int. J. Mol. Sci.* **23** 610

[44] Liu W, Du H, Zhang M, Liu K, Liu H, Xie H, Zhang X and Si C 2020 Bacterial cellulose-based composite scaffolds for biomedical applications: a review *ACS Sustain. Chem. Eng.* **8** 7536–62

[45] Zhang K, Barhoum A, Xiaoqing C, Haoyi L and Samyn P 2019 Cellulose nanofibers: fabrication and surface functionalization techniques *Handbook of Nanofibers* ed A Barhoum, M Bechelany and A Makhlouf (Cham: Springer International) pp 1–41

[46] Lam E and Hemraz U D 2021 Preparation and surface functionalization of carboxylated cellulose nanocrystals *Nanomaterials* **11** 1641

[47] Imiete I E, Giannini L, Tadiello L, Orlandi M and Zoia L 2023 The effect of sulfate half-ester groups on the mechanical performance of cellulose nanocrystal-natural rubber composites *Cellulose* **30** 8929–40

[48] Patoary M K, Islam S R, Farooq A, Rashid M A, Sarker S, Hossain M Y, Rakib M A N, Al-Amin M and Liu L 2023 Phosphorylation of nanocellulose: state of the art and prospects *Ind. Crops Prod.* **201** 116965

[49] Pierre G, Punta C, Delattre C, Melone L, Dubessay P, Fiorati A, Pastori N, Galante Y M and Michaud P 2017 TEMPO-mediated oxidation of polysaccharides: an ongoing story *Carbohydr. Polym.* **165** 71–85

[50] Zaman M, Xiao H, Chibante F and Ni Y 2012 Synthesis and characterization of cationically modified nanocrystalline cellulose *Carbohydr. Polym.* **89** 163–70

[51] Xiao G, Wang Y, Zhang H, Zhu Z and Fu S 2020 Dialdehyde cellulose nanocrystals act as multi-role for the formation of ultra-fine gold nanoparticles with high efficiency *Int. J. Biol. Macromol.* **163** 788–800

[52] Le Gars M, Roger P, Belgacem N and Bras J 2020 Role of solvent exchange in dispersion of cellulose nanocrystals and their esterification using fatty acids as solvents *Cellulose* **27** 4319–36

[53] Trinh B M and Mekonnen T 2018 Hydrophobic esterification of cellulose nanocrystals for epoxy reinforcement *Polymer* **155** 64–74

[54] Abushammala H and Mao J 2019 A review of the surface modification of cellulose and nanocellulose using aliphatic and aromatic mono- and di-isocyanates *Molecules* **24** 2782

[55] Cañas-Gutiérrez A, Martinez-Correa E, Suárez-Avendaño D, Arboleda-Toro D and Castro-Herazo C 2020 Influence of bacterial nanocellulose surface modification on calcium phosphates precipitation for bone tissue engineering *Cellulose* **27** 10747–63

[56] Arumughan V, Nypelö T, Hasani M and Larsson A 2021 Fundamental aspects of the non-covalent modification of cellulose via polymer adsorption *Adv. Colloid Interface Sci.* **298** 102529

[57] Dudefoi W, Dhuiège B, Capron I and Sèbe G 2022 Controlled hydrophobic modification of cellulose nanocrystals for tunable pickering emulsions *Carbohydr. Polym. Technol. Appl.* **3** 100210

[58] Hu Z, Berry R M, Pelton R and Cranston E D 2017 One-pot water-based hydrophobic surface modification of cellulose nanocrystals using plant polyphenols *ACS Sustain. Chem. Eng.* **5** 5018–26

[59] D'Acierno F and Capron I 2023 Modulation of surface properties of cellulose nanocrystals through adsorption of tannic acid and alkyl cellulose derivatives *Carbohydr. Polym.* **319** 121159

[60] Ni Y, Fan L and Sun Y 2020 Interfacial properties of cellulose nanoparticles with different lengths from ginkgo seed shells *Food Hydrocoll.* **109** 106121

[61] Jiménez A, Fabra M J, Talens P and Chiralt A 2012 Edible and biodegradable starch films: a review *Food Bioprocess Technol.* **5** 2058–76

[62] Arvanitoyannis I S 1999 Totally and partially biodegradable polymer blends based on natural and synthetic macromolecules: preparation, physical properties, and potential as food packaging materials *J. Macromol. Sci.* C **39** 205–71

[63] Amaraweera S M *et al* 2021 Development of starch-based materials using current modification techniques and their applications: a review *Molecules* **26** 6880

[64] Marichelvam M K, Jawaid M and Asim M 2019 Corn and rice starch-based bio-plastics as alternative packaging materials *Fibers* **7** 32

[65] Gadhave R V, Das A, Mahanwar P A and Gadekar P T 2018 Starch based bio-plastics: the future of sustainable packaging *Open J. Polym. Chem.* **08** 21–33

[66] Maurer H W 2009 Starch in the paper industry *Starch* (Amsterdam: Elsevier) pp 657–713

[67] Kalia S, Bhattacharya A, Prajapati S K and Malik A 2021 Utilization of starch effluent from a textile industry as a fungal growth supplement for enhanced α-amylase production for industrial application *Chemosphere* **279** 130554

[68] Sivamaruthi B S, Nallasamy P K, Suganthy N, Kesika P and Chaiyasut C 2022 Pharmaceutical and biomedical applications of starch-based drug delivery system: a review *J. Drug Deliv. Sci. Technol.* **77** 103890

[69] Watcharakitti J, Win E E, Nimnuan J and Smith S M 2022 Modified starch-based adhesives: a review *Polymers* **14** 2023

[70] Hossain M T, Shahid M A, Akter S, Ferdous J, Afroz K, Refat K R I, Faruk O, Jamal M S I, Uddin M N and Samad M A B 2024 Cellulose and starch-based bioplastics: a review of advances and challenges for sustainability *Polym.-Plast. Technol. Mater.* **63** 1329–49

[71] Torres F G, Troncoso O P, Torres C, Díaz D A and Amaya E 2011 Biodegradability and mechanical properties of starch films from Andean crops *Int. J. Biol. Macromol.* **48** 603–6

[72] Lourdin D, Putaux J-L, Potocki-Véronèse G, Chevigny C, Rolland-Sabaté A and Buléon A 2015 Crystalline structure in starch *Starch* ed Y Nakamura (Tokyo: Springer) pp 61–90

[73] Donmez D, Pinho L, Patel B, Desam P and Campanella O H 2021 Characterization of starch–water interactions and their effects on two key functional properties: starch gelatinization and retrogradation *Curr. Opin. Food Sci.* **39** 103–9

[74] Yu X, Chen L, Jin Z and Jiao A 2021 Research progress of starch-based biodegradable materials: a review *J. Mater. Sci.* **56** 11187–208

[75] Abidin M Z A Z, Julkapli N M, Juahir H, Azaman F, Sulaiman N H and Abidin I Z 2015 Fabrication and properties of chitosan with starch for packaging application *Malays. J. Anal. Sci.* **19** 1032–42

[76] Ebnesajjad S 2011 Characteristics of adhesive materials *Handbook of Adhesives and Surface Preparation* (Amsterdam: Elsevier) pp 137–83

[77] V. Gadhave R, Mahanwar P A and Gadekar P T 2017 Starch-based adhesives for wood/wood composite bonding: review *Op. J. Phys. Chem.* **07** 19–32

[78] Maulana M I *et al* 2022 Environmentally friendly starch-based adhesives for bonding high-performance wood composites: a review *Forests* **13** 1614

[79] Kumoro A C, Amalia R, Budiyati C S, Retnowati D S and Ratnawati R 2015 Preparation and characterization of physicochemical properties of glacial acetic acid modified Gadung (*Diocorea hispida* Dennst) flours *J. Food Sci. Technol.* **52** 6615–22

[80] Rosida D F, Yuliani R and Djajati S 2020 Modification of *Colocasia esculenta* starch with acetylation process *Nusantara Science and Technology Proc. 4th Int. Seminar of Research Month* (New York: Galaxy Science) pp 369–78

[81] Colussi R, Pinto V Z, El Halal S L M, Vanier N L, Villanova F A, Marques E Silva R, Da Rosa Zavareze E and Dias A R G 2014 Structural, morphological, and physicochemical properties of acetylated high-, medium-, and low-amylose rice starches *Carbohydr. Polym.* **103** 405–13

[82] Dapčević Hadnađev T R, Dokić L P, Hadnađev M S, Pojić M M and Torbica A M 2014 Rheological and breadmaking properties of wheat flours supplemented with octenyl succinic anhydride-modified waxy maize starches *Food Bioprocess Technol.* **7** 235–47

[83] Sun Y, Gu J, Tan H, Zhang Y and Huo P 2018 Physicochemical properties of starch adhesives enhanced by esterification modification with dodecenyl succinic anhydride *Int. J. Biol. Macromol.* **112** 1257–63

[84] Van Soest J J G and Vliegenthart J F G 1997 Crystallinity in starch plastics: consequences for material properties *Trends Biotechnol.* **15** 208–13

[85] Zhang Y, Liu X, Wang Y, Jiang P and Quek S 2016 Antibacterial activity and mechanism of cinnamon essential oil against *Escherichia coli* and *Staphylococcus aureus Food Control* **59** 282–9

[86] Wilpiszewska K and Czech Z 2022 An effect of carboxymethyl starch addition on adhesion to paper of water-soluble pressure-sensitive adhesive *Cellulose* **29** 5251–63

[87] Radley J A 1976 *Industrial Uses of Starch and its Derivatives* (Dordrecht: Springer)

[88] Kruger L and Lacourse N 1990 Starch based adhesives *Handbook of Adhesives* ed I Skeist (Boston, MA: Springer) pp 153–66

[89] Vanier N L, El Halal S L M, Dias A R G and da Rosa Zavareze E 2017 Molecular structure, functionality and applications of oxidized starches: a review *Food Chem.* **221** 1546–59

[90] Neitzel N, Hosseinpourpia R and Adamopoulos S 2023 A dialdehyde starch-based adhesive for medium-density fiberboards *BioRes* **18** 2155–71

[91] Sagnelli D *et al* 2017 Cross-linked amylose bio-plastic: a transgenic-based compostable plastic alternative *Int. J. Mol. Sci.* **18** 2075

[92] Dai L, Zhang J and Cheng F 2019 Effects of starches from different botanical sources and modification methods on physicochemical properties of starch-based edible films *Int. J. Biol. Macromol.* **132** 897–905

[93] Ghanbarzadeh B, Almasi H and Entezami A A 2011 Improving the barrier and mechanical properties of corn starch-based edible films: effect of citric acid and carboxymethyl cellulose *Ind. Crops Prod.* **33** 229–35

[94] Leceta I, Guerrero P and De La Caba K 2013 Functional properties of chitosan-based films *Carbohydr. Polym.* **93** 339–46

[95] Ma X, Yu J and Wang N 2007 Fly ash-reinforced thermoplastic starch composites *Carbohydr. Polym.* **67** 32–9

[96] Souza R C R and Andrade C T 2002 Investigation of the gelatinization and extrusion processes of corn starch *Adv. Polym. Technol.* **21** 17–24

[97] Ma X 2004 The plastcizers containing amide groups for thermoplastic starch *Carbohydr. Polym.* **57** 197–203

[98] Ma X F, Yu J G and Wan J J 2006 Urea and ethanolamine as a mixed plasticizer for thermoplastic starch *Carbohydr. Polym.* **64** 267–73

[99] Chen H, Xie F, Chen L and Zheng B 2019 Effect of rheological properties of potato, rice and corn starches on their hot-extrusion 3D printing behaviors *J. Food Eng.* **244** 150–8

[100] Haryńska A, Janik H, Sienkiewicz M, Mikolaszek B and Kucińska-Lipka J 2021 PLA–potato thermoplastic starch filament as a sustainable alternative to the conventional PLA filament: processing, characterization, and FFF 3D printing *ACS Sustain. Chem. Eng.* **9** 6923–38

[101] Ju Q, Tang Z, Shi H, Zhu Y, Shen Y and Wang T 2022 Thermoplastic starch based blends as a highly renewable filament for fused deposition modeling 3D printing *Int. J. Biol. Macromol.* **219** 175–84

[102] Bangar S P, Purewal S S, Trif M, Maqsood S, Kumar M, Manjunatha V and Rusu A V 2021 Functionality and applicability of starch-based films: an eco-friendly approach *Foods* **10** 2181

[103] Jiménez A, Fabra M J, Talens P and Chiralt A 2012 Effect of re-crystallization on tensile, optical and water vapour barrier properties of corn starch films containing fatty acids *Food Hydrocoll.* **26** 302–10

[104] Issa A, Schimmel K, Worku M, Shahbazi G, Ibrahim S and Tahergorabi R 2018 Sweet potato starch-based nanocomposites: development, characterization and biodegradability *Starch—Starke* **70** 1700273

[105] Merino D, Mansilla A Y, Casalongué C A and Alvarez V A 2019 Effect of nanoclay addition on the biodegradability and performance of starch-based nanocomposites as mulch films *J. Polym. Environ.* **27** 1959–70

[106] Fourati Y, Magnin A, Putaux J-L and Boufi S 2020 One-step processing of plasticized starch/cellulose nanofibrils nanocomposites via twin-screw extrusion of starch and cellulose fibers *Carbohydr. Polym.* **229** 115554

[107] Vaezi K, Asadpour G and Sharifi S H 2020 Bio nanocomposites based on cationic starch reinforced with montmorillonite and cellulose nanocrystals: fundamental properties and biodegradability study *Int. J. Biol. Macromol.* **146** 374–86

[108] Ravi Kumar M N V 2000 A review of chitin and chitosan applications *React. Funct. Polym.* **46** 1–27

[109] Pillai C K S, Paul W and Sharma C P 2009 Chitin and chitosan polymers: chemistry, solubility and fiber formation *Prog. Polym. Sci.* **34** 641–78

[110] Mogilevskaya E L, Akopova T A, Zelenetskii A N and Ozerin A N 2006 The crystal structure of chitin and chitosan *Polym. Sci.* A **48** 116–23

[111] Huang J, Zhong Y, Lu A, Zhang L and Cai J 2020 Temperature and time-dependent self-assembly and gelation behavior of chitin in aqueous KOH/urea solution *Giant* **4** 100038

[112] Morais E S, Lopes A M D C, Freire M G, Freire C S R, Coutinho J A P and Silvestre A J D 2020 Use of ionic liquids and deep eutectic solvents in polysaccharides dissolution and extraction processes towards sustainable biomass valorization *Molecules* **25** 3652

[113] Tolesa L D, Gupta B S and Lee M-J 2019 Chitin and chitosan production from shrimp shells using ammonium-based ionic liquids *Int. J. Biol. Macromol.* **130** 818–26

[114] Sun X, Wei Q, Yang Y, Xiao Z and Ren X 2022 In-depth study on the extraction and mechanism of high-purity chitin based on NADESs method *J. Environ. Chem. Eng.* **10** 106859

[115] Khajavian M, Vatanpour V, Castro-Muñoz R and Boczkaj G 2022 Chitin and derivative chitosan-based structures—preparation strategies aided by deep eutectic solvents: a review *Carbohydr. Polym.* **275** 118702

[116] Hong S, Yuan Y, Yang Q, Chen L, Deng J, Chen W, Lian H, Mota-Morales J D and Liimatainen H 2019 Choline chloride-zinc chloride deep eutectic solvent mediated preparation of partial O-acetylation of chitin nanocrystal in one step reaction *Carbohydr. Polym.* **220** 211–8

[117] Xu H, Zhang L, Zhang H, Luo J and Gao X 2021 Green fabrication of chitin/chitosan composite hydrogels and their potential applications *Macromol. Biosci.* **21** 2000389

[118] Li F, You X, Li Q, Qin D, Wang M, Yuan S, Chen X and Bi S 2021 Homogeneous deacetylation and degradation of chitin in NaOH/urea dissolution system *Int. J. Biol. Macromol.* **189** 391–7

[119] Kasprzak D and Galiński M 2021 Chitin and chitin-cellulose composite hydrogels prepared by ionic liquid-based process as the novel electrolytes for electrochemical capacitors *J. Solid State Electrochem.* **25** 2549–63

[120] King C, Shamshina J L, Gurau G, Berton P, Khan N F A F and Rogers R D 2017 A platform for more sustainable chitin films from an ionic liquid process *Green Chem.* **19** 117–26

[121] Barzic A I and Albu R M 2021 Optical properties and biointerface interactions of chitin *Polym. Bull.* **78** 6535–48

[122] Lv J, Lv X, Ma M, Oh D-H, Jiang Z and Fu X 2023 Chitin and chitin-based biomaterials: a review of advances in processing and food applications *Carbohydr. Polym.* **299** 120142

[123] Islam M M, Islam R, Mahmudul Hassan S M, Karim M R, Rahman M M, Rahman S, Nur Hossain M, Islam D, Aftab Ali Shaikh M and Georghiou P E 2023 Carboxymethyl chitin and chitosan derivatives: synthesis, characterization and antibacterial activity *Carbohydr. Polym. Technol. Appl.* **5** 100283

[124] Gheorghiţă D, Moldovan H, Robu A, Biţa A-I, Grosu E, Antoniac A, Corneschi I, Antoniac I, Bodog A D and Băcilă C I 2023 Chitosan-based biomaterials for hemostatic applications: a review of recent advances *Int. J. Mol. Sci.* **24** 10540

[125] May K L, Tangso K J, Hawley A, Boyd B J and Clulow A J 2020 Interaction of chitosan-based dietary supplements with fats during lipid digestion *Food Hydrocoll.* **108** 105965

[126] Fraunhofer IGB 2024 LaChiPur: Functionalized chitosan as a biobased flocculant for the treatment of complex wastewater *Press release* Fraunhofer Institute for Interfacial Engineering and Biotechnology IGB

[127] Kulka K and Sionkowska A 2023 Chitosan based materials in cosmetic applications: a review *Molecules* **28** 1817

[128] Rinaudo M 2006 Chitin and chitosan: properties and applications *Prog. Polym. Sci.* **31** 603–32

[129] Zhou M, Liu Z, Liu T, Zhu Y and Lin N 2022 Tradeoff between amino group and crystallinity of chitin nanocrystals as a functional component in fluorescent nail coatings *ACS Sustain. Chem. Eng.* **10** 10327–38

[130] Aranaz I, Alcántara A R, Civera M C, Arias C, Elorza B, Heras Caballero A and Acosta N 2021 Chitosan: an overview of its properties and applications *Polymers* **13** 3256

[131] Domard A 1987 pH and c.d. measurements on a fully deacetylated chitosan: application to Cu^{II}–polymer interactions *Int. J. Biol. Macromol.* **9** 98–104
[132] Raafat D and Sahl H 2009 Chitosan and its antimicrobial potential—a critical literature survey *Microb. Biotechnol.* **2** 186–201
[133] Jin T, Liu T, Lam E and Moores A 2021 Chitin and chitosan on the nanoscale *Nanoscale Horiz.* **6** 505–42
[134] Desbrières J and Babak V 2010 Interfacial properties of chitin and chitosan based systems *Soft Matter* **6** 2358
[135] Nilsen-Nygaard J, Strand S, Vårum K, Draget K and Nordgård C 2015 Chitosan: gels and interfacial properties *Polymers* **7** 552–79
[136] Payet L and Terentjev E M 2008 Emulsification and stabilization mechanisms of O/W emulsions in the presence of chitosan *Langmuir* **24** 12247–52
[137] Mai-ngam K 2006 Comblike poly(ethylene oxide)/hydrophobic C_6 branched chitosan surfactant polymers as anti-infection surface modifying agents *Colloids Surf.* B **49** 117–25
[138] Flórez M, Guerra-Rodríguez E, Cazón P and Vázquez M 2022 Chitosan for food packaging: recent advances in active and intelligent films *Food Hydrocoll.* **124** 107328
[139] Singh S, Nwabor O F, Syukri D M and Voravuthikunchai S P 2021 Chitosan-poly(vinyl alcohol) intelligent films fortified with anthocyanins isolated from *Clitoria ternatea* and *Carissa carandas* for monitoring beverage freshness *Int. J. Biol. Macromol.* **182** 1015–25
[140] Cazón P and Vázquez M 2020 Mechanical and barrier properties of chitosan combined with other components as food packaging film *Environ. Chem. Lett.* **18** 257–67
[141] Fernandes S C M, Freire C S R, Silvestre A J D, Pascoal Neto C, Gandini A, Berglund L A and Salmén L 2010 Transparent chitosan films reinforced with a high content of nano-fibrillated cellulose *Carbohydr. Polym.* **81** 394–401
[142] Bof M J, Bordagaray V C, Locaso D E and García M A 2015 Chitosan molecular weight effect on starch-composite film properties *Food Hydrocoll.* **51** 281–94
[143] Bonilla J and Sobral P J A 2016 Investigation of the physicochemical, antimicrobial and antioxidant properties of gelatin-chitosan edible film mixed with plant ethanolic extracts *Food Biosci.* **16** 17–25
[144] Valenzuela C, Abugoch L and Tapia C 2013 Quinoa protein–chitosan–sunflower oil edible film: mechanical, barrier and structural properties *LWT—Food Sci. Technol.* **50** 531–7
[145] Di Pierro P, Chico B, Villalonga R, Mariniello L, Damiao A E, Masi P and Porta R 2006 Chitosan–whey protein edible films produced in the absence or presence of transglutaminase: analysis of their mechanical and barrier properties *Biomacromolecules* **7** 744–9
[146] Baharlouei P and Rahman A 2022 Chitin and chitosan: prospective biomedical applications in drug delivery, cancer treatment, and wound healing *Mar. Drugs* **20** 460
[147] Shamshina J L, Oldham (Konak) T and Rogers R D 2019 Applications of chitin in agriculture *Sustainable Agriculture Reviews 36: Chitin and Chitosan: Applications in Food, Agriculture, Pharmacy, Medicine and Wastewater Treatment* ed G Crini and E Lichtfouse (Cham: Springer International) pp 125–46
[148] Bai L, Liu L, Esquivel M, Tardy B L, Huan S, Niu X, Liu S, Yang G, Fan Y and Rojas O J 2022 Nanochitin: chemistry, structure, assembly, and applications *Chem. Rev.* **122** 11604–74
[149] Ifuku S and Saimoto H 2012 Chitin nanofibers: preparations, modifications, and applications *Nanoscale* **4** 3308–18

[150] Carvalho L C R, Queda F, Santos C V A and Marques M M B 2016 Selective modification of chitin and chitosan: en route to tailored oligosaccharides *Chem. Asian J.* **11** 3468–81
[151] Gopalan Nair K, Dufresne A, Gandini A and Belgacem M N 2003 Crab shell chitin whiskers reinforced natural rubber nanocomposites. 3. Effect of chemical modification of chitin whiskers *Biomacromolecules* **4** 1835–42
[152] Liu P, Sehaqui H, Tingaut P, Wichser A, Oksman K and Mathew A P 2014 Cellulose and chitin nanomaterials for capturing silver ions (Ag^{+}) from water via surface adsorption *Cellulose* **21** 449–61
[153] Leuba J L and Stossel P 1986 Chitosan and other polyamines: antifungal activity and interaction with biological membranes *Chitin in Nature and Technology* ed R Muzzarelli, C Jeuniaux and G W Gooday (Boston, MA: Springer) pp 215–22
[154] Leonida M D, Belbekhouche S, Benzecry A, Peddineni M, Suria A and Carbonnier B 2018 Antibacterial hop extracts encapsulated in nanochitosan matrices *Int. J. Biol. Macromol.* **120** 1335–43
[155] Divya K, Smitha V and Jisha M S 2018 Antifungal, antioxidant and cytotoxic activities of chitosan nanoparticles and its use as an edible coating on vegetables *Int. J. Biol. Macromol.* **114** 572–7
[156] Chen W, Li Y, Yang S, Yue L, Jiang Q and Xia W 2015 Synthesis and antioxidant properties of chitosan and carboxymethyl chitosan-stabilized selenium nanoparticles *Carbohydr. Polym.* **132** 574–81
[157] Hamed I, Özogul F and Regenstein J M 2016 Industrial applications of crustacean by-products (chitin, chitosan, and chitooligosaccharides): a review *Trends Food Sci. Technol.* **48** 40–50
[158] Khajouei R A *et al* 2018 Extraction and characterization of an alginate from the Iranian brown seaweed *Nizimuddinia zanardini Int. J. Biol. Macromol.* **118** 1073–81
[159] Clementi F, Fantozzi P, Mancini F and Moresi M 1995 Optimal conditions for alginate production by *Azotobacter vinelandii Enzyme Microb. Technol.* **17** 983–8
[160] Hentati F, Ursu A V, Pierre G, Delattre C, Trica B, Abdelkafi S, Djelveh G, Dobre T and Michaud P 2019 Production, extraction and characterization of alginates from seaweeds *Handbook of Algal Technologies and Phytochemicals* (Boca Raton, FL: CRC Press)
[161] Grant G T, Morris E R, Rees D A, Smith P J C and Thom D 1973 Biological interactions between polysaccharides and divalent cations: the egg-box model *FEBS Lett.* **32** 195–8
[162] Agüero L, Zaldivar-Silva D, Peña L and Dias M L 2017 Alginate microparticles as oral colon drug delivery device: a review *Carbohydr. Polym.* **168** 32–43
[163] Qin Y, Jiang J, Zhao L, Zhang J and Wang F 2018 Applications of alginate as a functional food ingredient *Biopolymers for Food Design* (Amsterdam: Elsevier) pp 409–29
[164] Ahmad Raus R, Wan Nawawi W M F and Nasaruddin R R 2021 Alginate and alginate composites for biomedical applications *Asian J. Pharm. Sci.* **16** 280–306
[165] Nair M S, Tomar M, Punia S, Kukula-Koch W and Kumar M 2020 Enhancing the functionality of chitosan- and alginate-based active edible coatings/films for the preservation of fruits and vegetables: a review *Int. J. Biol. Macromol.* **164** 304–20
[166] Thakur S 2021 An overview on alginate based bio-composite materials for wastewater remedial *Mater. Today Proc.* **37** 3305–9
[167] Pereira R F, Carvalho A, Gil M H, Mendes A and Bártolo P J 2013 Influence of *Aloe vera* on water absorption and enzymatic *in vitro* degradation of alginate hydrogel films *Carbohydr. Polym.* **98** 311–20

[168] Eslami Z, Elkoun S, Robert M and Adjallé K 2023 A review of the effect of plasticizers on the physical and mechanical properties of alginate-based films *Molecules* **28** 6637

[169] da Silva M A, Bierhalz A C K and Kieckbusch T G 2009 Alginate and pectin composite films crosslinked with Ca^{2+} ions: effect of the plasticizer concentration *Carbohydr. Polym.* **77** 736–42

[170] Gao C, Pollet E and Avérous L 2017 Innovative plasticized alginate obtained by thermo-mechanical mixing: effect of different biobased polyols systems *Carbohydr. Polym.* **157** 669–76

[171] Pongjanyakul T and Puttipipatkhachorn S 2007 Alginate-magnesium aluminum silicate films: effect of plasticizers on film properties, drug permeation and drug release from coated tablets *Int. J. Pharm.* **333** 34–44

[172] El Miri N, Aziz F, Aboulkas A, El Bouchti M, Ben Youcef H and El Achaby M 2018 Effect of plasticizers on physicochemical properties of cellulose nanocrystals filled alginate bionanocomposite films *Adv. Polym. Tech.* **37** 3171–85

[173] Chen P, Xie F, Tang F and McNally T 2020 Unexpected plasticization effects on the structure and properties of polyelectrolyte complexed chitosan/alginate materials *ACS Appl. Polym. Mater.* **2** 2957–66

[174] Remuñán-lópez C and Bodmeier R 1996 Mechanical and water vapor transmission properties of polysaccharide films *Drug Dev. Ind. Pharm.* **22** 1201–9

[175] Zharkevich V, Melekhavets N, Savitskaya T and Hrynshpan D 2023 Enhancement of barrier properties regarding contaminants from recycled paperboard by coating packaging materials with starch and sodium alginate blends *Sustain. Chem. Pharm.* **32** 101001

[176] Kadzińska J, Bryś J, Ostrowska-Ligęza E, Estéve M and Janowicz M 2020 Influence of vegetable oils addition on the selected physical properties of apple–sodium alginate edible films *Polym. Bull.* **77** 883–900

[177] Senturk Parreidt T, Schott M, Schmid M and Müller K 2018 Effect of presence and concentration of plasticizers, vegetable oils, and surfactants on the properties of sodium-alginate-based edible coatings *Int. J. Mol. Sci.* **19** 742

[178] Baek S-K, Kim S and Song K B 2018 Characterization of *Ecklonia cava* alginate films containing cinnamon essential oils *Int. J. Mol. Sci.* **19** 3545

[179] Frank K, Garcia C V, Shin G H and Kim J T 2018 Alginate biocomposite films incorporated with cinnamon essential oil nanoemulsions: physical, mechanical, and anti-bacterial properties *Int. J. Polym. Sci.* **2018** 1–8

[180] Jansi R, Vinay B, Revathy M S, Sasikumar P, Marasamy L, Janani A, Haldhar R, Kim S-C, Almarhoon Z M and Hossain M K 2024 Synergistic blends of sodium alginate and pectin biopolymer hosts as conducting electrolytes for electrochemical applications *ACS Omega* **9** 13906–16

[181] Aadil K R, Prajapati D and Jha H 2016 Improvement of physcio-chemical and functional properties of alginate film by *Acacia* lignin *Food Packag. Shelf Life* **10** 25–33

[182] Ye Z, Ma P, Tang M, Li X, Zhang W, Hong X, Chen X and Chen D 2017 Interactions between calcium alginate and carrageenan enhanced mechanical property of a natural composite film for general packaging application *Polym. Bull.* **74** 3421–9

[183] Siddaramaiah , Swamy T M M, Ramaraj B and Lee J H 2008 Sodium alginate and its blends with starch: thermal and morphological properties *J. Appl. Polym. Sci.* **109** 4075–81

[184] Li K, Zhu J, Guan G and Wu H 2019 Preparation of chitosan-sodium alginate films through layer-by-layer assembly and ferulic acid crosslinking: film properties, characterization, and formation mechanism *Int. J. Biol. Macromol.* **122** 485–92
[185] Zhang X, Jiang Y, Han L and Lu X 2021 Biodegradable polymer hydrogel-based tissue adhesives: a review *Biosurf. Biotribol.* **7** 163–79
[186] Davidovich-Pinhas M and Bianco-Peled H 2011 Alginate–PEGAc: a new mucoadhesive polymer *Acta Biomater.* **7** 625–33
[187] Jeon O, Samorezov J E and Alsberg E 2014 Single and dual crosslinked oxidized methacrylated alginate/PEG hydrogels for bioadhesive applications *Acta Biomater.* **10** 47–55
[188] Xie Y, Liu C, Zhang J, Li Y, Li B and Liu S 2024 Crosslinking alginate at water-in-water pickering emulsions interface to control the interface structure and enhance the stress resistance of the encapsulated probiotics *J. Colloid Interface Sci.* **655** 653–63
[189] Gao Z, Gao C, Jiang W, Xu L, Hu B, Yao X, Li Y and Wu Y 2023 *In situ* crosslinking sodium alginate on oil–water interface to stabilize the O/W emulsions *Food Hydrocoll.* **135** 108233
[190] Ebringerová A, Hromádková Z and Heinze T 2005 Hemicellulose *Polysaccharides I Advances in Polymer Science* **vol 186** ed T Heinze (Berlin: Springer) pp 1–67
[191] Peterson A A, Vogel F, Lachance R P, Fröling M, Antal M J and Tester J W 2008 Thermochemical biofuel production in hydrothermal media: a review of sub- and supercritical water technologies *Energy Environ. Sci.* **1** 32
[192] Hosseinaei O, Wang S, Enayati A A and Rials T G 2012 Effects of hemicellulose extraction on properties of wood flour and wood–plastic composites *Composites* A **43** 686–94
[193] Terrett O M and Dupree P 2019 Covalent interactions between lignin and hemicelluloses in plant secondary cell walls *Curr. Opin. Biotechnol.* **56** 97–104
[194] Huang L-Z, Ma M-G, Ji X-X, Choi S-E and Si C 2021 Recent developments and applications of hemicellulose from wheat straw: a review *Front. Bioeng. Biotechnol.* **9** 690773
[195] Kent J 2024 Global hemicellulose market by type (polyxylose, polyglucose mannose), by application (alcohol, food), by geographic scope and forecast *Report* 522070 Verified Market Reports https://www.verifiedmarketreports.com/product/hemicellulose-market-size-and-forecast/#:%E2%88%BC:text=Hemicellulose%20Market%20size%20was%20valued,%2C%20distributed%2C%20and%20used%20worldwide (Accessed: 4 May 2024)
[196] Peng F, Bian J, Peng P, Xiao H, Ren J-L, Xu F and Sun R-C 2012 Separation and characterization of acetyl and non-acetyl hemicelluloses of *Arundo donax* by ammonium sulfate precipitation *J. Agric. Food Chem.* **60** 4039–47
[197] Pere J, Pääkkönen E, Ji Y and Retulainen E 2018 Influence of the hemicellulose content on the fiber properties, strength, and formability of handsheets *BioRes* **14** 251–63
[198] Lima D U, Loh W and Buckeridge M S 2004 Xyloglucan–cellulose interaction depends on the sidechains and molecular weight of xyloglucan *Plant Physiol. Biochem.* **42** 389–94
[199] Pękala P, Szymańska-Chargot M and Zdunek A 2023 Interactions between non-cellulosic plant cell wall polysaccharides and cellulose emerging from adsorption studies *Cellulose* **30** 9221–39
[200] Khodayari A, Thielemans W, Hirn U, Van Vuure A W and Seveno D 2021 Cellulose–hemicellulose interactions—a nanoscale view *Carbohydr. Polym.* **270** 118364

[201] Lahtinen M H, Valoppi F, Juntti V, Heikkinen S, Kilpeläinen P O, Maina N H and Mikkonen K S 2019 Lignin-rich PHWE hemicellulose extracts responsible for extended emulsion stabilization *Front. Chem.* **7** 871

[202] Zhao Y, Shakeel U, Saif Ur Rehman M, Li H, Xu X and Xu J 2020 Lignin-carbohydrate complexes (LCCs) and its role in biorefinery *J. Clean. Prod.* **253** 120076

[203] Carvalho D M D, Lahtinen M H, Lawoko M and Mikkonen K S 2020 Enrichment and identification of lignin–carbohydrate complexes in softwood extract *ACS Sustain. Chem. Eng.* **8** 11795–804

[204] Rao J, Lv Z, Chen G and Peng F 2023 Hemicellulose: structure, chemical modification, and application *Prog. Polym. Sci.* **140** 101675

[205] Mikkonen K S, Xu C, Berton-Carabin C and Schroën K 2016 Spruce galactoglucomannans in rapeseed oil-in-water emulsions: efficient stabilization performance and structural partitioning *Food Hydrocoll.* **52** 615–24

[206] Mikkonen K S, Merger D, Kilpeläinen P, Murtomäki L, Schmidt U S and Wilhelm M 2016 Determination of physical emulsion stabilization mechanisms of wood hemicelluloses via rheological and interfacial characterization *Soft Matter* **12** 8690–700

[207] Farhat W, Venditti R A, Hubbe M, Taha M, Becquart F and Ayoub A 2017 A review of water-resistant hemicellulose-based materials: processing and applications *ChemSusChem* **10** 305–23

[208] Lopez M, Bizot H, Chambat G, Marais M-F, Zykwinska A, Ralet M-C, Driguez H and Buléon A 2010 Enthalpic studies of xyloglucan−cellulose interactions *Biomacromolecules* **11** 1417–28

[209] Kabel M A, Van Den Borne H, Vincken J-P, Voragen A G J and Schols H A 2007 Structural differences of xylans affect their interaction with cellulose *Carbohydr. Polym.* **69** 94–105

[210] Kishani S, Vilaplana F, Ruda M, Hansson P and Wågberg L 2020 Influence of solubility on the adsorption of different xyloglucan fractions at cellulose–water interfaces *Biomacromolecules* **21** 772–82

[211] Bosmans T J, Stépán A M, Toriz G, Renneckar S, Karabulut E, Wågberg L and Gatenholm P 2014 Assembly of debranched xylan from solution and on nanocellulosic surfaces *Biomacromolecules* **15** 924–30

[212] Andrewartha K A, Phillips D R and Stone B A 1979 Solution properties of wheat-flour arabinoxylans and enzymically modified arabinoxylans *Carbohydr. Res.* **77** 191–204

[213] Olorunsola E O, Akpabio E I, Adedokun M O and Ajibola D O 2018 Emulsifying properties of hemicelluloses *Science and Technology Behind Nanoemulsions* ed S Karakuş (Rijeka: InTech)

[214] Hagiopol C 2021 Natural polyphenols applications *Natural Polyphenols from Wood* (Amsterdam: Elsevier) pp 259–314

[215] Fulcrand H, Rouméas L, Billerach G, Aouf C and Dubreucq E 2019 Advances in bio-based thermosetting polymers *Recent Advances in Polyphenol Research* ed H Halbwirth, K Stich, V Cheynier and S Quideau (New York: Wiley) pp 285–334

[216] Rezg R, El-Fazaa S, Gharbi N and Mornagui B 2014 Bisphenol A and human chronic diseases: current evidences, possible mechanisms, and future perspectives *Environ. Int.* **64** 83–90

[217] Vandenberg L N *et al* 2012 Hormones and endocrine-disrupting chemicals: low-dose effects and nonmonotonic dose responses *Endocr. Rev.* **33** 378–455

[218] Brenelli L B *et al* 2016 Acidification treatment of lignin from sugarcane bagasse results in fractions of reduced polydispersity and high free-radical scavenging capacity *Ind. Crops Prod.* **83** 94–103

[219] Vanholme R, De Meester B, Ralph J and Boerjan W 2019 Lignin biosynthesis and its integration into metabolism *Curr. Opin. Biotechnol.* **56** 230–9

[220] Borrega M, Päärnilä S, Greca L G, Jääskeläinen A-S, Ohra-aho T, Rojas O J and Tamminen T 2020 Morphological and wettability properties of thin coating films produced from technical lignins *Langmuir* **36** 9675–84

[221] Figueiredo P, Lintinen K, Hirvonen J T, Kostiainen M A and Santos H A 2018 Properties and chemical modifications of lignin: towards lignin-based nanomaterials for biomedical applications *Prog. Mater Sci.* **93** 233–69

[222] Christopher L P, Yao B and Ji Y 2014 Lignin biodegradation with laccase-mediator systems *Front. Energy Res.* **2**

[223] Lisý A, Ház A, Nadányi R, Jablonský M and Šurina I 2022 About hydrophobicity of lignin: a review of selected chemical methods for lignin valorisation in biopolymer production *Energies* **15** 6213

[224] Ruwoldt J, Heen Blindheim F and Chinga-Carrasco G 2023 Functional surfaces, films, and coatings with lignin—a critical review *RSC Adv.* **13** 12529–53

[225] Mahmood N, Yuan Z, Schmidt J and Xu C (C) 2016 Depolymerization of lignins and their applications for the preparation of polyols and rigid polyurethane foams: a review *Renew. Sustain. Energy Rev.* **60** 317–29

[226] Fagbemigun T K and Mai C 2023 Production and characterisation of self-blowing lignin-based foams *Eur. J. Wood Prod.* **81** 579–90

[227] Carlos De Haro J, Magagnin L, Turri S and Griffini G 2019 Lignin-based anticorrosion coatings for the protection of aluminum surfaces *ACS Sustain. Chem. Eng.* **7** 6213–22

[228] Zhou S-J, Wang H-M, Xiong S-J, Sun J-M, Wang Y-Y, Yu S, Sun Z, Wen J-L and Yuan T-Q 2021 Technical lignin valorization in biodegradable polyester-based plastics (BPPs) *ACS Sustain. Chem. Eng.* **9** 12017–42

[229] Lu X and Gu X 2023 A review on lignin-based epoxy resins: lignin effects on their synthesis and properties *Int. J. Biol. Macromol.* **229** 778–90

[230] Gao Z, Lang X, Chen S and Zhao C 2021 Mini-review on the synthesis of lignin-based phenolic resin *Energy Fuels* **35** 18385–95

[231] Gong X, Meng Y, Lu J, Tao Y, Cheng Y and Wang H 2022 A review on lignin-based phenolic resin adhesive *Macro Chem. Phys.* **223** 2100434

[232] Parit M and Jiang Z 2020 Towards lignin derived thermoplastic polymers *Int. J. Biol. Macromol.* **165** 3180–97

[233] Tanase-Opedal M, Espinosa E, Rodríguez A and Chinga-Carrasco G 2019 Lignin: a biopolymer from forestry biomass for biocomposites and 3D printing *Materials* **12** 3006

[234] Sameni J, Krigstin S, Santos Rosa D D, Leao A and Sain M 2013 Thermal characteristics of lignin residue from industrial processes *BioResources* **9** 725–37

[235] Kun D and Pukánszky B 2017 Polymer/lignin blends: interactions, properties, applications *Eur. Polym. J.* **93** 618–41

[236] Ariffin H, Sapuan S M and Hassan M A 2019 *Lignocellulose for Future Bioeconomy* (Amsterdam: Elsevier)

[237] Wang H, Pu Y, Ragauskas A and Yang B 2019 From lignin to valuable products—strategies, challenges, and prospects *Bioresour. Technol.* **271** 449–61

[238] Kienberger M, Maitz S, Pichler T and Demmelmayer P 2021 Systematic review on isolation processes for technical lignin *Processes* **9** 804

[239] Shorey R, Gupta A and Mekonnen T H 2021 Hydrophobic modification of lignin for rubber composites *Ind. Crops Prod.* **174** 114189

[240] Lizundia E, Sipponen M H, Greca L G, Balakshin M, Tardy B L, Rojas O J and Puglia D 2021 Multifunctional lignin-based nanocomposites and nanohybrids *Green Chem.* **23** 6698–760

[241] Pereira A, Hoeger I C, Ferrer A, Rencoret J, Del Rio J C, Kruus K, Rahikainen J, Kellock M, Gutiérrez A and Rojas O J 2017 Lignin films from spruce, eucalyptus, and wheat straw studied with electroacoustic and optical sensors: effect of composition and electrostatic screening on enzyme binding *Biomacromolecules* **18** 1322–32

[242] Tammelin T, Österberg M, Johansson L-S and Laine J 2006 Preparation of lignin and extractive model surfaces by using spincoating technique—application for QCM-D studies *Nordic Pulp Paper Res. J.* **21** 444–50

[243] Norgren M, Notley S M, Majtnerova A and Gellerstedt G 2006 Smooth model surfaces from lignin derivatives. I. Preparation and characterization *Langmuir* **22** 1209–14

[244] Belgacem M N, Blayo A and Gandini A 1996 Surface characterization of polysaccharides, lignins, printing ink pigments, and ink fillers by inverse gas chromatography *J. Colloid Interface Sci.* **182** 431–6

[245] Notley S M and Norgren M 2010 Surface energy and wettability of spin-coated thin films of lignin isolated from wood *Langmuir* **26** 5484–90

[246] Pasquini D, Balogh D T, Antunes P A, Constantino C J L, Curvelo A A S, Aroca R F and Oliveira O N 2002 Surface morphology and molecular organization of lignins in Langmuir –Blodgett films *Langmuir* **18** 6593–6

[247] Kopacic S, Ortner A, Guebitz G, Kraschitzer T, Leitner J and Bauer W 2018 Technical lignins and their utilization in the surface sizing of paperboard *Ind. Eng. Chem. Res.* **57** 6284–91

[248] Hao C, Liu T, Zhang S, Brown L, Li R, Xin J, Zhong T, Jiang L and Zhang J 2019 A High-lignin-content, removable, and glycol-assisted repairable coating based on dynamic covalent bonds *ChemSusChem.* **12** 1049–58

[249] Bode D, Wilson P and Craun G P 2015 Lignin based coating compositions *Patent* Worldwide WO2014095800A1

[250] Wang H, Qiu X, Liu W, Fu F and Yang D 2017 A novel lignin/ZnO hybrid nanocomposite with excellent UV-absorption ability and its application in transparent polyurethane coating *Ind. Eng. Chem. Res.* **56** 11133–41

[251] Griffini G, Passoni V, Suriano R, Levi M and Turri S 2015 Polyurethane coatings based on chemically unmodified fractionated lignin *ACS Sustain. Chem. Eng.* **3** 1145–54

[252] Hajirahimkhan S, Xu C C and Ragogna P J 2018 Ultraviolet curable coatings of modified lignin *ACS Sustain. Chem. Eng.* **6** 14685–94

[253] Henn K A, Forsman N, Zou T and Österberg M 2021 Colloidal lignin particles and epoxies for bio-based, durable, and multiresistant nanostructured coatings *ACS Appl. Mater. Interfaces* **13** 34793–806

[254] Zhao C, Huang J, Yang L, Yue F and Lu F 2019 Revealing structural differences between alkaline and kraft lignins by HSQC NMR *Ind. Eng. Chem. Res.* **58** 5707

[255] Gong Z, Yang G, Huang L, Chen L, Luo X and Shuai L 2023 Phenol-assisted depolymerisation of condensed lignins to mono-/poly-phenols and bisphenols *Chem. Eng. J.* **455** 140628

[256] Alonso M V, Rodríguez J J, Oliet M, Rodríguez F, García J and Gilarranz M A 2001 Characterization and structural modification of ammonic lignosulfonate by methylolation *J. Appl. Polym. Sci.* **82** 2661–8

[257] Zhao L, Griggs B, Chen C-L, Gratzl J and Hse C-Y 1994 Utilization of softwood kraft lignin as adhesive for the manufacture of reconstituted wood *J. Wood Chem. Tech.* **14** 127–45

[258] Du X, Li J and Lindström M E 2014 Modification of industrial softwood kraft lignin using Mannich reaction with and without phenolation pretreatment *Ind. Crops Prod.* **52** 729–35

[259] Matsushita Y and Yasuda S 2003 Reactivity of a condensed-type lignin model compound in the Mannich reaction and preparation of cationic surfactant from sulfuric acid lignin *J. Wood. Sci.* **49** 166–71

[260] Podschun J, Stücker A, Buchholz R I, Heitmann M, Schreiber A, Saake B and Lehnen R 2016 Phenolated lignins as reactive precursors in wood veneer and particleboard adhesion *Ind. Eng. Chem. Res.* **55** 5231–7

[261] Jiang X, Liu J, Du X, Hu Z, Chang H and Jameel H 2018 Phenolation to improve lignin reactivity toward thermosets application *ACS Sustain. Chem. Eng.* **6** 5504–12

[262] Khan T A, Lee J-H and Kim H-J 2019 Lignin-based adhesives and coatings *Lignocellulose for Future Bioeconomy* (Amsterdam: Elsevier) pp 153–206

[263] Huang C, Peng Z, Li J, Li X, Jiang X and Dong Y 2022 Unlocking the role of lignin for preparing the lignin-based wood adhesive: a review *Ind. Crops Prod.* **187** 115388

[264] Yang G, Gong Z, Luo X, Chen L and Shuai L 2023 Bonding wood with uncondensed lignins as adhesives *Nature* **621** 511–5

[265] Zhang W, Qiu X, Wang C, Zhong L, Fu F, Zhu J, Zhang Z, Qin Y, Yang D and Xu C C 2022 Lignin derived carbon materials: current status and future trends *Carbon Res.* **1** 14

[266] Guizani C, Widsten P, Siipola V, Paalijärvi R, Berg J, Pasanen A, Kalliola A and Torvinen K 2024 New insights into the chemical activation of lignins and tannins using K_2CO_3—a combined thermoanalytical and structural study *Carbon Lett.* **34** 371–86

[267] Xi Y, Yang D, Qiu X, Wang H, Huang J and Li Q 2018 Renewable lignin-based carbon with a remarkable electrochemical performance from potassium compound activation *Ind. Crops Prod.* **124** 747–54

[268] Pérez-Rodríguez S, Pinto O, Izquierdo M T, Segura C, Poon P S, Celzard A, Matos J and Fierro V 2021 Upgrading of pine tannin biochars as electrochemical capacitor electrodes *J. Colloid Interface Sci.* **601** 863–76

[269] Jia G, Yu Y, Wang X, Jia C, Hu Z, Yu S, Xiang H and Zhu M 2023 Highly conductive and porous lignin-derived carbon fibers *Mater. Horiz.* **10** 5847–58

[270] Puziy A M, Poddubnaya O I and Sevastyanova O 2018 Carbon materials from technical lignins: recent advances *Top. Curr. Chem.* **376** 33

[271] Arbenz A and Avérous L 2015 Chemical modification of tannins to elaborate aromatic biobased macromolecular architectures *Green Chem.* **17** 2626–46

[272] Dhawale P V, Vineeth S K, Gadhave R V, Jabeen Fatima M J, Vijay Supekar M, Kumar Thakur V and Raghavan P 2022 Tannin as a renewable raw material for adhesive applications: a review *Mater. Adv.* **3** 3365–88

[273] Shirmohammadli Y, Efhamisisi D and Pizzi A 2018 Tannins as a sustainable raw material for green chemistry: a review *Ind. Crops Prod.* **126** 316–32

[274] Arbenz A and Avérous L 2014 Synthesis and characterization of fully biobased aromatic polyols—oxybutylation of condensed tannins towards new macromolecular architectures *RSC Adv.* **4** 61564–72

[275] Santos S C R, Bacelo H A M, Boaventura R A R and Botelho C M S 2019 Tannin-adsorbents for water decontamination and for the recovery of critical metals: current state and future perspectives *Biotechnol. J.* **14** 1900060

[276] Gao M, Wang Z, Yang C, Ning J, Zhou Z and Li G 2019 Novel magnetic graphene oxide decorated with persimmon tannins for efficient adsorption of malachite green from aqueous solutions *Colloids Surf.* A **566** 48–57

[277] Hagerman A E and Butler L G 1981 The specificity of proanthocyanidin-protein interactions *J. Biol. Chem.* **256** 4494–7

[278] Yang S, Wu W, Jiao Y, Cai Z and Fan H 2017 Preparation of NBR/tannic acid composites by assembling a weak IPN structure *Compos. Sci. Technol.* **153** 40–7

[279] Zhai Y, Wang J, Wang H, Song T, Hu W and Li S 2018 Preparation and characterization of antioxidative and UV-protective larch bark tannin/PVA composite membranes *Molecules* **23** 2073

[280] Shutova T G, Agabekov V E and Lvov Y M 2007 Reaction of radical cations with multilayers of tannic acid and polyelectrolytes *Russ. J. Gen. Chem.* **77** 1494–501

[281] Yen K-C, Mandal T K and Woo E M 2008 Enhancement of bio-compatibility via specific interactions in polyesters modified with a bio-resourceful macromolecular ester containing polyphenol groups *J. Biomed. Mater. Res.* A **86** 701–12

[282] Pizzi A 2009 Polyflavonoid tannins self-condensation adhesives for wood particleboard *J. Adhes.* **85** 57–68

[283] Morris R, Black K A and Stollar E J 2022 Uncovering protein function: from classification to complexes *Essays Biochem.* **66** 255–85

[284] Miserez A, Yu J and Mohammadi P 2023 Protein-based biological materials: molecular design and artificial production *Chem. Rev.* **123** 2049–111

[285] Abascal N C and Regan L 2018 The past, present and future of protein-based materials *Open Biol.* **8** 180113

[286] Sato K, Oide M and Nakasako M 2023 Prediction of hydrophilic and hydrophobic hydration structure of protein by neural network optimized using experimental data *Sci Rep.* **13** 2183

[287] Sun J, Su J, Ma C, Göstl R, Herrmann A, Liu K and Zhang H 2020 Fabrication and mechanical properties of engineered protein-based adhesives and fibers *Adv. Mater.* **32** 1906360

[288] Huang F F and Rha C 1974 Protein structures and protein fibers—a review *Polym. Eng. Sci.* **14** 81–91

[289] Qi Y, Wang H, Wei K, Yang Y, Zheng R-Y, Kim I S and Zhang K-Q 2017 A review of structure construction of silk fibroin biomaterials from single structures to multi-level structures *Int. J. Mol. Sci.* **18** 237

[290] Zhao C, Xiao Y, Ling S, Pei Y and Ren J 2021 Structure of collagen *Fibrous Proteins: Design, Synthesis, and Assembly* ed S Ling (New York: Springer) pp 17–25

[291] Schreiber G 2020 Protein–protein interaction interfaces and their functional implications *Protein—Protein Interaction Regulators* ed S Roy and H Fu (London: The Royal Society of Chemistry) pp 1–24

[292] Schmidt D R, Waldeck H and Kao W J 2009 Protein adsorption to biomaterials *Biological Interactions on Materials Surfaces* ed D A Puleo and R Bizios (New York: Springer) pp 1–18

[293] Du N, Ye F, Sun J and Liu K 2022 Stimuli-responsive natural proteins and their applications *ChemBioChem.* **23** e202100416

[294] Dong C and Lv Y 2016 Application of collagen scaffold in tissue engineering: recent advances and new perspectives *Polymers* **8** 42

[295] Schiller T and Scheibel T 2024 Bioinspired and biomimetic protein-based fibers and their applications *Commun. Mater.* **5** 1–18

[296] Calva-Estrada S J, Jiménez-Fernández M and Lugo-Cervantes E 2019 Protein-based films: advances in the development of biomaterials applicable to food packaging *Food Eng. Rev.* **11** 78–92

[297] Yue H, Mai L, Xu C, Yang C, Shuttleworth P S and Cui Y 2023 Recent advancement in bio-based adhesives derived from plant proteins for plywood application: a review *Sustain. Chem. Pharm.* **33** 101143

[298] Gough C R, Callaway K, Spencer E, Leisy K, Jiang G, Yang S and Hu X 2021 Biopolymer-based filtration materials *ACS Omega* **6** 11804–12

[299] Gosline J M, Denny M W and DeMont M E 1984 Spider silk as rubber *Nature* **309** 551–2

[300] Tao H, Kaplan D L and Omenetto F G 2012 Silk materials—a road to sustainable high technology *Adv. Mater.* **24** 2824–37

[301] Cho S Y, Yun Y S, Jang D, Jeon J W, Kim B H, Lee S and Jin H-J 2017 Ultra strong pyroprotein fibres with long-range ordering *Nat. Commun.* **8** 74

[302] Sanchez Ramirez D O, Vineis C, Cruz-Maya I, Tonetti C, Guarino V and Varesano A 2023 Wool keratin nanofibers for bioinspired and sustainable use in biomedical field *J. Funct. Biomater.* **14** 5

[303] Guo C, Zhang J, Jordan J S, Wang X, Henning R W and Yarger J L 2018 Structural comparison of various silkworm silks: an insight into the structure–property relationship *Biomacromolecules* **19** 906–17

[304] Zhao G, Qing H, Huang G, Genin G M, Lu T J, Luo Z, Xu F and Zhang X 2018 Reduced graphene oxide functionalized nanofibrous silk fibroin matrices for engineering excitable tissues *NPG Asia Mater.* **10** 982–94

[305] Zhang C, Wang X, Fan S, Lan P, Cao C and Zhang Y 2021 Silk fibroin/reduced graphene oxide composite mats with enhanced mechanical properties and conductivity for tissue engineering *Colloids Surf.* B **197** 111444

[306] Wang R, de Kort B J, Smits A I P M and Weiss A S 2020 Elastin in vascular grafts *Tissue-Engineered Vascular Grafts* ed B H Walpoth, H Bergmeister, G L Bowlin, D Kong, J I Rotmans and P Zilla (Cham: Springer International) pp 379–410

[307] Gough C R, Rivera-Galletti A, Cowan D A, Salas-de la Cruz D and Hu X 2020 Protein and polysaccharide-based fiber materials generated from ionic liquids: a review *Molecules* **25** 3362

[308] Chen T, Duan M, Shi P and Fang S 2017 Ultrathin nanoporous membranes derived from protein-based nanospheres for high-performance smart molecular filtration *J. Mater. Chem.* A*5 20208–16*

[309] Lubasova D, Netravali A, Parker J and Ingel B 2014 Bacterial filtration efficiency of green soy protein based nanofiber air filter *J. Nanosci. Nanotechnol.* **14** 4891–8

[310] Li K, Li C, Tian H, Yuan L, Xiang A, Wang C, Li J and Rajulu A V 2020 Multifunctional and efficient air filtration: a natural nanofilter prepared with zein and polyvinyl alcohol *Macromol. Mater. Eng.* **305** 2000239

[311] Kadam V, Truong Y B, Schutz J, Kyratzis I L, Padhye R and Wang L 2021 Gelatin/β-cyclodextrin bio-nanofibers as respiratory filter media for filtration of aerosols and volatile organic compounds at low air resistance *J. Hazard. Mater.* **403** 123841

[312] Wang Z, Cui Y, Feng Y, Guan L, Dong M, Liu Z and Liu L 2021 A versatile silk fibroin based filtration membrane with enhanced mechanical property, disinfection and biodegradability *Chem. Eng. J.* **426** 131947

[313] Yang F and Yang P 2021 Protein-based separation membranes: state of the art and future trends *Adv. Energy Sustain. Res.* **2** 2100008

[314] Wu M-B, Yang F, Yang J, Zhong Q, Körstgen V, Yang P, Müller-Buschbaum P and Xu Z-K 2020 Lysozyme membranes promoted by hydrophobic substrates for ultrafast and precise organic solvent nanofiltration *Nano Lett.* **20** 8760–7

[315] Bolisetty S and Mezzenga R 2016 Amyloid–carbon hybrid membranes for universal water purification *Nat. Nanotech.* **11** 365–71

[316] White A P, Gibson D L, Collinson S K, Banser P A and Kay W W 2003 Extracellular polysaccharides associated with thin aggregative fimbriae of *Salmonella enterica* Serovar Enteritidis *J. Bacteriol.* **185** 5398–407

[317] Bolisetty S, Coray N M, Palika A, Prenosil G A and Mezzenga R 2020 Amyloid hybrid membranes for removal of clinical and nuclear radioactive wastewater *Environ. Sci.: Water Res. Technol.* **6** 3249–54

[318] Bolisetty S, Arcari M, Adamcik J and Mezzenga R 2015 Hybrid amyloid membranes for continuous flow catalysis *Langmuir* **31** 13867–73

[319] Lin S, Liu W, Fu X, Luo M, Liu H and Zhong W-H 2023 An all-protein aerogel with a nanofiber/foam structure for versatile air filtration *Mater. Today Chem.* **34** 101760

[320] Souzandeh H, Johnson K S, Wang Y, Bhamidipaty K and Zhong W-H 2016 Soy-protein-based nanofabrics for highly efficient and multifunctional air filtration *ACS Appl. Mater. Interfaces* **8** 20023–31

[321] Souzandeh H, Molki B, Zheng M, Beyenal H, Scudiero L, Wang Y and Zhong W-H 2017 Cross-linked protein nanofilter with antibacterial properties for multifunctional air filtration *ACS Appl. Mater. Interfaces* **9** 22846–55

[322] Christopherson D A, Yao W C, Lu M, Vijayakumar R and Sedaghat A R 2020 High-Efficiency particulate air filters in the era of COVID-19: function and efficacy *Otolaryngol.–Head Neck Surg.* **163** 1153–5

[323] Kandasamy S, Yoo J, Yun J, Kang H-B, Seol K-H, Kim H-W and Ham J-S 2021 Application of whey protein-based edible films and coatings in food industries: an updated overview *Coatings* **11** 1056

[324] Jaski A C, Schmitz F, Horta R P, Cadorin L, Da Silva B J G, Andreaus J, Paes M C D, Riegel-Vidotti I C and Zimmermann L M 2022 Zein—a plant-based material of growing importance: new perspectives for innovative uses *Ind. Crops Prod.* **186** 115250

[325] Masanabo M A, Ray S S and Emmambux M N 2022 Properties of thermoplastic maize starch–zein composite films prepared by extrusion process under alkaline conditions *Int. J. Biol. Macromol.* **208** 443–52

[326] Wang Z, Hu S and Wang H 2017 Scale-up preparation and characterization of collagen/sodium alginate blend films *J. Food Qual.* **2017** 4954259

[327] Peng X *et al* 2022 High-strength collagen-based composite films regulated by water-soluble recombinant spider silk proteins and water annealing *ACS Biomater. Sci. Eng.* **8** 3341–53

[328] Cho S Y, Lee S Y and Rhee C 2010 Edible oxygen barrier bilayer film pouches from corn zein and soy protein isolate for olive oil packaging *LWT—Food Sci. Technol.* **43** 1234–9

[329] Vanden Braber N L, Di Giorgio L, Aminahuel C A, Díaz Vergara L I, Martín Costa A O, Montenegro M A and Mauri A N 2021 Antifungal whey protein films activated with low quantities of water soluble chitosan *Food Hydrocoll.* **110** 106156

[330] Galus S and Kadzińska J 2016 Whey protein edible films modified with almond and walnut oils *Food Hydrocoll.* **52** 78–86

[331] Bhaskar R, Zo S M, Narayanan K B, Purohit S D, Gupta M K and Han S S 2023 Recent development of protein-based biopolymers in food packaging applications: a review *Polym. Test.* **124** 108097

[332] Pushp P, Bhaskar R, Kelkar S, Sharma N, Pathak D and Gupta M K 2021 Plasticized poly (vinylalcohol) and poly(vinylpyrrolidone) based patches with tunable mechanical properties for cardiac tissue engineering applications *Biotechnol. Bioeng.* **118** 2312–25

[333] Mohamed S A A, El-Sakhawy M and El-Sakhawy M A-M 2020 Polysaccharides, protein and lipid -based natural edible films in food packaging: a review *Carbohydr. Polym.* **238** 116178

[334] Song T, Qian S, Lan T, Wu Y, Liu J and Zhang H 2022 Recent advances in bio-based smart active packaging materials *Foods* **11** 2228

[335] Song H, Choi I, Choi Y J, Yoon C S and Han J 2020 High gas barrier properties of whey protein isolate-coated multi-layer film at pilot plant facility and its application to frozen marinated meatloaf packaging *Food Packag. Shelf Life* **26** 100599

[336] Shubha A, Sharmita G and Anita L 2024 Production and characterization of human hair keratin bioplastic films with novel plasticizers *Sci. Rep.* **14** 1186

[337] Silverman H G and Roberto F F 2007 Understanding marine mussel adhesion *Mar. Biotechnol.* **9** 661–81

[338] Heinritz C, Ng X J and Scheibel T 2024 Bio-inspired protein-based and activatable adhesion systems *Adv. Funct. Mater.* **34** 2303609

[339] Castillo J J, Shanbhag B K and He L 2017 Comparison of natural extraction and recombinant mussel adhesive proteins approaches *Food Bioactives: Extraction and Biotechnology Applications* ed M Puri (Cham: Springer International) pp 111–35

[340] Oni O V, Lawrence M A, Zappi M E and Chirdon W M 2023 A review of strategies to enhance the water resistance of green wood adhesives produced from sustainable protein sources *Sustainability* **15** 14779

[341] Chen N, Lin Q, Zheng P, Rao J, Zeng Q and Sun J 2019 A sustainable bio-based adhesive derived from defatted soy flour and epichlorohydrin *Wood Sci. Technol.* **53** 801–17

[342] Zhu X, Song C, Sun X, Wang D, Cai D, Wang Z, Chen Y and Chen X 2021 Improved water resistance of TA-modified soy adhesive: effect of complexation *Int. J. Adhes. Adhes.* **108** 102858

[343] Zhang W, Sun H, Zhu C, Wan K, Zhang Y, Fang Z and Ai Z 2018 Mechanical and water-resistant properties of rice straw fiberboard bonded with chemically-modified soy protein adhesive *RSC Adv.* **8** 15188–95

[344] Li J, Zhang B, Li X, Yi Y, Shi F, Guo J and Gao Z 2019 Effects of typical soybean meal type on the properties of soybean-based adhesive *Int. J. Adhes. Adhes.* **90** 15–21
[345] Qi G, Li N, Wang D and Sun X S 2016 Development of high-strength soy protein adhesives modified with sodium montmorillonite clay *J. Am. Oil Chem. Soc.* **93** 1509–17
[346] Schwarzenbrunner R, Barbu M C, Petutschnigg A and Tudor E M 2020 Water-resistant casein-based adhesives for veneer bonding in biodegradable ski cores *Polymers* **12** 1745
[347] Pojanavaraphan T, Magaraphan R, Chiou B-S and Schiraldi D A 2010 Development of biodegradable foamlike materials based on casein and sodium montmorillonite clay *Biomacromolecules* **11** 2640–6
[348] Yamamoto H and Takimoto T 1991 Synthesis and conformational study of cuticle collagen models and application as a bioadhesive *J. Mater. Chem.* **1** 947–54
[349] Li X, Zhou Y, Li J, Li K and Li J 2021 Perm-inspired high-performance soy protein isolate and chicken feather keratin-based wood adhesive without external crosslinker *Macromol. Mater. Eng.* **306** 2100498
[350] Ostendorf K, Reuter P and Euring M 2020 Manufacturing medium-density fiberboards and wood fiber insulation boards using a blood albumin adhesive on a pilot scale *BioRes.* **15** 1531–46
[351] Gunasekaran S and Lin H 2012 Glue from slaughterhouse animal blood *Patent* USA US8092584B2
[352] Esparza Y, Bandara N, Ullah A and Wu J 2018 Hydrogels from feather keratin show higher viscoelastic properties and cell proliferation than those from hair and wool keratins *Mater. Sci. Eng.* C **90** 446–53
[353] Nordqvist P, Lawther M, Malmström E and Khabbaz F 2012 Adhesive properties of wheat gluten after enzymatic hydrolysis or heat treatment—a comparative study *Ind. Crops Prod.* **38** 139–45
[354] Wei Y, Yao J, Shao Z and Chen X 2020 Water-resistant zein-based adhesives *ACS Sustain. Chem. Eng.* **8** 7668–79
[355] Rathi S, Saka R, Domb A J and Khan W 2019 Protein-based bioadhesives and bioglues *Polym. Adv. Technol.* **30** 217–34
[356] Tran H A, Hoang T T, Maraldo A, Do T N, Kaplan D L, Lim K S and Rnjak-Kovacina J 2023 Emerging silk fibroin materials and their applications: new functionality arising from innovations in silk crosslinking *Mater. Today* **65** 244–59
[357] Zhou Y, Zeng G, Zhang F, Luo J, Li X, Li J and Fang Z 2021 Toward utilization of agricultural wastes: development of a novel keratin reinforced soybean meal-based adhesive *ACS Sustain. Chem. Eng.* **9** 7630–7
[358] Schmidt G, Christ P E, Kertes P E, Fisher R V, Miles L J and Wilker J J 2023 Underwater bonding with a biobased adhesive from tannic acid and zein protein *ACS Appl. Mater. Interfaces* **15** 32863–74
[359] Ball P 2021 Ned Seeman (1945–2021) *Nature* **600** 605
[360] Kulkarni J A, Witzigmann D, Thomson S B, Chen S, Leavitt B R, Cullis P R and van der Meel R 2021 The current landscape of nucleic acid therapeutics *Nat. Nanotechnol.* **16** 630–43
[361] Hu Q, Li H, Wang L, Gu H and Fan C 2019 DNA nanotechnology-enabled drug delivery systems *Chem. Rev.* **119** 6459–506
[362] Li B L, Zhang H, Li N B, Qian H and Leong D T 2022 Materialistic interfaces with nucleic acids: principles and their impact *Adv. Funct. Mater.* **32** 2201172

[363] Xie W Y, Huang W T, Li N B and Luo H Q 2011 Design of a dual-output fluorescent DNA logic gate and detection of silver ions and cysteine based on graphene oxide *Chem. Commun.* **48** 82–4

[364] Xia N, Cheng J, Tian L, Zhang S, Wang Y and Li G 2023 Hybridization chain reaction-based electrochemical biosensors by integrating the advantages of homogeneous reaction and heterogeneous detection *Biosensors* **13** 543

[365] Li Q *et al* 2020 A poly(thymine)-melamine duplex for the assembly of DNA nanomaterials *Nat. Mater.* **19** 1012–8

[366] Chen X *et al* 2021 Covalent bisfunctionalization of two-dimensional molybdenum disulfide *Angew. Chem. Int. Ed.* **60** 13484–92

[367] Li B L, Zou H L, Lu L, Yang Y, Lei J L, Luo H Q and Li N B 2015 Size-dependent optical absorption of layered MoS_2 and DNA oligonucleotides induced dispersion behavior for label-free detection of single-nucleotide polymorphism *Adv. Funct. Mater.* **25** 3541–50

[368] Zhu C, Zeng Z, Li H, Li F, Fan C and Zhang H 2013 Single-layer MoS_2-based nanoprobes for homogeneous detection of biomolecules *J. Am. Chem. Soc.* **135** 5998–6001

[369] Wang D *et al* 2021 Functionalizing DNA nanostructures with natural cationic amino acids *Bioact. Mater.* **6** 2946–55

[370] Dutta K, Das R, Medeiros J and Thayumanavan S 2021 Disulfide bridging strategies in viral and nonviral platforms for nucleic acid delivery *Biochemistry* **60** 966–90

[371] Gandioso A, Massaguer A, Villegas N, Salvans C, Sánchez D, Brun-Heath I, Marchán V, Orozco M and Terrazas M 2017 Efficient siRNA-peptide conjugation for specific targeted delivery into tumor cells *Chem. Commun.* **53** 2870–3

[372] Duan R, Li T, Duan Z, Huang F and Xia F 2020 Near-infrared light activated nucleic acid cascade recycling amplification for spatiotemporally controllable signal amplified mRNA imaging *Anal. Chem.* **92** 5846–54

[373] Nehzati S, Summers A O, Dolgova N V, Zhu J, Sokaras D, Kroll T, Pickering I J and George G N 2021 Hg(II) binding to thymine bases in DNA *Inorg. Chem.* **60** 7442–52

[374] Jouha J and Xiong H 2021 DNAzyme-functionalized nanomaterials: recent preparation, current applications, and future challenges *Small* **17** 2105439

[375] Fang Z, Li M and Zuo X 2023 DNA composites and applications in bioanalysis *Adv. Sens. Res.* **2** 2300002

[376] Cutler J I, Auyeung E and Mirkin C A 2012 Spherical nucleic acids *J. Am. Chem. Soc.* **134** 1376–91

[377] Bhatt N, Huang P-J J, Dave N and Liu J 2011 Dissociation and degradation of thiol-modified DNA on gold nanoparticles in aqueous and organic solvents *Langmuir* **27** 6132–7

[378] Banga R J, Chernyak N, Narayan S P, Nguyen S T and Mirkin C A 2014 Liposomal spherical nucleic acids *J. Am. Chem. Soc.* **136** 9866–9

[379] Banga R J, Meckes B, Narayan S P, Sprangers A J, Nguyen S T and Mirkin C A 2017 Cross-linked micellar spherical nucleic acids from thermoresponsive templates *J. Am. Chem. Soc.* **139** 4278–81

[380] Ge J, Hu Y, Deng R, Li Z, Zhang K, Shi M, Yang D, Cai R and Tan W 2020 Highly sensitive microRNA detection by coupling nicking-enhanced rolling circle amplification with MoS_2 quantum dots *Anal. Chem.* **92** 13588–94

[381] Hu Y *et al* 2019 Carbon-nanotube reinforcement of DNA–silica nanocomposites yields programmable and cell-instructive biocoatings *Nat. Commun.* **10** 5522

[382] Song P *et al* 2020 Programming bulk enzyme heterojunctions for biosensor development with tetrahedral DNA framework *Nat. Commun.* **11** 838

[383] Kamaly N, Yameen B, Wu J and Farokhzad O C 2016 Degradable controlled-release polymers and polymeric nanoparticles: mechanisms of controlling drug release *Chem. Rev.* **116** 2602–63

[384] Li F, Li J, Dong B, Wang F, Fan C and Zuo X 2021 DNA nanotechnology-empowered nanoscopic imaging of biomolecules *Chem. Soc. Rev.* **50** 5650–67

[385] Rothemund P 2006 Folding DNA to create nanoscale shapes and patterns *Nature* **440** 297–302

[386] Hong F, Zhang F, Liu Y and Yan H 2017 DNA origami: scaffolds for creating higher order structures *Chem. Rev.* **117** 12584–640

[387] Wu T C, Rahman M and Norton M L 2014 From nonfinite to finite 1D arrays of origami tiles *Acc. Chem. Res.* **47** 1750–8

[388] Dey S *et al* 2021 DNA origami *Nat. Rev. Methods Primers* **1** 1–24

[389] Woo S and Rothemund P W K 2011 Programmable molecular recognition based on the geometry of DNA nanostructures *Nat. Chem.* **3** 620–7

[390] Oktay E, Alem F, Hernandez K, Girgis M, Green C, Mathur D, Medintz I L, Narayanan A and Veneziano R 2023 DNA origami presenting the receptor binding domain of SARS-CoV-2 elicit robust protective immune response *Commun. Biol.* **6** 1–11

[391] Chikkaraddy R *et al* 2018 Mapping nanoscale hotspots with single-molecule emitters assembled into plasmonic nanocavities using DNA origami *Nano Lett.* **18** 405–11

[392] Zhan P *et al* 2023 Recent advances in DNA origami-engineered nanomaterials and applications *Chem. Rev.* **123** 3976–4050

[393] Majikes J M and Liddle J A 2022 Synthesizing the biochemical and semiconductor worlds: the future of nucleic acid nanotechnology *Nanoscale* **14** 15586–95

[394] Hu Y and Fan C 2022 Nanocomposite DNA hydrogels emerging as programmable and bioinstructive materials systems *Chem.* **8** 1554–66

[395] Li J, Song W and Li F 2023 Polymeric DNA hydrogels and their applications in drug delivery for cancer therapy *Gels* **9** 239

[396] Morán M C, Miguel M G and Lindman B 2007 DNA gel particles: particle preparation and release characteristics *Langmuir* **23** 6478–81

[397] Zhang K and Yam V W-W 2020 Platinum(II) non-covalent crosslinkers for supramolecular DNA hydrogels *Chem. Sci.* **11** 3241–9

[398] Jian X, Feng X, Luo Y, Li F, Tan J, Yin Y and Liu Y 2021 Development, preparation, and biomedical applications of DNA-based hydrogels *Front. Bioeng. Biotechnol.* **9** 661409

[399] Morya V, Walia S, Mandal B B, Ghoroi C and Bhatia D 2020 Functional DNA based hydrogels: development, properties and biological applications *ACS Biomater. Sci. Eng.* **6** 6021–35

[400] Wang C and Zhang J 2022 Recent advances in stimuli-responsive DNA-based hydrogels *ACS Appl. Bio Mater.* **5** 1934–53

[401] Lu S, Wang S, Zhao J, Sun J and Yang X 2018 A pH-controlled bidirectionally pure DNA hydrogel: reversible self-assembly and fluorescence monitoring *Chem. Commun.* **54** 4621–4

Chapter 2

Bio-based surface coatings

Yonggui Wang, Xiangyu Tang, Fanjun Yu and Xinyan Fan

2.1 Introduction

Recent studies have thoroughly addressed the main properties of coating materials. In today's society, material production is increasingly supported by tailored surface coatings designed to meet diverse needs, including protection, design enhancement, functionality, durability, and the crucial considerations of environmental and health impacts. The surface of materials is constantly subjected to various damaging factors, which include ultraviolet light, changes in humidity, and mechanical damage as well as the action of chemicals, oxygen, microorganisms, etc. Surface modification using coating technology has been developed to protect surfaces and enhance their resistance to various factors. Moreover, adding suitable functional/reinforcement additives to enhance the barrier properties of coatings, improve their overall performance, and make them more durable is crucial [1–3]. This not only extends the lifespan of the coatings, reducing the need for frequent replacements and thus minimizing waste generation, but also contributes to the broader sustainability goals of resource conservation and environmental protection. Coating materials, including polymers, inorganic materials, and composite polymer/inorganic materials, are commonly used because of their wide application prospects. In addition to their inherent sustainable protective qualities, the current focus on utilizing renewable resources to replace petrochemical-based polymers has heightened attention to the importance of incorporating suitable functional additives into coatings. Renewable feedstocks are available as biopolymers such as cellulose, lignin, chitin, essential oils (EOs), and so on. Alternatively, one can source molecular building blocks of materials from renewable resources, allowing the design of various bio-based, biomimetic, and bio-inspired coating materials via biopolymer modification or monomer engineering.

Sustainable bio-based resources, as fascinating and potential materials, have innumerable applications as composite coating materials. In this review, we describe how recent advances in bio-based coating research have helped us develop highly

doi:10.1088/978-0-7503-6184-2ch2

advanced functional coatings. Numerous bio-based coatings are available for different applications. In this review, we primarily chose three main types of bio-based coatings: superwetting, flame-retardant, and antibacterial coatings. The functional mechanism, preparation, and application of each bio-based coating are described herein.

2.2 Bio-based superwetting coatings

Special wettability in nature has attracted attention due to phenomena such as the special mobility of liquid droplets on the surface of lotus leaves, the anisotropic wettability of butterfly wings, and the surface sliding liquid used by hogweed to catch insects [4]. These characteristics have inspired the scientific study of wettability in nature, giving rise to a distinct research field dedicated to investigating wettability in various research domains. Thomas Young described the water contact angle (CA) of liquids in detail in 1804 and defined the surface wettability based on the size of the contact angle [5]. Solid surfaces are generally classified as hydrophilic ($10° < \theta < 90°$), hydrophobic ($90° < \theta < 150°$), superhydrophilic ($\theta < 10°$) and superhydrophobic ($150° < \theta < 180°$) [6].

The three models of superhydrophobicity—Wenzel, Cassie–Baxter, and the Lotus effect—describe surface wetting states influenced by roughness and air pockets, leading to varying degrees of water repellency. The thermodynamic equilibrium among the three coexisting phases, namely the solid phase, the liquid phase, and the vapor phase, predominantly dictates the apparent contact angle. Simultaneously, this equilibrium is affected by the scale ratio between the measured liquid droplet and hierarchical structure, influencing the observed contact angle. Under ideal conditions, where the solid surface exhibits perfect planar geometry and chemical homogeneity (figure 2.1(a)), the CA can be determined using the well-established Young's equation. Here, γ_{SL}, γ_{SV}, and γ_{LV} represent the interfacial energies of the solid–liquid, solid-gas, and liquid-gas interfaces. This equation illustrates that the three interface energies at the position of the three-phase contact can achieve a state of mutual equilibrium. In accordance with Young's equation, surfaces can be classified in terms of their wettability is classified as follows: hydrophilic surfaces ($\theta < 90°$); hydrophobic surfaces ($\theta > 90°$).

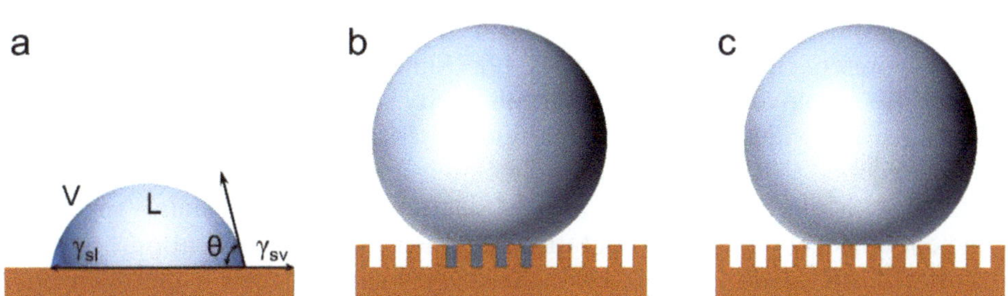

Figure 2.1. Typical wetting behavior of a droplet on rough solid substrates. (a) Young's mode. (b) Wenzel's mode. (c) Cassie–Baxter's mode.

$$\cos\theta = \frac{\gamma_{SV} - \gamma_{SL}}{\gamma_{LV}}. \tag{2.1}$$

However, in practical scenarios, solid surfaces having varying degrees of roughness, deviating from the fundamental assumption of Young's equation. In 1936, Wenzel introduced a representative wetting model for a rough-surfaced solid [7]. According to Wenzel's model, the liquid fully infiltrates the microscale/nanoscale structure of the rough surface (figure 2.1(b)), effectively elucidating the role of surface roughness in dictating surface wetting behavior. Here, θ_W represents the apparent contact angle on the rough surface, θ is the Young's contact angle on the ideal surface, and r denotes the roughness factor. Given that the roughness factor (r) for rough surfaces consistently exceeds 1, it is evident that surface roughness serves to enhance the intrinsic wettability.

$$\cos\theta_W = r\cos\theta. \tag{2.2}$$

In addition, the Cassie model, presented in 1944 by Cassie and Baxter, is another typical wetting model [8]. Their findings highlight that the presence of air pockets between liquids and the substrate can profoundly influence the unique wettability characteristics of surfaces (figure 2.1(c)). For complex surfaces with porous structures, the Cassie–Baxter model is often used to interpret them. Where f_{SL} represents the proportion to the liquid contact surface

$$\cos\theta_{CB} = f_{SL}\cos\theta + (1 - f_{SL})\cos\pi = f_{SL}\cos\theta + f_{SL}. \tag{2.3}$$

In the Cassie–Baxter state, the porous structured surface can capture air, allowing liquid to rapidly wet or roll on the surface. Liquid droplets, compared to the Wenzel state, exhibit greater mobility in this state. Additionally, research has revealed that under certain specific environmental conditions, certain surfaces can undergo a mutual transition between the Wenzel and Cassie–Baxter states.

2.2.1 Bio-based superhydrophobic coatings

To date, numerous bio-inspired artificial superhydrophobic surfaces were fabricated by replicating the micro/nanostructures found in natural superhydrophobic surfaces. These surfaces are primarily constructed by techniques such as nanofabrication, photolithography, or wet chemistry, with the aim of mimicking the intricate structures present in natural surfaces [9]. Simultaneously, low surface energy materials are employed to confer hydrophobicity. However, in practical industrial applications, the widespread use of perfluorinated compounds has raised environmental concerns [10]. Furthermore, the environmental impact and high costs associated with the raw materials and fabrication processes of many superhydrophobic coatings have imposed limitations on their broader application.

Bio-based materials have several advantages in preparing superhydrophobic coatings. These materials are derived from renewable plant-based resources, aligning with sustainability principles and reducing the environmental impact during processing. Furthermore, their biocompatibility reduces the risk of adverse reactions in biomedical applications [11]. Bio-based materials contain diverse functional groups, such as hydroxyl and ester groups, which allow for the introduction of various functional moieties in coating preparation. This enables precise control over the coating's properties, including adhesion, hydrophobicity, durability, and responsiveness to environmental stimuli, allowing for tailored performance to meet specific application requirements. Additionally, the versatile structures of bio-based materials enable the modulation of surface morphology through appropriate processing methods, which can influence wetting characteristics. Furthermore, the lower cost of bio-based materials contributes to cost-effectiveness in the production of superhydrophobic coatings. Additionally, certain bio-based materials possess inherent antibacterial properties, which can enhance the functionality of the coating in specific applications [12]. In conclusion, the use of bio-based materials in creating superhydrophobic coatings not only promotes environmental preservation and sustainable development but also provides various functionalities and wide-ranging application potential to the coatings.

Cellulose is the most abundant natural high-molecular-weight compound in the biosphere and serves as a primary component of renewable biomass resources. It is widely distributed in various natural plants, constituting a crucial component of cell walls in plants such as trees, cotton, and straw [13]. Additionally, cellulose is found in the cell walls of algae, fungi, bacteria, and certain marine organisms like tunicates [14]. Although cellulose has many advantages, its physical and chemical properties limit its use in the field of superhydrophobicity. When preparing hydrophobic coatings using cellulose, its high hydrophilicity can make it difficult to achieve the desired effects. To address this issue, chemical modifications of the -OH groups of cellulose are often conducted according to the specific requirements of practical applications. Consequently, there are two main approaches for crafting superhydrophobic coatings based on cellulose: Surface modification of cellulose-based materials, or directly modifying cellulose on a molecular level to create superhydrophobic properties [15].

The deposition of modified materials onto cellulose surfaces represents a straightforward way for the fabrication of superhydrophobic cellulose materials. For instance, the deposition of octadecyltrimethoxysilane-modified cellulose nanocrystals (CNCs) onto cotton fabric surfaces enables the preparation of superhydrophobic cotton textiles [16]. The superhydrophobic CNCs are co-immersed with cured epoxy resin, serving as a bonding agent, to intimately adhere to cotton fibers. In an alternative approach, by employing a surface sol–gel method, a titanium dioxide (TiO_2) gel film is first coated onto each cellulose layer of a bulk cellulose material, such as laboratory filter paper, to obtain TiO_2-precoated cellulose material. After further self-assembly of an octyltrimethoxysilane (OTMS) monolayer on the TiO_2 surface, the originally superhydrophilic filter paper transforms into a superhydrophobic state, with the resulting composite sheet exhibiting a water contact angle of 154.8°. This composite sheet is highly stable in strong acidic or basic aqueous solutions and retains the corresponding physical characteristics of the primary filter paper, such as flexibility and mechanical performance. Notably, the OTMS monolayer cannot self-assemble directly onto the filter paper surface without the TiO_2 pre-coating, indicating that the TiO_2 coating is a crucial factor for the subsequent assembly of the silane monolayer [17].

A commonly used method for preparing cellulose-based superhydrophobic coatings involves modifying the cellulose first, then directly spraying it onto the substrate surface [18]. For instance, the utilization of 10-undecenoyl chloride-modified cellulose microcrystals, followed by solvent-driven self-assembly, enables the fabrication of superhydrophobic coatings on various surfaces, such as glass, paper, wood and so on [19]. The selected 10-undecenoyl chloride serves as a low surface energy material, imparting hydrophobic alkyl chains. Notably, the terminal double bonds undergo self-crosslinking upon heating, thereby enhancing the mechanical properties of the coating. Chemical vapor deposition (CVD) is also a common method for preparing superhydrophobic cellulose coatings. In this approach, pre-prepared nanocellulose particles are modified with methyltrimethoxysilane through chemical vapor deposition [20]. Subsequently, the modified particles are proportionally mixed with polydimethylsiloxane (PDMS) to formulate a coating, resulting in the attainment of a superhydrophobic surface.

Lignin, a by-product of industrial pulp and paper production, has gained significant attention within the scientific community due to its potential use as a sustainable and renewable source for chemicals, fuels, materials, and drug carriers. The molecular structure of lignin encompasses diverse functional groups, such as phenolic hydroxyl and carboxyl groups, providing ample chemical reaction sites for the introduction of functional moieties and facilitating the construction of superhydrophobic properties [21]. Wood lignin exhibits favorable biocompatibility within living organisms, making it suitable for various biomedical applications, including drug delivery systems and medical coatings [22]. Its wide availability from different plant sources and variations in structure resulting from diverse growth environments contribute to the potential for tailoring superhydrophobic coatings with distinct properties. Additionally, the antioxidative properties of wood lignin enhance the stability and durability of lignin-based superhydrophobic coatings [23]. In summary, wood lignin, as a material for superhydrophobic coating fabrication, possesses natural origins, multifunctionality, and tunable characteristics, providing robust support for the development of superhydrophobic surfaces with broad applications.

As an example chemical modification of lignin can be used to create superhydrophobic coatings through a spray-coating technique. The modification of alkaline lignin involves the use of 1H,1H,2H,2H-perfluorodecyltriethoxysilane (PFTEOS) to replace hydrophilic groups with PFTEOS, resulting in modified lignin-PFTEOS [23]. The application of a spray-coating composed of lignin-PFTEOS on different substrates results in the creation of superhydrophobic lignin-based surfaces that exhibit a contact angle of up to 169°. These surfaces also display significant resistance to friction and corrosion. However, the introduction of fluorine-containing functional groups contradicts the initial goal of using biomass materials. As a result, researchers are actively seeking greener reagents to functionalize lignin with superhydrophobic properties. By physically blending lignin, beeswax, and cotton, it is possible to create a porous material derived from biomass that possesses both superhydrophobic and superoleophilic properties [24].

Table 2.1. Bio-based materials for superhydrophobic surfaces: surface coating methods and corresponding applications.

Materials	Method	Application	References
MCC	Spray-coating	Self-cleaning	[19, 29, 30]
CNC	Dip-coating	Oil–water separation	[31]
CNC	Spray-coating	Oil–water separation	[32]
CNC	CVD	Fast water-removing	[33, 34]
CNF	Dip-coating	Self-cleaning	[35]
CNF	Spray-coating	Fast water-removing	[36–38]
CNF	CVD	Self-cleaning	[39]
Cellulose acetate	Electrospinning	Oil–water separation	[40, 41]
Lignin	Spray-coating	Self-cleaning	[42–44]
Lignin	CVD	Oil–water separation	[45, 46]
Chitosan	Spray-coating	Self-cleaning	[47, 48]
Chitosan	Dip-coating	Oil–water separation	[49, 50]
Biowax	Spray-coating	Self-cleaning	[51]
Biowax	Thermal evaporation	Self-cleaning	[52, 53]
Starch	Dip-coating	Oil–water separation	[54]

The low surface energy of beeswax and the micro/nanostructures provided by lignin contribute to this effect. Applying a mixture of beeswax and lignin onto cotton enhances its properties, resulting in a material with superhydrophobic and superoleophilic characteristics for oil–water separation. Hydroxyethylated lignin undergoes hydrophobic modification using oleic acid. The resulting lignin-based superhydrophobic material can be spray-coated onto food packaging to remove fluid food residues. In addition, various methods have been reported for preparing lignin into superhydrophobic coatings, including self-assembly [25], layer-by-layer addition [26], electrospinning [27], and physical blending [28].

In addition to cellulose and lignin, various other bio-based materials have found extensive applications in the preparation of superhydrophobic coatings. Examples include chitosan, biowaxes, plant extracts, starch, and more. Table 2.1 provides specific methods and applications of superhydrophobic coatings derived from biomass materials.

2.2.2 Bio-based superhydrophilic coatings

The term 'superhydrophilic' was coined by Fujishima *et al* [55] to describe surfaces with a micro-nano hierarchical structure where the roughness factor exceeds 1. On such surfaces, water can completely spread out, forming a flat film. The wettability of surfaces depends on both their surface free energy and geometric structure. A superhydrophilic surface is characterized by a water contact angle of less than 10°. Superhydrophilic surfaces are prevalent in nature and serve various functional purposes. For example, tear fluid spreads into a film over the surface of the cornea, ensuring proper eye function. The superhydrophilicity of shark skin reduces drag, aiding in their efficient movement underwater. Similarly, the superhydrophilic surface of fish scales provides them with ultra-oleophobic properties, enabling fish to navigate through oily waters with ease [56]. Research into these biological surfaces has spurred the development of synthetic superhydrophilic materials. Advances in biomimicry and nanotechnology have led to the discovery of new properties and potential applications for superhydrophilic materials. These materials hold promise for diverse fields such as construction materials, industrial anti-corrosion, oil–water separation, and biomedical applications [57]. This intersection of natural inspiration and technological innovation marks a new phase in the evolution of superhydrophilic materials, expanding their utility and efficiency across various industries.

Superhydrophilic materials are typically composed of polymers, inorganic metal materials, and inorganic hybrid materials [58]. Conventional polymer materials exhibit poor formability, susceptibility to contamination, low mechanical strength, and difficulty in modification. On the other hand, inorganic hybrid materials boast higher mechanical strength and superior corrosion resistance, albeit remaining challenging to process and expensive. Notably, bio-based superhydrophilic materials are derived from renewable resources, such as plant extracts, biopolymers, and other naturally occurring substances. This contrasts sharply with conventional superhydrophilic materials, which often rely on petrochemical derivatives and synthetic processes that can be environmentally harmful [59]. The use of bio-based sources ensures a sustainable supply chain, reducing dependency on finite resources and mitigating the ecological footprint associated with material production. One of the most significant advantages of bio-based superhydrophilic materials is their biodegradability. These materials can naturally decompose into non-toxic by-products when disposed of, significantly reducing the environmental impact compared to non-biodegradable synthetic materials [59]. This property is particularly valuable in applications such as packaging and disposable medical devices, where the material's end-of-life disposal is a critical consideration.

Oil–water separation is a critical process in addressing environmental pollution caused by oil spills, industrial effluents, and wastewater. Conventional methods for oil–water separation often involve complex and costly processes. In recent years, cellulose-based superhydrophilic materials have emerged as a promising solution due to their natural abundance, biodegradability, and excellent water affinity. Cellulose is inherently biodegradable, breaking down into non-toxic byproducts in natural environments. This property is particularly advantageous in oil–water separation, where the materials used must not contribute to secondary pollution. The use of cellulose-based materials ensures that the separation process is environmentally friendly from start to finish. Cellulose can be chemically modified to enhance its superhydrophilicity and oleophobicity. Superhydrophilic surfaces attract water while repelling oil, allowing for efficient separation of oil and water mixtures. This dual functionality is crucial for applications where rapid and effective separation is required. Hydrogels, which exhibit a water-rich structure composed of cross-linked hydrophilic polymer chains, play a significant role in this field.

Cellulose-based hydrogels are particularly common in coating materials. Coating a cellulose-based hydrogel onto a metal mesh enables gravity-driven oil–water separation [60]. In this process, cellulose dissolves in a solvent composed of LiCl and dimethylacetamide. The steel mesh is immersed in the solution, treated with citric acid to accelerate gelation. The modified cellulose enhances the hydrophilicity of the metal mesh, with a water contact angle (WCA) of 0° [61]. Additionally, the modified mesh exhibits superamphiphilic properties. The hydrophobic polyvinylidene difluoride (PVDF) membrane is immersed in an alkaline solution containing cellulose at room temperature. The cellulose in the solution forms strong hydrogen bonds with the membrane surface, thereby enhancing hydrophilicity [62]. The resulting PVDF membrane exhibits superhydrophilic and underwater superoleophobic surfaces, demonstrating excellent anti-fouling and anti-adhesion properties when used in crude oil treatment. The cellulose-modified PVDF membrane is also effective in separating various water-in-oil emulsions, achieving a high oil removal rate (>99%). Table 2.2 shows in detail the preparation method of cellulose superhydrophilic materials.

Lignin, in addition to cellulose, is frequently used to prepare superhydrophilic coatings. Lignin-based superhydrophilic coatings are often used in the field of adsorption and oil–water separation. For instance, the preparation of lignin-based polyurethane foams was achieved by utilizing lignin polyols in combination with conventional polyurethane foaming processes, incorporating carbon nanotubes [74]. These novel lignin-based polyurethane foam adsorbents exhibit outstanding

Table 2.2. Preparation method of cellulose superhydrophilic materials.

Materials	Methods	References
Cellulose@sodium alginate/$CaCO_3$ membrane	Alternating soaking method	[63]
Cotton fabric/polydopamine	*In situ* surface deposition and a dip-coating method	[64]
Cellulose/polydopamine/polyethyleneimine nanofibrous membrane	Electrospinning method	[65]
Dopamine methacrylamide modified cotton fabric	Free-radical polymerization reaction	[66]
Cellulose/$CaCO_3$/epichlorohydrin sponge	Vacuum freeze-drying method	[67]
Cellulose solution-coated cotton fabric	Surface coating method	[68]
Oxidized-CNFs aerogel microspheres	Atomization freeze-drying method	[69]
Graphene oxide@CNF membrane	Electrospinning and freeze-drying method	[70]
Cellulose-coated nylon membrane	Chemical crosslinking method	[71]
Cellulose nanofibers/guar gum/multiwalled carbon nanotubes	Physical doping method	[72]
CNF/ vinyltrimethoxysilane/acrylic acid/ 1-(3-(dimethylamino)propyl)-3-ethyl carbodiimide hydrochloride (EDC)	HRP-catalyzed method	[73]

environmental friendliness, high oil absorption efficiency, and excellent performance in repetitive cycling. The foams produced using this method exhibit exceptional solar thermal conversion and adsorption–desorption properties. They effectively absorb sunlight, increase surface temperatures, and improve oil recovery rates. Additionally, the polyurethane foams based on lignin, prepared in this way, can degrade in alkaline environments, demonstrating their environmental friendliness. After synthesizing lignin-based polyurethane foams, polydopamine particles were deposited on the foam surface through *in situ* polymerization under weakly alkaline conditions to increase surface roughness [75]. Next, the foam was modified with phytic acid to achieve surface superhydrophilicity and underwater superoleophobicity. Oil–water separation tests were conducted using PDA- and PA-modified lignin-based foams with mixtures of water and n-hexane, cyclohexane, toluene, and pump oil. The results showed separation efficiencies in excess of 99% for the tested mixture types.

Chitosan, a natural polysaccharide derived from chitin, exhibits several unique properties that make it highly valuable in the field of superhydrophilicity. Chitosan is non-toxic, non-irritating, and biodegradable, making it suitable for various biomedical applications. The amino and hydroxyl groups in chitosan allow for various chemical modifications, enhancing its functionality and range of applications. Chitosan can absorb a significant amount of water molecules, forming hydrogels, which is crucial for its performance in superhydrophilic materials. The water contact angle (WCA) of neat chitosan films, prepared using casting technology, was measured at 66.1°. When chitosan-sodium phytate films were coated with SiO_2 nanoparticles (NPs), its WCA was reduced from 100°–3°, and the film surface was changed from hydrophobic to superhydrophilic. Furthermore, the oxygen transmission rate (OTR) was significantly reduced, and the mechanical properties of the film were improved [76]. Chitosan can adsorb heavy metal ions and organic pollutants from water, purifying it effectively [77]. Its superhydrophilic nature enhances its ability to capture and remove contaminants. Due to its biocompatibility and superhydrophilicity, chitosan is used in wound dressings and tissue engineering scaffolds, promoting wound healing and cell growth [78]. Chitosan can be used as a coating material on various substrates to enhance their hydrophilicity, antibacterial properties, and biocompatibility [79]. For example, chitosan coatings can improve the performance of medical devices and biosensors. The superhydrophilic nature of chitosan makes it suitable for producing highly absorbent materials, such as diapers, sanitary napkins, and other personal care products. The amines in chitosan can be protonated under acidic conditions, conferring a positive charge and film-forming properties to chitosan. Based on electrostatic interactions and ionic crosslinking, a durable superhydrophilic chitosan-alginate hydrogel can be prepared for oil–water separation [79]. By integrating polysaccharide-based superhydrophobic surfaces and the hierarchical micro-/nanostructures, the as-fabricated CS-ALG hydrogel coated mesh exhibits excellent underwater superoleophobicity and anti-oil-fouling performance.

In conclusion, bio-based materials have made a great impact in the field of superhydrophilic coatings due to their environmental friendliness and biocompatibility, as well as their renewability and ease of modification, and provide directions that can be learnt from the development of bio-based coatings.

2.2.3 Bio-based superslippery coatings

A superslippery surface refers to a surface characterized by an extremely low surface friction coefficient and outstanding lubrication performance. Such surfaces typically exhibit the rapid movement of water droplets on them, akin to surfaces coated with lubricants. Superslippery surfaces, characterized by their exceptional ability to repel water and reduce friction, feature high water repellency with contact angles exceeding 150°, causing water droplets to bead up and roll off effortlessly [80]. In contrast, superhydrophobic surfaces emphasize the repellency of water, characterized by high contact angles, causing water droplets to bead up and roll off rapidly. Such surfaces find applications in waterproofing, anti-pollution, and oil–water separation, relying on the increased repellency to water for these functionalities. In summary, superslippery surfaces focus on reducing friction and enhancing lubrication, while superhydrophobic surfaces emphasize water repellency and anti-pollution characteristics. The design and fabrication of these two surface features typically involve different surface engineering and coating technologies [81]. The special structure of *Nepenthes alata* inspires researchers to adopt liquid lubricants to smooth the surface of a substrate. Tuteja and co-workers proposed a superslippery surface by designing overhangs on the fluorinated rough surfaces. The surface overhangs can slow down the wetting of liquid droplets toward the surface by confinement of air inside the local surface curvature [82].

Since friction strongly depends on the smoothness of substrates, lubricants, such as air, liquid lubricants, and solid lubricants, are widely adopted to smooth the surface of substrates. When designing a superslippery surface, two key factors need to be considered: the interaction between droplets and lubricants, and the roughness of the surface [83]. The interaction between the droplets and the lubricants should not be too strong to allow droplets move freely. However, there are variations in the selection criteria for lubricants, taking into account the working conditions of the lubricating oil. The chosen lubricant aims to mitigate solid–solid interactions, preventing atomic-level contact between two opposing surfaces. A surface is typically considered superslippery if the coefficient of friction (COF) is less than 0.01 [80].

Biomass materials offer significant advantages in the development of superslippery surfaces due to their sustainability, biodegradability, and inherent biocompatibility [83]. These materials, derived from renewable sources such as plants and microorganisms, can be engineered to exhibit exceptional water repellency and low friction. Traditional methods for preparing superslippery surfaces typically involve introducing fluorinating agents into polytetrafluoroethylene (PTFE) nanofibers to create a smooth liquid-filled porous surface [84]. The selected liquid lubricant exhibits extremely low surface tension, providing excellent repellency to liquids with higher surface tension. When preparing superslippery coatings based on biomass materials, attention must be given not only to the effects of elastic fluid dynamics but also to the relationship between the macromolecular chemical structure and lubrication behavior. Specifically, cellulose, due to its strong hygroscopicity, readily forms a water film on the substrate surface, making it an ideal choice for superslippery coating applications. For instance, modifying CNF through *in situ* polymerization of trichlorovinylsilane, followed by attaching transparent and reactive porous trichlorovinylsilane-functionalized nanofibrillated cellulose nanofibers (TCVS-SNFs) to the surface of CNF films, introduces nanoscale roughness and reactive vinyl groups. Functionalization can be achieved through a photo-click thiol–ene reaction. Subsequent lubricant injection results in a superslippery CNF film that exhibits perfect repellency to both low surface tension aqueous and organic liquids [85].

Utilizing cellulose derivatives as a substrate to prepare superslippery surfaces is also a viable strategy. Fabricating porous films by combining cellulose acetate with carbon nanotubes, followed by the injection of paraffin wax as a lubricant, results in a superslippery coating with excellent anti-icing/de-icing performance and self-healing capabilities, showcasing outstanding photothermal conversion properties [86]. Chitosan has exceptional film-forming properties and can be used to produce films and coating materials. Furthermore, an electrostatic and hydrogen bonding layer-by-layer self-assembly approach is employed due to the strong electrostatic interactions between chitosan and alginates. The bottom layer is constructed using chitosan, sodium alginate, and polyvinylpyrrolidone (PVPON). To create porosity in the structure, pH-sensitive cross-linked chitosan and alginate films are immersed in an alkaline buffer solution to remove PVPON through hydrogen bonding. Then, the film surface is coated with biocompatible almond oil to achieve a superslippery surface [87].

Bio-based materials can serve not only as substrates for superslippery coatings but also as lubricants. Under identical test conditions, the friction coefficient of cellulose treated with alkali and added to castor oil is significantly lower than that without additives [88]. Additionally, cellulose esters with modified fatty chains serve as additives in lubricants for superslippery surfaces [89]. The combination of lignin and ionic liquids enhances the lubrication mechanism of the lubricant. Lubricants comprising choline and amino acid with added lignin demonstrate improved frictional behavior due to strong hydrogen bonding interactions [90]. Likewise, the performance of lignin is enhanced in lubricants composed of ethylene glycol and polyethylene glycol, attributed to strong hydrogen bonding interactions [91]. The synthesis of acylated chitosan schiff bases is employed as an additive for multifunctional bio-based lubricants, displaying antioxidant and lubricating properties in butyl palmitate/hardened esters [18].

In summary, superslippery surfaces are generally applied to regulate the motion behavior of droplets on solid surfaces, while the purpose of lubricants is to reduce the frictional force of droplets on the corresponding surface by smoothing the contact interface. The application of superslippery surfaces currently faces challenges, with the most prominent being the issue of lubricant consumption. Therefore, in future developments, it is crucial to develop suitable substrates to anchor the lubricant and lubricants with low evaporation rates and excellent weather resistance to manufacture high-performance superslippery surfaces.

2.3 Bio-based flame-retardant coatings

Materials that are commonly used in daily life, such as textiles, fibers, and other polymers, are particularly susceptible to fire owing to their organic structural characteristics, and the rapid spread of flames poses a great threat to human lives and property. Traditionally, materials have been made less flammable by incorporating flame retardants. However, the bulk incorporation of flame retardants can often degrade the mechanical properties of the material [92–94]. Hence, researchers are exploring a strategy to render coatings flame-retardant, while maintaining inherent substrate properties. The earliest flame retardants contain halogenated compounds as active ingredients [95, 96]. The flame retardance mechanism of halogens involves releasing halogen radicals during combustion, which then interfere with the chemical reactions in the flame, effectively inhibiting the flame's propagation and reducing the overall burning rate. This mechanism is called 'radical quenching' or 'radical trapping'. However, these retardants are toxic and bioaccumulative, and halogenated coating systems have gradually come under scrutiny [97, 98]. To avoid the aforementioned shortcomings of halogenated coatings, researchers have been actively pursuing halogen-free flame-retardant technology based on flame retardants containing phosphorus [99], nitrogen [100], silicone [101], boron [102], and the metal-family [103]. Furthermore, bio-based flame-retardant coatings have attracted the attention of researchers because of their renewable and ecofriendly characteristics.

Bio-based flame-retardant coatings predominantly consist of intumescent flame retardants (IFRs) because biomass molecules naturally offer the necessary sources of carbon, acid, and gas required for IFR functionality. The three compounds together play a synergistic role in improving the flame retardancy. High-carbon biomass molecules such as chitosan (CS), lignin, and starch can be used as carbon sources to form an expanded carbon layer on the substrate surface. Phytic acid [104], citric acid [105] and tannic acid [106] are excellent acid source reserves. Glycine [107] and urea [108] are excellent gas source reserves. Furthermore, deoxyribonucleic acid (DNA) can form a complete IFR system suitable for flame-retardant applications [109]. These bio-based molecules can effectively enhance the flame retardancy of materials when used as flame-retardant coatings on the substrate via layer-by-layer self-assembly (LbL), sol-gel treatment, plasma-aided surface modification, and polyelectrolyte complex technologies. These methods involve mild conditions, simple equipment, and flexible preparation processes. The components are applied on the substrate surface, creating a protective layer that intercepts the transfer of combustible materials and heat upon contact with flames, effectively interrupting the combustion process and achieving flame retardation.

2.3.1 Mechanism for bio-based flame-retardant coatings

The flame-retardant mechanism of bio-based flame-retardant coatings is similar to that of conventional flame retardants. The combustion process of the substrate can be interrupted through condensed-phase action, gas-phase action, or a synergistic combination of both. Flame-retardant coatings increase the time available for people to evacuate safely by delaying substrate ignition and inhibiting the spread of flames and toxic fumes. During the combustion process, flame retardants effectively inhibit or prevent the physical or chemical change of the substrate in different ways to achieve the reaction of heat absorption, covering and dilution of combustible gases.

2.3.1.1 The flame-retardant mechanism in the condensed-phase action

In the condensed phase, the addition of flame-retardant coatings can change the substrate pyrolysis pathway, creating a dense, stable charcoal layer on the surface of the promoted substrate. First, an acid source promotes the production of inorganic acid and catalyzes the carbon source, triggering the substrate dehydration and charring. Then, the gas source releases non-combustible gas during combustion conditions, causing the charcoal layer to expand. This charcoal layer typically acts as a physical barrier to the transfer of oxygen and heat between the substrate and gas phase. In addition, the porous structure of the charcoal layer allows it to adsorb toxic fumes released during combustion [110]. Currently, many studies have been conducted on the use of biomass materials as condensed-phase flame-retardant materials. The preparation of coatings containing phosphorylated chitosan on the surface of cotton fabrics has been used to improve the persistent flame retardancy of the fabrics [111]. Phytic acid, D-sorbitol, and glycine were used as flame retardants synthesized from acid, carbon, and gas sources, respectively, and applied to the surface coating of polyurethane foam using UV curing technology. The coating created a dense charcoal layer that protected the foam structure after combustion [107].

2.3.1.2 The flame-retardant mechanism in the gas-phase action

Gas-phase flame retardants primarily function in the gas-phase environment by slowing down the combustion chain reaction or inhibiting the final stages of combustion, thereby achieving flame retardation. The retardation effect relies on two mechanisms, one is the releases of a large number of non-combustible gases upon thermal decomposition of the coating, these gases dilute the concentration of combustible gases as well as oxygen, slowing down the spread of combustion [110]. Nitrogen-containing flame retardants such as β-cyclodextrin phosphate ammonium salt [112], phytate ammonium salt [113], and protein [114] can release nitrogen, ammonia, etc. These gases also can facilitate the expansion of the charcoal layer, ultimately forming a porous expanded charcoal layer that insulates the charcoal from oxygen and heat. Thus they achieve the effect of gas-phase and condensed-phase flame retardancy. The second mechanism interferes with the large number of H· and HO· radicals generated during a fire which maintain the combustion. The decomposition of phosphorus-containing flame retardants generates free radicals such as P· and PO· to capture the H· and HO· radicals, interrupting the chain reaction of combustion to achieve flame retardancy [112]. Overall, the mechanism of N/P intumescent flame-retardant coatings is shown in figure 2.2.

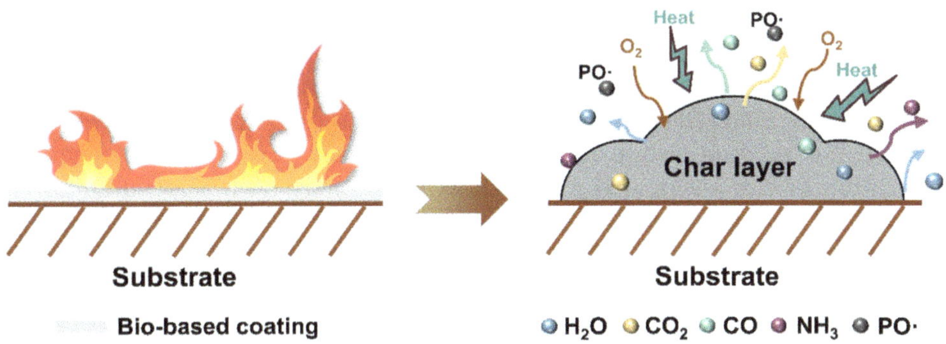

Figure 2.2. N/P synergistic intumescent flame-retardant mechanism for coatings.

2.3.2 Bio-based flame-retardant coatings through compound formation

Due to the structural constraints of biomass molecules, individual components within certain biomass molecules can typically fulfill only one or two roles within the intumescent flame-retardant (IFR) system, leading to a limited flame-retardant effect. Typically, enhancing flame-retardant efficiency involves compounding flame retardants or chemically modifying biomass molecules to introduce multiple flame-retardant elements. This results in coatings with several flame-retardant elements coexisting, which can synergistically strengthen their participation in the IFR system and enhance the flame-retardant efficiency of the single bio-based molecule structure.

Chitosan is one of the bio-based materials commonly used for flame-retardant coatings, due to the fact that, in an IFR system, the hydroxyl groups enriched in CS can be dehydrated to carbon when heated. At the same time, the amino groups in CS releases NH_3 and N_2 to dilute the gas-phase combustible gases concentration. In the following we will discuss several examples of electrostatically bonded CS complexes with other materials, aimed at creating flame-retardant surface coatings. Ammonium polyphosphate (APP), which is widely used at present, is a polyanionic electrolyte with high nitrogen and phosphorous contents. As an efficient inorganic nitrogen–phosphorus expansion-type flame-retardant, it functions via the condensed and gas phases. Phosphate anions combine with the positive charge of ammonium (NH_4^+) in CS to form an ionic bond under acidic conditions, forming a stable coating on PET or cotton fabrics under the interaction of positive and negative charges [115, 116]. Considering the typical CS and APP two-component flame-retardant coatings, researchers further added inorganic materials such as diatomite [117], graphene oxide (GO) [118], and SiO_2 [119] to the system. The combination of inorganic materials with phosphate ester formed via pyrolysis creates a more stable and stronger carbon structure on the substrate surface. The CS/GO/APP system demonstrates a significant reduction in both the peak heat release rate (PHRR) and total heat release (THR) by 45% each after depositing fifteen layers on the surface of wood. The excellent flame-retardant performance is attributed to the synergistic effects of CS, GO and APP.

In addition to inorganic phosphorus flame retardants, chitosan can interact with other bio-based molecules on an electrostatic base to create superior flame-retardant coatings. Lignin is a macromolecular polymer composed of three phenylpropane structural units, characterized by a complex benzene ring structure. This structure imparts a high residual carbon content during thermal decomposition. For instance, when cotton fabrics are treated with an LbL technology using lignosulfonate (LS) and CS, a 25.2% increase in weight percent can be observed and the residual charcoal rate increases from 3.5% (without treatment) to 16.1% after a cone calorimetry test (CCT) [120]. Differing to the previous examples, the chemical structure of DNA can be considered an IFR system in itself. Since DNA contains phosphoric acid, polyphosphoric acid precursors, polyhydroxy carbon sources (deoxyribose) and nitrogenous bases, it combines several IFR properties within itself. When DNA is combined with CS on a cotton surface with up to 20 layers, the PHRR and THR were reduced by 40% and 30%, respectively [121].

The phosphorus contents in coatings has an important influence on flame-retardant efficiency. Phytic acid (PA), a bio-based small molecule with a high phosphorus content (28%), is mainly extracted from natural plants such as grains and seeds. The structure has six highly reactive phosphoric acid groups that provide a strong dehydrating ability. The CS/PA system is a typical representative of fully bio-based flame-retardant coatings. This system reduces the PHRR and THR of cotton fabrics by 60% and 61%, respectively, and increases the residual carbon rate from 0% to 20%. Upon adding a certain amount of biochar to this system, the flame-retardant efficiency is further enhanced, and the PHRR and THR decrease by 88.66% and 88.69%, respectively, and the residual charcoal is increased to 91% [122].

PA, a polyanionic electrolyte, can be electrostatically bound to not only CS but also to other positively charged compounds such as urea, melamine, and polyethyleneimine (PEI). Furthermore, PA has a unique ability to chelate metal ions. The combination of PA with other compounds through electrostatic action can improve the durability and mechanical properties of PA flame-retardant coatings. One of the advantages of the coating systems is their straightforward production, depending on the availability of the process. Benefiting from the formation of an insoluble coating on the wool surface, PA/PEI-treated wool fibers exhibit self-extinguishing properties even after ten washes, indicating the excellent flame-retardant capability and washing durability of the modified fibers. The researchers attribute the durability to ionic and amide covalent reactions between the amino groups of PEI and carboxyl groups of wool fiber [123]. By adding metal ions as synergists to the CS/PA system, an efficient flame-retardant coating can be created on cotton fabrics. In addition to the excellent catalytic carbonization ability of metal ions (Ba^{2+}), phosphoric acid promotes the formation of a charcoal layer on the surface of cotton fabrics, thereby preventing the transfer of heat and reducing the PHRR (−61%) and THR (−59%) remarkably [124]. Table 2.3 summarizes the strategic effects of a representative LbL assembly and non-covalent complex coatings.

Table 2.3. Flame-retardant performances for representative LbL assembly and non-covalent complex coatings.

			CCT results		
Substrate	Coating	Number	PHRR	THR	References
Wool	PA/PEI	15BL	−40%	−44%	[123]
Polyethylene terephthalate	PA/BPEI/APP	3BL	−64%	−43%	[125]
Polyethylene terephthalate	PA/polysiloxane	5BL	−65%	−59%	[126]
Cotton	CS/APP	15BL	−23%	−56%	[127]
Cotton	CS/titanate nanotubes	20BL	−38%	−14%	[128]
PUF	CS/phosphorylated cellulose nanofibers	5BL	−31%	—	[129]
PUF	CS/APP	20BL	−66%	−11%	[130]
Polyamide 66	PA/oxidized sucrose	3HL	−36%	−12%	[131]

BL refers to bilayer.
HL refers to hexa-layer.

2.3.3 Bio-based flame-retardant coatings through chemical modification formation

Compounding flame retardants to form a coating is a simple process and easy to operate, but because the effective flame-retardant components and the biomass substrate are only physically interacting with each other, their lasting flame retardancy is always a difficult problem to solve. One of the effective ways to obtain durable flame retardancy is chemical modification. Proper design of bio-based molecular structures by introducing flame-retardant elements (typically in the form of phosphonic-acid groups and amino groups) through chemical grafting covalent bonding can help improve the flame retardancy of coatings.

Although the covalent bonding is durable, it usually requires relatively complex processing means of compounding. Grafting PA onto cotton reduces the PHRR of cotton by nearly 95%, while also providing washing durability, with an increase of less than 20% in weight. The possible reaction between APA and cotton fabric is presented in figures 2.3(a) and (b). This is due to the catalytic effect of dicyandiamide, the amine salt of PA (APA) and the hydroxyl group of cotton fibers to form a stable P–O–C covalent bond, which is difficult to break during the washing cycle, so that the cotton can maintain flame retardancy [132]. Similarly, to develop bio-based flame-retardant coatings with high efficiency and good durability, researchers can leverage the active sites (hydroxyl, carboxyl, and amino groups) present on materials such as chitosan, cellulose, and lignin. These active sites can be utilized to graft elements such as nitrogen and phosphorus into the material structure via covalent bonding to obtain bio-based flame-retardant coatings. This is typically achieved through chemical modifications such as esterification and etherification [102, 111, 131]. The direct application of phosphorylated CS as a coating on cotton fabrics resulted in reductions

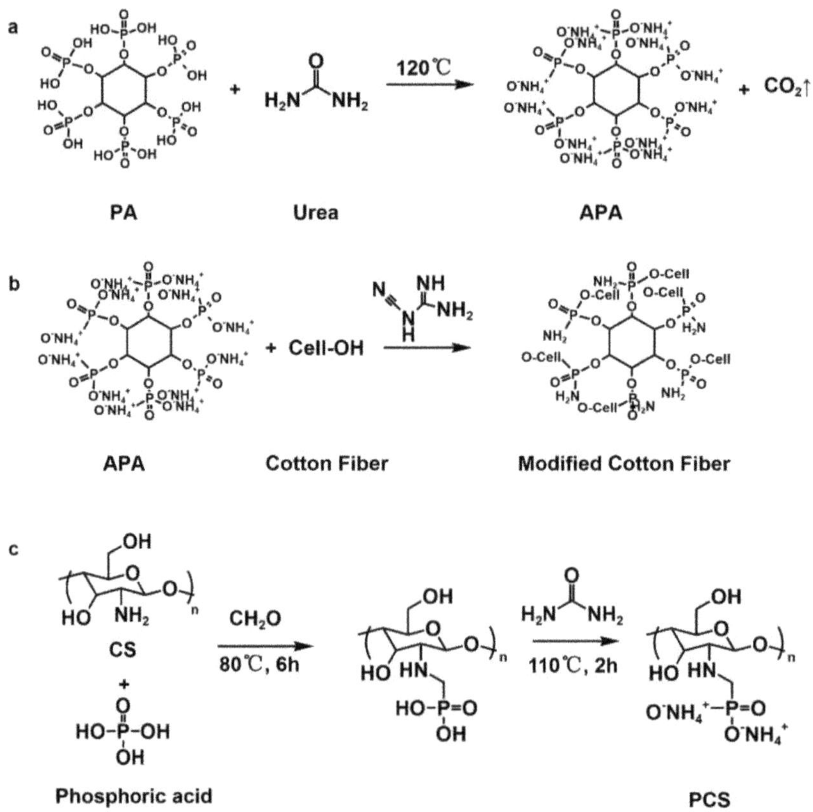

Figure 2.3. (a) The synthesis of APA, (b) crosslinking reaction between cotton fibers and APA, and (c) the preparation of PCS.

of 88.2% and 59.4% in PHRR and THR, respectively, compared with pure cotton. Importantly, even after ten washing cycles, the flame retardancy of the treated cotton fabrics was still better than that of the control owing to partial chemical grafting of phosphorylated CS (PCS) on the cellulose [111]. The preparation of PCS is shown in figure 2.3(c). Flame-retardant small molecules can also be grafted into other biomass molecules, including phosphorylated chitosan, and then constructed on the surface of fibers through a dip–dry–cure process. The cotton fibers treated with phosphorylated chitosan still behave satisfied flame retardancy with an LOI value of 23.5%, which are superior to the control sample [111].

In addition to grafting flame-retardant functional groups onto the substrate, it is also feasible to synthesize reactive flame-retardant molecules that react with biomass-based materials. Cheng *et al* synthesized an environmentally friendly active flame-retardant based on phytic acid, pentaerythritol and 1,2,3,4-butanetetracarboxylic acid, whose carboxyl group can form ester and amide bonds with the hydroxyl group and amino group of wool, which can provide better washing durability to the treated wool. The wool fabric treated with the flame-retardant at 0.14 mol l^{-1} remained self-extinguishing even after 20 washes [133]. Similarly, tartaric acid phytate ester, prepared via esterification with the carboxyl group in the tartaric acid molecule introduced into the phytic acid molecule, can likewise be bonded to silk fabrics via ester and amide covalent bonds [134]. Combining two flame-retardant coating modification strategies, non-covalent and covalent, will endow coatings with even better flame retardancy. Hypophosphorous acid-modified CS and PEI combined further significantly reduced the PHRR (−73%) and THR (−80%) [135].

Bio-based flame-retardant coatings offer an environmentally friendly alternative to chemical flame-retardant materials by utilizing materials such as CS, lignin, PA and so on. These coatings provide the ability to act through a gas-phase mechanism, releasing non-combustible gases to dilute flammable gases, and through a condensed phase to form a charcoal layer that isolates and protects the underlying material. These biodegradable, non-toxic coatings are suitable for textiles, wood products and electronics, addressing safety and environmental concerns. Ongoing research aims to improve the performance, durability and scalability of these biomass-based coatings to meet industry standards for a sustainable future.

2.4 Bio-based antibacterial coatings

Bacterial infection and colonization are severe public health concerns. The prevalence of bacterial infections in nosocomial facilities has dramatically increased the morbidity and mortality of patients. Moreover, bacterial adhesion and subsequent deposition on medical instruments and devices, including implants, ventilators, catheters, and so forth, and bio-fouling are major challenges faced by the biomedical industry at present. A series of coatings was developed to address these problems, including coatings infused with bactericidal reagents and incorporated with bactericidal moieties, thereby achieving ‘contract killing’. Some special structures are constructed for antibacterial activity such as hydrophobic or superhydrophobic surfaces to achieve anti-fouling (more details about superhydrophobic surfaces are described in section 2.2.1) [136–138]. Recently, bio-based resources have attracted much attention because of their intrinsic remarkable biocompatibility and have been developed to prepare coatings for sterilization. Among them, some resources (e.g. EOs, antimicrobial peptides) have intrinsic antibacterial properties because of their chemical structure. Although polysaccharides themselves lack antibacterial activities, they can still serve as templates in potential antibacterial coatings through strategies using their functional groups and microstructures. In this chapter, we review the technology of bio-based antibacterial coatings from different perspectives, including the antibacterial agent, preparation methods, and application, as illustrated in figure 2.4.

2.4.1 Antibacterial agents

During the long evolutionary processes of plants and animals on the Earth, they have developed sophisticated systems to survive invasions of viruses and bacteria. Antibacterial agents derived from plants and animals can be classified into the

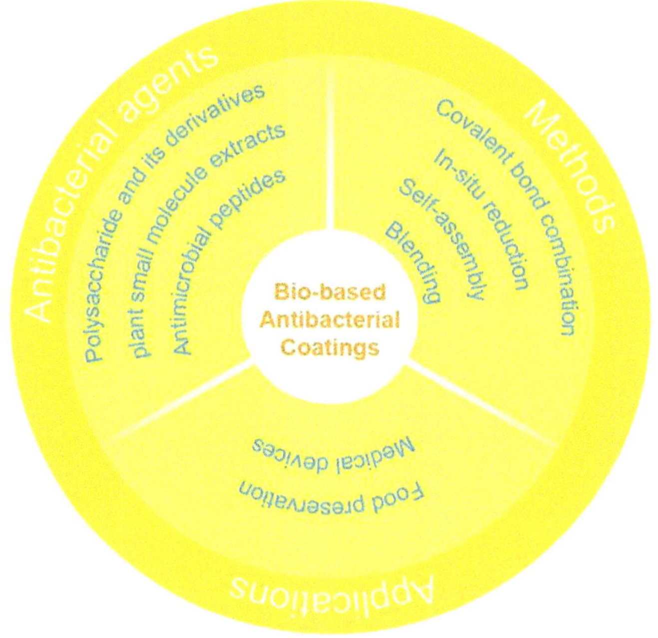

Figure 2.4. Schematic representation of the technology of bio-based antibacterial coatings in different perspectives.

following types: (i) plant small molecule extracts, (ii) antimicrobial peptides, and (iii) polysaccharides and some of their derivatives.

2.4.1.1 Plant small molecule extracts

Plant extracts consist of phenolic compounds that are responsible for their antibacterial properties. The chemical structures of phenolic compounds play an important role in antibacterial properties. There is at least one hydroxyl group in one phenol structure, which disrupts the cell membrane of the pathogenic microorganisms and causes leakage of their components [139]. Plant extracts, such as tannin acid (figure 2.5(a)), curcumin (figure 2.5(b)), and EOs (figure 2.5(c)), are commonly used to prevent bacterial invasion. Wang *et al* [140] designed a tannin acid-based water polyurethane coating with ZnO/GO, which exhibits great mechanical properties, self-healing efficiency, and antibacterial capabilities. Tannic acid not only interacts with the proteins and lipids present in bacterial cell membranes, but also inhibits key enzyme activity within bacteria. In addition, a combination of tannic acid and ZnO/GO causes the leakage of cell contents. Qu *et al* [141] reported a curcumin-based antibacterial coating on a titanium surface through surface polymerization of KH570. Due to curcumin, the cell membrane polarity of bacteria was increased, their cell membrane permeability was disrupted, cell lysis occurred, and the bacteria (*Escherichia coli* and *Staphylococcus aureus*) were killed.

EOs from plants and spices and aromatic and volatile liquids extracted from plant materials are usually used in food packages and massage treatments [142].

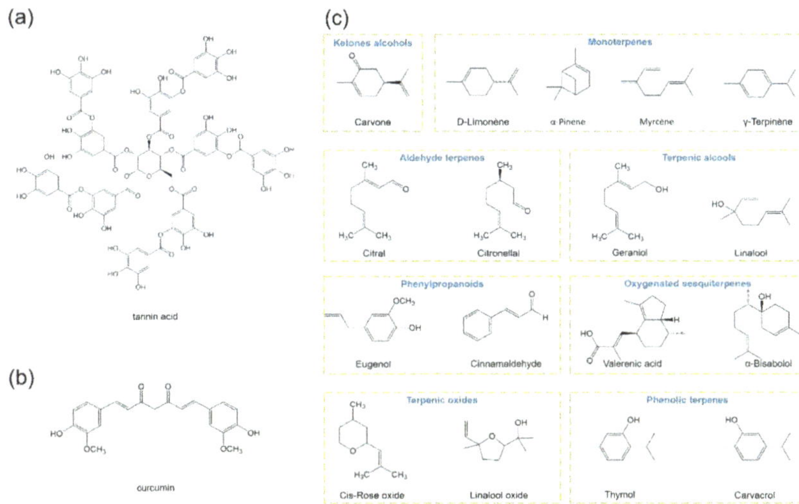

Figure 2.5. Some small molecule plant extracts: (a) tannin acid, (b) curcumin, and (c) a few components of essential oils.

EOs contain a mixture of lipophilic and small volatile compounds such as aldehydes, phenols, terpenes, ketones, alcohol, and others which are responsible for their biological properties (figure 2.5(c)). Moreover, EOs have attracted increasing attention owing to their excellent broad-spectrum bactericidal properties and antioxidant activities. Because EOs contain phenol groups, they show outstanding antibacterial properties. Phenol groups can destroy the structure and permeability of cell membranes, and hydroxyl groups can inhibit microorganism enzyme activity [143, 144]. In addition, EOs contain functional groups such as alcohols, ethers, ketones, and aldehydes that contribute to their antioxidant activities [145]. Mu *et al* [144] coated octadecyl trichlorosilane-modified mesoporous silica nanoparticles on an Al surface and infused cinnamon EOs into the nanoparticles with periodic nanopores. As a result, planktonic bacteria were killed via diffusing of cinnamon EOs.

2.4.1.2 Antimicrobial peptides

A peptide is a short chain made up of two or more amino acids. Antimicrobial peptides are potent natural compounds that are produced by living organisms such as humans, bacteria, fungi, and actinomycetes, and form an important part of innate immunity. Such active peptides are generally cationic short peptides consisting of 5–100 amino acid residues with molecular weights between 1 and 5 kDa (figure 2.6). Antimicrobial peptides exhibit a broad-spectrum antibacterial function and an amphiphilic structure. Owing to the presence of cations, antimicrobial peptides can destroy the bacterial cell membrane and kill bacteria by preferentially combining with the negative parts of the cell membrane and inserting the hydrophobic end into the phospholipid molecular layer [146]. In addition, mechanisms such as inhibition of protein synthesis and stimulation of the host defense system confer

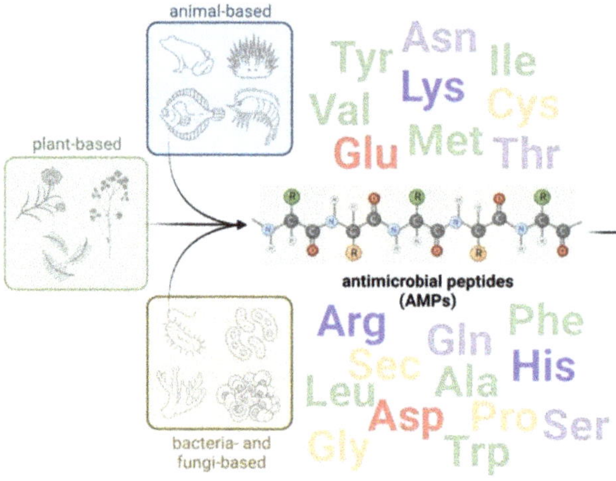

Figure 2.6. Occurrence of antimicrobial peptides and their application to inactivate disease-related microorganisms. (Reproduced with permission from [146]. Copyright 2022 Elsevier.)

virucidal capacity to several antimicrobial peptides [147]. Wang *et al* [148] designed a coating consisting of antimicrobial peptides, antibiotics, and polymyxin B. This coating was used on different biomedical device surfaces such as polystyrene, silicone, polyurethane, and titanium, and showed high bactericidal activity, but low cytotoxicity and hemolytic activity.

2.4.1.3 Polysaccharides and their derivatives

Polysaccharides contain various chemical components such as cellulose, starch, galactomannans, pectin, chitosan, alginate, pullulan, and so on. Among them, because of their antibacterial activity, chitosan and hyaluronic acid have been used directly as antibacterial agents for various biomedical and pharmaceutical applications. Other polysaccharides without antibacterial activities still function as templates for potential antibacterial coatings. Therefore, in the following, we want to focus on (i) chitosan (with intrinsic antibacterial properties) and (ii) cellulose and their derivatives (without intrinsic antibacterial properties) as representatives of polysaccharides for antibacterial coating approaches.

Chitosan is derived from chitin extracted from crustacean shells and plays an important role in the biomedical field as an antibacterial agent. The inhibition activity of chitosan depends on the type, concentration, molecular weight, and degree of deacetylation. Positively charged NH_3^+ in chitosan can interact with the negatively charged bacterial membranes, which may cause the leaching of low-molecular-weight material, nucleic acids, proteins, etc [149]. Huang *et al* [150] developed chitosan-based coated fabrics through a padding–drying–curing process, showing excellent antibacterial properties against *E. coli* and *S. aureus* reaching 95.1% and 99.9%, respectively.

Cellulose is the most abundant polysaccharide on Earth and comprises linear glucan chains linked by β-1,4-glycosidic bonds. It acts as an antibacterial platform because of the presence of flexible and adjustable chemical groups and microstructures. Nanocellulose such as cellulose nanocrystals (CNCs) and cellulose nanofibrils (CNFs) or nanofibrillated cellulose (NFC) can be isolated by chemical, mechanical, enzymatical or combination methods from raw materials containing cellulose. Nanocellulose has been applied in various filed such as coating, packing, biomedicine, energy storage devices, and so on [151]. The surface of nanocellulose has abundant hydrogen bonds and lacks antibacterial properties. Nanocellulose has been chemically modified through esterification, silylation, amidation, phosphorylation, and carboxymethylation to enhance its antibacterial properties [152–154]. Various quaternary ammonium salts with different alkyl chain lengths were grafted onto CNC through a nucleophilic addition reaction [155]. Hongrattanavichit and Aht-Ong [156] developed antibacterial cotton fabric through immersing the cotton fabric in an organosilane (amino silane (3-aminopropyl) and octadecyl dimethyl (3-trimethoxysilylpropyl) ammonium chloride) modified CNF dispersion. 0.25% of octadecyl dimethyl (3-trimethoxysilylpropyl) ammonium chloride-modified CNF coating is the ideal, showing good antibacterial properties (99.99% bacterial reduction) and water-repellent performance (140° water contact angle). Similarly, dissolved cellulose was modified using protoporphyrin IX (PpIX) and quaternary ammonium salt groups, exhibiting a markedly enhanced reactive oxygen species (ROS) yield [157]. Benefiting from porphyrin-based photosensitizers (PpIX), the inhibition efficiencies of cellulose-based photosensitizers (CPS) toward drug-resistant *E. coli* and *S. aureus* are 84% and 98%, respectively, after white-light irradiation for 1 min at a flux rate of 40 mW cm^{-2}.

2.4.2 Preparation methods

Almost all functional coatings are deposited on a substrate using common methods such as self-assembly [3, 158], dip-coating/impregnation [150, 159], spin coating [160], and spraying coating [161]. Biomass ingredients play different key roles in the coating; for example, some ingredients have antibacterial properties while others function as templates for active agents. We discuss the preparation methods of antibacterial coatings by introducing the antibacterial agents in coatings and their interaction with biomass ingredients.

2.4.2.1 Self-assembly

Self-assembly is a process where molecules organize into stable structures without external guidance, driven by their intrinsic properties such as shape, charge, and hydrophobicity. This concept can be used to create functional antibacterial coatings by leveraging natural processes. Cyclodextrins (α-CDs) are cyclic oligosaccharides containing six α-cyclodextrin, seven β-cyclodextrin, eight γ-cyclodextrin or more glucopyranose units linked by α-(1,4) bonds. The most remarkable feature of α-CDs is their ability to form solid inclusion complexes (host–guest complexes) with various solid, liquid, and gaseous compounds through molecular complexation [162]. Inspired by the mechanism of articular cartilage lubrication, a soft/hard combination strategy was proposed by Shi's group for producing composite ceramic coatings [163]. Here, the host–guest interaction-induced self-assembly between α-CDs and

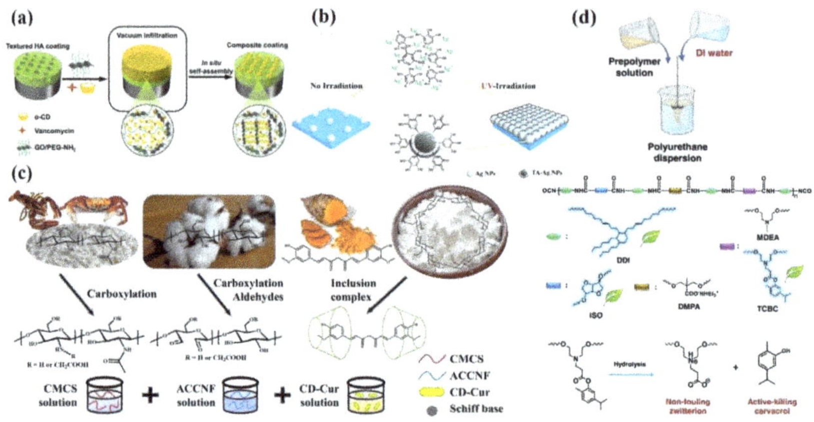

Figure 2.7. (a) Self-assembly. (Reproduced with permission from [164]. Copyright 2022 American Chemical Society.) (b) *In situ* reduction. (Reproduced with permission from [3]. Copyright 2021 American Chemical Society.) (c) Blending. (Reproduced with permission from [173]. Copyright 2023 Elsevier.) (d) Covalent bond combination. (Reproduced with permission from [168]. Copyright 2022 American Chemical Society.)

poly(ethylene glycol)-modified Ag NPs results in the formation of Ag NP-hybrid supramolecular hydrogel *in situ* in the holes of a textured Y_2O_3-stabilized ZrO_2 ceramic coating. This coating provided a new chance of designing antibacterial and self-lubricating ceramic coatings for artificial joint applications. Recently, they also developed a novel hydroxyapatite (HA) composite coating by incorporating GO-hybrid pseudopolyrotaxane supramolecular hydrogels prepared via the host–guest interaction between α-CD and PEG chains in the holes of textured HA coating through a vacuum infiltration process (figure 2.7(a)). The HA composite coating can be endowed with an antibacterial function after loading vancomycin in the GO hydrogel [164]. These composite coatings provide new insights for designing antibacterial and self-lubricating ceramic coatings for artificial joint applications. Layer-by-layer self-assembly is a simple method based on electrostatic interactions between polyanions and polycations. A coating was prepared by synthesizing a composite comprising poly-L-lysine that could interact with sodium alginate-modified quaternary ammonium and Ag NPs [165]. This coating promoted mineralization and improved cytocompatibility.

2.4.2.2 In situ *reduction*

The *in situ* reduction process usually involves reductant and oxidant or oxidizing and reducing agents generating material. As a common component in antibacterial coating, Ag NPs can be prepared through one redox reaction. During this process, biobased compounds, such as dopamine and tannic acid, play the important role of reducing agents generating material. Mussels strongly adhere to all types of materials in the sea. This adhesion is attributed to multiple interactions, such as π–π stacking, metal complexation, hydrogen bonding, and covalent crosslinking, between the catechol structure in mussel proteins and substrates. Inspired by this adhesion, an effective approach was reported in which dopamine, tannic acid, and urushiol with catechol structures were introduced into a coating for reducing Ag^+ [166]. Xu's group [3] functionalized the substrates of Ti, Si, polystyrene, glass, and PDMS with tannic acid–Ag NPs under a UV-assisted one-step deposition (figure 2.7 (b)). Because of the existence of Ag NPs that release Ag^+ ions, the modified substrate exhibited excellent antibacterial activity in the inhibition zone test. Moreover, the functionalized surfaces exhibited low cytotoxicity toward L929 mouse fibroblasts and showed no side effects on the major organs of Sprague–Dawley rats after implantation. Jo *et al* [167] leveraged a coating by synthesizing recombinant mussel adhesive proteins fused with functional silver-binding peptides for reducing Ag NPs on the fused protein-coated surface. Ag NPs were fully synthesized on various surfaces, including metal, plastic, and glass, by spin coating and demonstrated excellent antibacterial efficacy against gram-positive and gram-negative bacteria and good cytocompatibility with mammalian cells.

2.4.2.3 Covalent bond combination

There are functional groups (hydroxyl and aldehyde groups) in the chemical structures of EOs; therefore, chemical modification based on covalent bond combination is a common method for preparing antibacterial coatings. Carvacrol, one of the EOs, shows a relatively low minimum inhibitory concentration against both gram-positive and gram-negative bacteria. Huang *et al* [168] prepared a bio-based polyurethane coating with long-lasting antibacterial properties using spin coating. This coating contains a novel diol, wherein carvacrol is introduced as a pendant group into the nonfouling zwitterionic moieties, which can be continuously hydrolyzed in an aqueous environment (figure 2.7(d)). Consequently, the released carvacrol actively kills the bacteria, and carboxy betaine zwitterionic moieties eliminate future bacterial attachment. Similarly, Liu *et al* [169] reported on two types of antimicrobial peptides (WRWRWR-NH_2 and RLARIVVIRVAR-NH_2)) with different numbers of amino acid residues; these AMPs were embedded in the hydrogel coating via a chemical grafting method to achieve antibacterial properties. Yan *et al* [170] prepared a tung oil/*Eucommia* rubber composite coating (ERTO) based on conjugated C–C bonds via the Diels–Alder reaction. As a composite protective coating, ERTO integrated the characteristics of both rubber and tung oil, exhibiting outstanding corrosion resistance and excellent antibacterial properties.

2.4.2.4 Blending

Blending two different ingredients through physical blending is a wise alternative to create a new coating with desired properties showing a synergistic effect of the individual materials. Compared with other methods involving complex or tedious chemical reactions, physical blending is the direct and simple method of combining antibacterial agents and biomass materials for certain applications [171, 172]. Zhou *et al* [173] mixed three biomass materials (carboxymethyl chitosan, aldehyde carboxylate cellulose nanofibers, and hydroxypropyl γ-cyclodextrin/curcumin) in an aqueous solution. This mixture could be easily applied to different substrates to form conformal coatings/films with tunable thickness via spraying and dipping approaches. These films are useful for the keeping food fresh (figure 2.7(c)).

Inspired by piezoelectricity and the porous structure of living bone, Wei *et al* [2] developed a coating with barium titanate (BTO) NPs and chitosan. This coating had an excellent antibacterial property with an antibacterial rate of 90.41%. The spontaneous polarization derived from the BTO and the mechanical force is essential for destroying the membrane of *Fusobacterium nucleatum*.

2.4.3 Applications

2.4.3.1 Food preservation

Globally, foodborne diseases are among the most severe and expensive public concerns, and bacteria tend to be the most widespread cause of these types of poisonings and infections [174]. Some of these microorganisms produce toxins that may pose serious health risks to consumers. In addition to microbial spoilage, endogenous enzyme activity and chemical changes such as lipid oxidation and nonenzymatic browning (the Maillard reaction) are responsible for the deterioration of stored fruits and vegetables. Coating is a technique used to preserve food for long periods by controlling water transport, surface ablation, and reducing the permeability of oxygen and carbon dioxide in to the volume of food [175].

During the twelfth to thirteenth centuries, hot-melt paraffin waxes were used as coatings for lemons and oranges. Subsequently, more polymeric materials derived from oxidized polyethylene, organic solvents, surfactants, and stabilizers became popular coating materials. However, nondegradable polymers affect human health and do not meet global environmental concerns. Additionally, traditional preservation methods no longer meet the needs; therefore, there is an urgent need to explore new and safe food preservation methods [143].

Currently, biomass coatings are predominantly popular in the field of food preservation (e.g. fruits, vegetables, and meats). Fruits and vegetables tend to spoil because of water loss, senescence processes, and microbial growth. Zhou *et al* [176] developed a coating with excellent antimicrobial and antioxidant properties for preserving passion fruit. The coating solution comprises quaternary ammonium chitosan and tannic acid (TA), endowing with antimicrobial and antioxidant properties, respectively. The coating is applied on the surface of the passion fruit by dip-coating. The resulting coating effectively reduces weight loss, delays the softening of the fruit, and prolongs the storage period. Mondal *et al* [177] extracted bioactive compounds 'crude algae ethanolic extract' (CAEE) that were rich in antioxidants and various bioactive compounds, predominantly carotenoids, proteins, and polysaccharides from *Dunaliella tertiolecta*. The bio-composite coating solution prepared by mixing CS and CAEE was applied to green chillies, and it provided superior antioxidant activity, prolonging the shelf life of the coated green chilies. Meats contain a large amount of protein, which is of high biological value and provides amino acids (especially essential amino acids), benefiting human health and well-being [178].

Meats are susceptible to chemical deterioration and microbiological contamination, which can cause health problems. Ghasemi *et al* [179] prepared an EO-based composite coating consisting of ZnO NPs, nanostructured lipid carriers of okra mucilage, and carboxymethyl cellulose, which can control the release of antimicrobial EOs.

This coating was applied on the surface of beef so that their effect continues during the storage period rather than directly adding EOs. The coatings showed a significant decrease in the growth rate of aerobic mesophilic bacteria and *S. aureus* ($P < 0.05$).

2.4.3.2 Medical devices

Implant-associated infections have increased dramatically in recent years, leading to reduced service life and secondary surgical procedures for implant replacement and even implant failure [180]. Hydrogel coatings were synthesized using a cationic quaternary ammonium-containing gelatin-based polymer, which showed good mechanical properties, biodegradability, excellent bactericidal activity against various types of bacteria, and high cytocompatibility with mammalian cells [181]. Periprosthetic joint infection is one of the common complications after arthroplasty. It is mainly attributed to *S. aureus*, a gram-positive bacterium that accounts for nearly half of the infection issues and leads to the ultimate failure of artificial joints [182, 183]. Vancomycin is a high-risk antibiotic used to treat severe infections. It is loaded in a hydrogel coating that serves as a drug carrier to exhibit sustained antibacterial effect against *S. aureus* [164]. Dressing of blood-contacting devices with robust and synergistic antibacterial agents has been explored for several decades, and hydrogel coatings with AMPs have shown excellent bactericidal and anti-adhesion properties against gram-positive and gram-negative bacteria [169]. The *in vivo* anti-infection performance was verified using the microbiological and histological results of animal experiments, showing promising clinical application potential in preventing bacterial infection and thrombosis for blood-contacting medical devices and related implants.

2.5 Other bio-based coatings

Some advanced bio-based coatings (e.g. radiative cooling and anti-corrosion) have drawn the attention of researchers for other applications. To achieve passive cooling, surface coating with high reflectivity is essential for directly reflecting solar irradiation to the atmosphere without consuming extra energy. Some NPs with high whiteness and a high electron energy gap, including TiO_2 (3.2 eV) and $BaSO_4$ (~6 eV) NPs, have been used to increase the reflectivity of passive cooling coatings [184]. Chen *et al* [185] prepared jatropha oil-based polyurethane (PU) prepolymer coatings with different proportions of $BaSO_4$ and TiO_2. Because the TiO_2 is smaller in size than $BaSO_4$, TiO_2 acts as a nanofiller and fills in the gaps between $BaSO_4$ particles. The reflectance increased with increasing TiO_2 proportion in the hybrid NPs, and the highest reflectance was achieved for the $Ba_{20}Ti_{20}$ sample. The reflectance of the optimal coating reached 97.30% in the UV region (200–380 nm), and the maximum temperature difference in surfaces before and after loading the coating reached 20.8 °C under infrared irradiation for 60 min. Some coatings are susceptible to mechanical damage and chemical corrosion under extreme conditions. Liu *et al* [186] prepared a coating based on biogas residue biochar/MoS_2 with excellent anti-icing performance. Because MoS_2 loading act as a barrier between the substrate and corrosive medium, the barrier ability of the coating increased substantially, thereby reducing the content of the corrosive medium at the coating–substrate interface. In the past two decades, self-healing materials with dynamic chemistry have attracted increasing attention. If the morphology and property of coating can be renewed after its deterioration, its service life will be greatly extended. Yu *et al* [187] prepared UV-curable coatings with self-healing properties from CNF-stabilized tung oil (TO)-based oligomer (TMHT)-based Pickering emulsions after drying and UV light-curing processes. CNFs considerably improved the storage stability of Pickering emulsions. Additionally, because of the presence of dynamic ester bonds in TMHT, the coatings exhibited excellent self-healing performance (78.05%–56.34%) at 150 °C without any catalyst or external force. More importantly, impedance at 0.01 Hz determined from Bode plots ($|Z|_{0.01\,Hz}$) for self-healing coating was higher than that for scratched coating.

2.6 Summary

This chapter provides a summary of bio-based coatings: superwetting, flame-retardant, and antibacterial coatings. Bio-based coating technology, as an advancement in the field of innovative material science, has shown tremendous potential across various applications. With the development of aim of carbon neutrality in the world, ecofriendly feedstocks are the most promising candidates for designing various biodegradable coatings, which is beneficial to the environment. These coatings harness the unique properties of natural or bio-based materials, imparting multifunctional characteristics such as superwetting, flame retardancy, and antibacterial effects to target surfaces. The superwetting property enables complete spreading or repelling of liquids, which is crucial for self-cleaning and anti-fog applications. Flame retardancy significantly enhances material safety, particularly in the construction and textile industries. The antibacterial function inhibits microbial growth, creating more hygienic environments, which is especially important for medical facilities and food processing industries. By integrating these multifunctional properties, biomass coatings not only promote sustainability but also greatly expand the application scope of bio-based materials.

However, challenges and issues for practical applications remain. We mention the following challenges for future advances in bio-based coatings. First, some of functional coatings involve complex steps and specific experimental instrumentation, which are far from a practical application. Therefore, it is essential to regulate and change the fabrication process to improve the efficiency and scalability of the chemical modification process, and expand the range of applicable biomass sources. Second, it is urgent to design a smart functional coating, and new designs are needed to maintain long-term stable superwetting, flame-retardant and antibacterial capabilities. Finally, most antibacterial coatings are effective for specific bacteria and not applied to others, which cannot satisfy practical applications.

Bibliography

[1] Canama G J C, Delco M C L, Talandron R A and Tan N P 2023 Synthesis of chitosan-silver nanocomposite and its evaluation as an antibacterial coating for mobile phone glass protectors *ACS Omega* **8** 17699–711

[2] Wei Y, Hu X, Shao J, Wang S, Zhang Y, Xie W Z, Wu Y X, Zeng X T and Zhang L L 2024 Daily sonic toothbrush triggered biocompatible $BaTiO_3$/chitosan multiporous coating with enhanced piezocatalysis for intraoral antibacterial activity *Mater. Today Commun.* **38** 107715
[3] He X, Gopinath K, Sathishkumar G, Guo L, Zhang K, Lu Z, Li C, Kang E T and Xu L 2021 UV-assisted deposition of antibacterial Ag–tannic acid nanocomposite coating *ACS Appl. Mater. Interfaces* **13** 20708–17
[4] Singh A K, Mishra S and Singh J K 2019 Underwater superoleophobic biomaterial based on waste potato peels for simultaneous separation of oil/water mixtures and dye adsorption *Cellulose* **26** 5497–511
[5] Bhushan B 2011 Biomimetics inspired surfaces for drag reduction and oleophobicity/philicity *Beilstein J. Nanotechnol.* **2** 66–84
[6] Fei L, He Z, Lacoste J D, Nguyen T H and Sun Y 2020 A mini review on superhydrophobic and transparent surfaces *Chem. Rec.* **20** 1257–68
[7] Wenzel R N 1936 Resistance of solid surfaces to wetting by water *Ind. Eng. Chem.* **28** 988–94
[8] Cassie A and Baxter S 1944 Wettability of porous surfaces *Trans. Faraday Soc.* **40** 546–51
[9] Bhushan B and Jung Y C 2011 Natural and biomimetic artificial surfaces for superhydrophobicity, self-cleaning, low adhesion, and drag reduction *Prog. Mater Sci.* **56** 1–108
[10] Lyu J, Wu B, Wu N, Peng C, Yang J, Meng Y and Xing S 2021 Green preparation of transparent superhydrophobic coatings with persistent dynamic impact resistance for outdoor applications *Chem. Eng. J.* **404** 126456
[11] Peng J, Wu L, Zhang H, Wang B, Si Y, Jin S and Zhu H 2022 Research progress on eco-friendly superhydrophobic materials in environment, energy and biology *Chem. Commun.* **58** 11201–19
[12] Chang M, Wang X, Lin Q, Li R, Zhao L, Ren J and Zhang F 2022 Formic acid–hydrogen peroxide treatment of furfural residue for production of nanocellulose, lignin, and nanoscale lignin *Green Chem.* **24** 6232–40
[13] Verma J, Petru M and Goel S 2024 Cellulose based materials to accelerate the transition towards sustainability *Ind. Crops Prod.* **210** 118078
[14] Nechyporchuk O, Belgacem M N and Bras J 2016 Production of cellulose nanofibrils: a review of recent advances *Ind. Crops Prod.* **93** 2–25
[15] Liu H, Gao S-W, Cai J-S, He C-L, Mao J-J, Zhu T-X, Chen Z, Huang J-Y, Meng K and Zhang K-Q 2016 Recent progress in fabrication and applications of superhydrophobic coating on cellulose-based substrates *Materials* **9** 124
[16] Putro J N, Ismadji S, Gunarto C, Yuliana M, Santoso S P, Soetaredjo F E and Ju Y H 2019 The effect of surfactants modification on nanocrystalline cellulose for paclitaxel loading and release study *J. Mol. Liq.* **282** 407–14
[17] Li S, Wei Y and Huang J 2010 Facile fabrication of superhydrophobic cellulose materials by a nanocoating approach *Chem. Lett.* **39** 20–1
[18] Singh R K, Kukrety A, Chatterjee A K, Thakre G D, Bahuguna G M, Saran S, Adhikari D K and Atray N 2014 Use of an acylated chitosan Schiff base as an ecofriendly multifunctional biolubricant additive *Ind. Eng. Chem. Res.* **53** 18370–9
[19] Tang X, Huang W, Xie Y, Xiao Z, Wang H, Liang D, Li J and Wang Y 2021 Superhydrophobic hierarchical structures from self-assembly of cellulose-based nanoparticles *ACS Sustain. Chem. Eng.* **9** 14101–11

[20] Zheng X and Fu S 2019 Reconstructing micro/nano hierarchical structures particle with nanocellulose for superhydrophobic coatings *Colloids Surf.* A **560** 171–9

[21] Ma B, Xiong F, Wang H, Qing Y, Chu F and Wu Y Tailorable and scalable production of eco-friendly lignin micro-nanospheres and their application in functional superhydrophobic coating *Chem. Eng. J.* **457** 141309

[22] Figueiredo P, Lintinen K, Hirvonen J T, Kostiainen M A and Santos H A 2018 Properties and chemical modifications of lignin: towards lignin-based nanomaterials for biomedical applications *Prog. Mater Sci.* **93** 233–69

[23] Liu X, Gao C, Fu C, Xi Y, Fatehi P, Zhao J R, Wang S, Gibril M E and Kong F 2022 Preparation and performance of lignin-based multifunctional superhydrophobic coating *Molecules* **27** 1440

[24] Zhang Y, Zhang Y, Cao Q, Wang C, Yang C, Li Y and Zhou J 2020 Novel porous oil-water separation material with super-hydrophobicity and super-oleophilicity prepared from beeswax, lignin, and cotton *Sci. Total Environ.* **706** 135807

[25] Yu S, Wang M, Xie Y, Qian W, Bai Y and Feng Q 2023 Lignin self-assembly and auto-adhesion for hydrophobic cellulose/lignin composite film fabrication *Int. J. Biol. Macromol.* **233** 123598

[26] Ren C, Li M, Huang W, Zhang Y and Huang J 2022 Superhydrophobic coating with excellent robustness and UV resistance fabricated using hydrothermal treated lignin nanoparticles by one-step spray *J. Mater. Sci.* **57** 18356–69

[27] Cao M, Hu Y, Cheng W, Huan S, Bai T, Niu Z, Zhao Y, Yue G, Zhao Y and Han G 2022 Lignin-based multi-scale cellular aerogels assembled from co-electrospun nanofibers for oil/water separation and energy storage *Chem. Eng. J.* **436** 135233

[28] Bang J, Kim J, Kim Y, Oh J-K and Kwak H W 2022 Preparation and characterization of hydrophobic coatings from carnauba wax/lignin blends *J. Korean Wood Sci. Technol.* **50** 149–58

[29] Sun S, Xu P, Xiao Q, Qiang X and Shi X 2022 One-step solvent-free fabrication of superhydrophobic cellulose powder with reversible wettability *Prog. Org. Coat.* **173** 107170

[30] Wang Y, Wang X, Heim L, Breitzke H, Buntkowsky G and Zhang K 2015 Superhydrophobic surfaces from surface-hydrophobized cellulose fibers with stearoyl groups *Cellulose* **22** 289–99

[31] Xu C, Wang S, Zhou L, Bi Y, Yang G, Wu J and Zhang X 2022 Facile fabrication of superhydrophobic cellulose/Fe_2O_3-STA film with nanoflower morphologies for heavy oil removal *Fibers Polym.* **23** 2692–8

[32] Huang J, Wang S, Lyu S and Fu F 2018 Preparation of a robust cellulose nanocrystal superhydrophobic coating for self-cleaning and oil-water separation only by spraying *Ind. Crops Prod.* **122** 438–47

[33] Zhu Z, Fu S and Basta A H 2020 A cellulose nanoarchitectonic: multifunctional and robust superhydrophobic coating toward rapid and intelligent water-removing purpose *Carbohydrate Polym.* **243** 116444

[34] Cheng Q, Guan C, Wang M, Li Y and Zeng J 2018 Cellulose nanocrystal coated cotton fabric with superhydrophobicity for efficient oil/water separation *Carbohydr. Polym.* **199** 390–6

[35] Ye M, Wang S, Ji X, Tian Z, Dai L and Si C 2022 Nanofibrillated cellulose-based superhydrophobic coating with antimicrobial performance *Adv. Compos. Hybrid Mater.* **6** 30

[36] Roy S, Zhai L, Kim J W, Kim H C and Kim J 2020 A novel approach of developing sustainable cellulose coating for self-cleaning-healing fabric *Prog. Org. Coat.* **140** 105500
[37] Wang T and Zhao Y 2021 Fabrication of thermally and mechanically stable superhydrophobic coatings for cellulose-based substrates with natural and edible ingredients for food applications *Food Hydrocoll.* **120** 106877
[38] Huang J, Cai P, Li M, Wu Q, Li Q and Wang S 2020 Preparation of CNF/PDMS superhydrophobic coatings with good abrasion resistance using a one-step spray method *Materials* **13** 5380
[39] Huang J, Lyu S, Fu F, Chang H and Wang S 2016 Preparation of superhydrophobic coating with excellent abrasion resistance and durability using nanofibrillated cellulose *RSC Adv.* **6** 106194–200
[40] Fang S K, Li H R, Feng S D, Wang P X, Yu Y, Zhang H and Guo J 2024 Cellulose acetate superhydrophobic coatings for efficient oil–water separation using a combination of electrostatic spraying and chemical vapor deposition *Polym. Eng. Sci.* **64** 254–63
[41] Arslan O, Aytac Z and Uyar T 2016 Superhydrophobic, hybrid, electrospun cellulose acetate nanofibrous mats for oil/water separation by tailored surface modification *ACS Appl. Mater. Interfaces* **8** 19747–54
[42] Huang H, Huang C, Xu C and Liu R 2022 Development and characterization of lotus-leaf-inspired bionic antibacterial adhesion film through beeswax *Food Packag. Shelf Life* **33** 100906
[43] Li M, Huang W, Ren C, Wu Q, Wang S and Huang J 2022 Preparation of lignin nanospheres based superhydrophobic surfaces with good robustness and long UV resistance *RSC Adv.* **12** 11517–25
[44] Meng H, Zhao Y, Wang S, Wang Y, Xiao Z, Wang H, Liang D and Xie Y 2022 Turning the morphology and wetting ability of self-assembled hierarchical structures from lignin stearoyl esters *Ind. Crops Prod.* **183** 114969
[45] Jiang Y, Zhang Y, Gao C, An Q, Xiao Z and Zhai S 2022 Superhydrophobic aerogel membrane with integrated functions of biopolymers for efficient oil/water separation *Sep. Purif. Technol.* **282** 120138
[46] Wu J, Ma X, Gnanasekar P, Wang F, Zhu J, Yan N and Chen J 2023 Superhydrophobic lignin-based multifunctional polyurethane foam with SiO_2 nanoparticles for efficient oil adsorption and separation *Sci. Total Environ.* **860** 160276
[47] Wang S, Sha J, Wang W, Qin C, Li W and Qin C 2018 Superhydrophobic surfaces generated by one-pot spray-coating of chitosan-based nanoparticles *Carbohydr. Polym.* **195** 39–44
[48] Roy S, Goh K, Verma C, Ghosh B D, Sharma K and Maji P K 2022 A facile method for processing durable and sustainable superhydrophobic chitosan-based coatings derived from waste crab shell *ACS Sustain. Chem. Eng.* **10** 4694–704
[49] Xue Q, Wu J, Lv Z, Lei Y, Liu X and Huang Y 2023 Photothermal superhydrophobic chitosan-based cotton fabric for rapid deicing and oil/water separation *Langmuir* **39** 9912–23
[50] Wang W, Lin J-H, Guo J, Sun R, Han G, Peng F, Chi S and Dong T 2023 Biomass chitosan-based tubular/sheet superhydrophobic aerogels enable efficient oil/water separation *Gels* **9** 346
[51] Zhang W, Lu P, Qian L and Xiao H 2014 Fabrication of superhydrophobic paper surface via wax mixture coating *Chem. Eng. J.* **250** 431–6

[52] Bhushan B, Jung Y C, Niemietz A and Koch K 2009 Lotus-like biomimetic hierarchical structures developed by the self-assembly of tubular plant waxes *Langmuir* **25** 1659–66
[53] Niemietz A, Wandelt K, Barthlott W and Koch K 2009 Thermal evaporation of multi-component waxes and thermally activated formation of nanotubules for superhydrophobic surfaces *Prog. Org. Coat.* **66** 221–7
[54] Wang F, Ma R and Tian Y 2022 Superhydrophobic starch-based adsorbent with honeycomb coral-like surface fabricated via facile immersion process for removing oil from water *Int. J. Biol. Macromol.* **207** 549–58
[55] Fujishima A, Rao T N and Tryk D A 2000 Titanium dioxide photocatalysis *J. Photochem. Photobiol.* C **1** 1–21
[56] Zhang L, Zhao N and Xu J 2014 Fabrication and application of superhydrophilic surfaces: a review *J. Adhes. Sci. Technol.* **28** 769–90
[57] Drelich J and Marmur A 2014 Physics and applications of superhydrophobic and superhydrophilic surfaces and coatings *Surf. Innov.* **2** 211–27
[58] Otitoju T, Ahmad A and Ooi B 2017 Superhydrophilic (superwetting) surfaces: a review on fabrication and application *J. Ind. Eng. Chem.* **47** 19–40
[59] Fan Q, Lu T, Deng Y, Zhang Y, Ma W, Xiong R and Huang C 2022 Bio-based materials with special wettability for oil–water separation *Sep. Purif. Technol.* **297** 121445
[60] Liu Y, Wen J, Gao Y, Li T, Wang H, Yan H, Niu B and Guo R 2018 Antibacterial graphene oxide coatings on polymer substrate *Appl. Surf. Sci.* **436** 624–30
[61] Ao C, Hu R, Zhao J, Zhang X, Li Q, Xia T, Zhang W and Lu C 2018 Reusable, salt-tolerant and superhydrophilic cellulose hydrogel-coated mesh for efficient gravity-driven oil/water separation *Chem. Eng. J.* **338** 271–7
[62] Deerattrakul V, Sakulaue P, Bunpheng A, Kraithong W, Pengsawang A, Chakthranont P, Iamprasertkun P and Itthibenchapong V 2023 Introducing hydrophilic cellulose nanofiber as a bio-separator for 'water-in-salt' based energy storage devices *Electrochim. Acta* **453** 142355
[63] Yang J, Cui J, Xie A, Dai J, Li C and Yan Y 2021 Facile preparation of superhydrophilic/underwater superoleophobic cellulose membrane with $CaCO_3$ particles for oil/water separation *Colloids Surf.* A **608** 125583
[64] Wang M, Peng M, Zhu J, Li Y and Zeng J-B 2020 Mussel-inspired chitosan modified superhydrophilic and underwater superoleophobic cotton fabric for efficient oil/water separation *Carbohydr. Polym.* **244** 116449
[65] Wang Q, Xie D, Chen J, Liu G and Yu M 2020 Facile fabrication of superhydrophobic and photoluminescent TEMPO-oxidized cellulose-based paper for anticounterfeiting application *ACS Sustain. Chem. Eng.* **8** 13176–84
[66] Liang L, Wang C, Wang H, Zhan H and Meng X 2018 Bioinspired fabric with superhydrophilicity and superoleophobicity for efficient oil/water separation *Fibers Polym.* **19** 1828–34
[67] Meng X, Dong Y, Zhao Y and Liang L 2020 Preparation and modification of cellulose sponge and application of oil/water separation *RSC Adv.* **10** 41713–9
[68] Zhang Y, Chen J, Hao B, Wang R and Ma P 2020 Preparation of cellulose-coated cotton fabric and its application for the separation of emulsified oil in water *Carbohydr. Polym.* **240** 116318
[69] Zhang F, Ren H, Dou J, Tong G and Deng Y 2017 Cellulose nanofibril based-aerogel microreactors: a high efficiency and easy recoverable W/O/W membrane separation system *Sci. Rep.* **7** 40096

[70] Ao C, Yuan W, Zhao J, He X, Zhang X, Li Q, Xia T, Zhang W and Lu C 2017 Superhydrophilic graphene oxide@electrospun cellulose nanofiber hybrid membrane for high-efficiency oil/water separation *Carbohydr. Polym.* **175** 216–22

[71] Zhang X, Wang C, Liu X, Wang J, Zhang C and Wen Y 2018 A durable and high-flux composite coating nylon membrane for oil–water separation *J. Clean. Prod.* **193** 702–8

[72] Li J, Xing G, Qiao M, Liu Z, Sun H, Jiao R, Li L, Zhang J and Li A 2023 Guar gum-based macroporous hygroscopic polymer for efficient atmospheric water harvesting *Langmuir* **39** 18161–70

[73] Fan B, Lin C, Li Z, Cui L, Xu B, Yu Y, Wang Q and Wang P 2023 Peroxidase-catalyzed fabrication of a mechanically robust nanofibrillated cellulose-based aerogel with antibacterial activity for efficient oil–water separation *ACS Appl. Polym. Mater.* **5** 10494–505

[74] Ma X *et al* 2021 Mechanically robust, solar-driven, and degradable lignin-based polyurethane adsorbent for efficient crude oil spill remediation *Chem. Eng. J.* **415** 128956

[75] Liao Y, Wang C, Huang C, Hussain Abdalkarim S Y, Wang L, Chen Z and Yu H-Y 2023 Robust cellulose/carboxymethyl chitosan composite films with high transparency and antibacterial ability for fresh fruit preservation *ACS Sustain. Chem. Eng.* **11** 5908–17

[76] Yu M, Zhao S, Yang L, Ji N, Wang Y, Xiong L and Sun Q 2021 Preparation of a superhydrophilic SiO_2 nanoparticles coated chitosan–sodium phytate film by a simple ethanol soaking process *Carbohydr. Polym.* **271** 118422

[77] Li X, Yu Z, Chen Q, Yang S, Li F and Yang Y 2019 Chitosan-coated filter paper with superhydrophilicity for treatment of oily wastewater in acidic and alkaline environments *Mater. Technol.* **34** 213–23

[78] Kankariya Y and Chatterjee B 2023 Biomedical application of chitosan and chitosan derivatives: a comprehensive review *Curr. Pharm. Design* **29** 1311–25

[79] Li Y, Zhang H, Ma C, Yin H, Gong L, Duh Y and Feng R 2019 Durable, cost-effective and superhydrophilic chitosan–alginate hydrogel-coated mesh for efficient oil/water separation *Carbohydr. Polym.* **226** 115279

[80] Zheng Z, Guo Z, Liu W and Luo J 2023 Low friction of superslippery and superlubricity: a review *Friction* **11** 1121–37

[81] Ji X, Shuai S, Liu S, Weng Y and Zheng F 2023 Silicon-based superslippery/superhydrophilic striped surface for highly efficient fog harvesting *Materials* **16** 5423

[82] Tuteja A, Choi W, Ma M, Mabry J M, Mazzella S A, Rutledge G C, Mckinley G H and Cohen R E 2007 Designing superoleophobic surfaces *Science* **318** 1618–22

[83] Tysoe W T and Spencer N D 2023 Super-slippery cellulose surfaces *Tribol. Lubr. Technol.* **79** 110–1

[84] Wong T-S, Kang S H, Tang S K, Smythe E J, Hatton B D, Grinthal A and Aizenberg J 2011 Bioinspired self-repairing slippery surfaces with pressure-stable omniphobicity *Nature* **477** 443–7

[85] Guo J, Fang W, Welle A, Feng W, Filpponen I, Rojas O J and Levkin P A 2016 Superhydrophobic and slippery lubricant-infused flexible transparent nanocellulose films by photoinduced thiol–ene functionalization *ACS Appl. Mater. Interfaces* **8** 34115–22

[86] Tan S, Han X, Cheng S, Guo P, Wang X, Che P, Jin R, Jiang L and Heng L 2023 Photothermal solid slippery surfaces with rapid self-healing, improved anti/de-icing and excellent stability *Macromol. Rapid Commun.* **44** 2200816

[87] Manabe K, Kyung K and Shiratori S 2015 Biocompatible slippery fluid-infused films composed of chitosan and alginate via layer-by-layer self-assembly and their antithrombogenicity *ACS Appl. Mater. Interfaces* **7** 4763–71
[88] Martín-Alfonso J E, López-Beltrán F, Valencia C and Franco J M 2018 Effect of an alkali treatment on the development of cellulose pulp-based gel-like dispersions in vegetable oil for use as lubricants *Tribol. Int.* **123** 329–36
[89] Da Cruz M G A, Budnyak T M, Rodrigues B V, Budnyk S and Slabon A 2021 Biocoatings and additives as promising candidates for ultralow friction systems *Green Chem. Lett. Rev.* **14** 358–81
[90] Jiang C, Li W, Nian J, Lou W and Wang X 2018 Tribological evaluation of environmentally friendly ionic liquids derived from renewable biomaterials *Friction* **6** 208–18
[91] Mu L, Wu J, Matsakas L, Chen M, Vahidi A, Grahn M, Rova U, Christakopoulos P, Zhu J and Shi Y 2018 Lignin from hardwood and softwood biomass as a lubricating additive to ethylene glycol *Molecules* **23** 537
[92] Cheng H *et al* 2023 Molecular design and properties of intrinsic flame-retardant P-N synergistic epoxy resin *J. Appl. Polym. Sci.* **141** e54885
[93] Li P, Wang J, Wang C, Xu C and Ni A 2024 The flame retardant and mechanical properties of the epoxy modified by an efficient DOPO-based flame retardant *Polymers* **16** 631
[94] Zhao B, Yue X, Tian Q, Qiu F, Li Y and Zhang T 2022 Bio-inspired BC aerogel/PVA hydrogel bilayer gel for enhanced daytime sub-ambient building cooling *Cellulose* **29** 7775–87
[95] Antia F K, Cullis C F and Hirschler M M 1981 The combined action of aluminium oxides and halogen compounds as flame retardants *Eur. Polym. J.* **17** 451–5
[96] Touré B, Lopez Cuesta J-M, Longerey M and Crespy A 1996 Incorporation of natural flame retardant fillers in an ethylene-propylene copolymer, in combination with a halogen-antimony systemv *Polym. Degrad. Stab.* **54** 345–52
[97] Guo L C *et al* 2023 Associations between serum polychlorinated biphenyls, halogen flame retardants, and renal function indexes in residents of an e-waste recycling area *Sci. Total Environ.* **858** 159746
[98] Trowbridge J, Gerona R, Mcmaster M, Ona K, Clarity C, Bessonneau V, Rudel R, Buren H and Morello-Frosch R 2022 Organophosphate and organohalogen flame-retardant exposure and thyroid hormone disruption in a cross-sectional study of female firefighters and office workers from San Francisco *Environ. Sci. Technol.* **56** 440–50
[99] Huo S, Song P, Yu B, Ran S, Chevali V S, Liu L, Fang Z and Wang H 2021 Phosphorus-containing flame retardant epoxy thermosets: recent advances and future perspectives *Prog. Polym. Sci.* **114** 101366
[100] Lu S, Chen S, Luo L, Yang Y, Wang J, Chen Y, Yang Y, Yuan Z and Chen X 2023 Molecules featuring the azaheterocycle moiety toward the application of flame-retardant polymers *ACS Chem. HealthSaf.* **30** 343–61
[101] Hamdani S, Longuet C, Perrin D, Lopez-Cuesta J-M and Ganachaud F 2009 Flame retardancy of silicone-based materials *Polym. Degrad. Stab.* **94** 465–95
[102] Cheng X-W, Wu Y-X, Huang Y-T, Jiang J-R, Xu J-T and Guan J-P 2020 Synthesis of a reactive boron-based flame retardant to enhance the flame retardancy of silk *React. Funct. Polym.* **156** 104731
[103] Yang J, Hong W, Zhang J, Liu M, Fu Z, Zhang Y, Guo Q, Li Y, Cai R and Qian K 2023 Wearable, biodegradable, and antibacterial multifunctional $Ti_3C_2T_x$ MXene/cellulose

paper for electromagnetic interference shielding and passive and active dual-thermal management *ACS Appl. Mater. Interfaces* **15** 23653–61

[104] Mokhena T C, Sadiku E R, Ray S S, Mochane M J, Matabola K P and Motloung M 2022 Flame retardancy efficacy of phytic acid: an overview *J. Appl. Polym. Sci.* **139** e52495

[105] Mengal N, Syed U, Malik S A, Ali Sahito I and Jeong S H 2016 Citric acid based durable and sustainable flame retardant treatment for Lyocell fabric *Carbohydr. Polym.* **153** 78–88

[106] Wang X, Yang G and Guo H 2023 Tannic acid as biobased flame retardants: a review *J. Anal. Appl. Pyrolysis* **174** 106111

[107] Yu J, Sun L, Ding L, Cao Y, Liu X, Ren Y and Li Y 2024 A UV-curable coating constructed from bio-based phytic acid, d-sorbitol and glycine for flame retardant modification of rigid polyurethane foam *Polym. Degrad. Stab.* **227** 110892

[108] Shi Y, Wang L, Wu M, Wang Y and Li H 2023 An eco-friendly B/P/N flame retardant for its fabrication of high-effective and durable flame-retardant cotton fabric *Cellulose* **30** 6621–38

[109] Costes L, Laoutid F, Brohez S and Dubois P 2017 Bio-based flame retardants: when nature meets fire protection *Mater. Sci. Eng.* R*117 1–25*

[110] Zhan W, Li L, Chen L, Kong Q, Chen M, Chen C, Zhang Q and Jiang J 2024 Biomaterials in intumescent fire-retardant coatings: a review *Prog. Org. Coat.* **192** 108483

[111] Li X L, Shi X H, Chen M J, Liu Q Y, Li Y M, Li Z, Huang Y H and Wang D Y 2022 Biomass-based coating from chitosan for cotton fabric with excellent flame retardancy and improved durability *Cellulose* **29** 5289–303

[112] Zhou X, Su X, Zhao J, Liu Y, Ren Y, Xu Z and Liu X 2024 Preparation of biomass-based green cotton fabrics with flame retardant, hydrophobic and self-cleaning properties *Cellulose* **31** 3871–92

[113] LI C-B, Wang F, Sun R-Y, Nie W-C, Song F and Wang Y-Z 2022 A multifunctional coating towards superhydrophobicity, flame retardancy and antibacterial performances *Chem. Eng. J.* **450** 138031

[114] Song F, Zhao Q, Zhu T, Bo C, Zhang M, Hu L, Zhu X, Jia P and Zhou Y 2022 Biobased coating derived from fish scale protein and phytic acid for flame-retardant cotton fabrics *Mater. Des.* **221** 110925

[115] Fang Y, Liu X and Tao X 2019 Intumescent flame retardant and anti-dripping of pet fabrics through layer-by-layer assembly of chitosan and ammonium polyphosphate *Prog. Org. Coat.* **134** 162–8

[116] Jimenez M, Guin T, Bellayer S, Dupretz R, Bourbigot S and Grunlan J C 2016 Microintumescent mechanism of flame-retardant water-based chitosan–ammonium polyphosphate multilayer nanocoating on cotton fabric *J. Appl. Polym. Sci.* **133** 43783

[117] Chen Z, Jiang J, Yu Y, Zhang Q, Chen T and Ni L 2020 Layer-by-layer assembled diatomite based on chitosan and ammonium polyphosphate to increase the fire safety of unsaturated polyester resins *Powder Technol.* **364** 36–48

[118] Yan Y, Dong S, Jiang H, Hou B, Wang Z and Jin C 2022 Efficient and durable flame-retardant coatings on wood fabricated by chitosan, graphene oxide, and ammonium polyphosphate ternary complexes via a layer-by-layer self-assembly approach *ACS Omega* **7** 29369–79

[119] Alongi J, Carosio F and Malucelli G 2012 Layer by layer complex architectures based on ammonium polyphosphate, chitosan and silica on polyester–cotton blends: flammability and combustion behaviour *Cellulose* **19** 1041–50

[120] Li P, Liu C, Xu Y J, Jiang Z-M, Liu Y and Zhu P 2020 Novel and eco-friendly flame-retardant cotton fabrics with lignosulfonate and chitosan through LBL: flame retardancy, smoke suppression and flame-retardant mechanism *Polym. Degrad. Stab.* **181** 109302
[121] Carosio F, Di Blasio A, Alongi J and Malucelli G 2013 Green DNA-based flame retardant coatings assembled through layer by layer *Polymer* **54** 5148–53
[122] Cheng X, Shi L, Fan Z, Yu Y and Liu R 2022 Bio-based coating of phytic acid, chitosan, and biochar for flame-retardant cotton fabrics *Polym. Degrad. Stab.* **199** 109898
[123] Cheng X W, Tang R C, Yao F and Yang X H 2019 Flame retardant coating of wool fabric with phytic acid/polyethyleneimine polyelectrolyte complex *Prog. Org. Coat.* **132** 336–42
[124] Zhang Z, Ma Z, Leng Q and Wang Y 2019 Eco-friendly flame retardant coating deposited on cotton fabrics from bio-based chitosan, phytic acid and divalent metal ions *Int. J. Biol. Macromol.* **140** 303–10
[125] Wang Y, Lan Y, Shi X, Sheng Y, Yang Y, Peng S and Xu J 2020 Highly efficient fabrication of self-extinguished flame-retardant and underwater superoleophobic coatings through layer-by-layer method *Mater. Chem. Phys.* **256** 123590
[126] Jiang Z, Wang C, Fang S, Ji P, Wang H and Ji C 2018 Durable flame-retardant and antidroplet finishing of polyester fabrics with flexible polysiloxane and phytic acid through layer-by-layer assembly and sol–gel process *J. Appl. Polym. Sci.* **135** 46414
[127] Wang W, Guo J, Liu X, Li H, Sun J, Gu X, Wang J, Zhang S and Li W 2020 Constructing eco-friendly flame retardant coating on cotton fabrics by layer-by-layer self-assembly *Cellulose* **27** 5377–89
[128] Pan H, Wang W, Pan Y, Zeng W, Zhan J, Song L, Hu Y and Liew K M 2015 Construction of layer-by-layer assembled chitosan/titanate nanotubes based nanocoating on cotton fabrics: flame retardant performance and combustion behavior *Cellulose* **22** 911–23
[129] Carosio F, Ghanadpour M, Alongi J and Wagberg L 2018 Layer-by-layer-assembled chitosan/phosphorylated cellulose nanofibrils as a bio-based and flame protecting nano-exoskeleton on pu foams *Carbohydr. Polym.* **202** 479–87
[130] Cain A A, Nolen C R, Li Y C, Davis R and Grunlan J C 2013 Phosphorous-filled nanobrick wall multilayer thin film eliminates polyurethane melt dripping and reduces heat release associated with fire *Polym. Degrad. Stab.* **98** 2645–52
[131] Kundu C K, Song L and Hu Y 2021 Sucrose derivative as a cross-linking agent in enhancing coating stability and flame retardancy of polyamide 66 textiles *Prog. Org. Coat.* **159** 106438
[132] Feng Y, Zhou Y, Li D, He S, Zhang F and Zhang G 2017 A plant-based reactive ammonium phytate for use as a flame-retardant for cotton fabric *Carbohydr. Polym.* **175** 636–44
[133] Cheng X-W, Guan J-P, Kiekens P, Yang X-H and Tang R-C 2019 Preparation and evaluation of an eco-friendly, reactive, and phytic acid-based flame retardant for wool *React. Funct. Polym.* **134** 58–66
[134] Cheng X-W, Zhang C, Jin W-J, Huang Y-T and Guan J-P 2021 Facile preparation of a sustainable and reactive flame retardant for silk fabric using plant extracts *Ind. Crops Prod.* **171** 113966
[135] Pan Y, Liu L, Zhang Y, Song L, Hu Y, Jiang S and Zhao H 2019 Effect of genipin crosslinked layer-by-layer self-assembled coating on the thermal stability, flammability and wash durability of cotton fabric *Carbohydr. Polym.* **206** 396–402

[136] Choudhary P and Das S K 2019 Bio-reduced graphene oxide as a nanoscale antimicrobial coating for medical devices *ACS Omega* **4** 387–97
[137] Li C, Xie C, Ou J, Xue M, Wang F, Lei S, Fang X, Zhou H and Li W 2018 ZnO superhydrophobic coating via convenient spraying and its biofouling resistance *Surf. Interface Anal.* **50** 1278–85
[138] Zhang W, Li S, Wei D, Zheng Z, Han Z and Liu Y 2023 Fabrication of a fluorine-free photocatalytic superhydrophobic coating and its long-lasting anticorrosion and excellent antibacterial abilities *Prog. Org. Coat.* **184** 107806
[139] Ong G, Kasi R and Subramaniam R 2021 A review on plant extracts as natural additives in coating applications *Prog. Org. Coat.* **151** 106091
[140] Wang Y, Yu Z, Zhang P, Wang Y, Chang Z, Jiang P, Xia J, Gao X and Bao Y 2024 Development of a bio-based self-healing waterborne polyurethane with dynamic phenol-carbamate network for enhanced antimicrobial and antiseptic performance *J. Polym. Sci.* **62** 4277–88
[141] Qu L, Li X, Zhou J, Cao K, Xie Q, Zhou P, Qian W and Yang Y 2023 A novel dual-functional coating based on curcumin/APEG polymer with antibacterial and antifouling properties *Appl. Surf. Sci.* **627** 157224
[142] Pavlátková L, Sedlaříková J, Pleva P, Peer P, Uysal-Unalan I and Janalíková M 2022 Bioactive zein/chitosan systems loaded with essential oils for food-packaging applications *J. Sci. Food Agric.* **103** 1097–104
[143] Li X L, Shen Y, Hu F, Zhang X-X, Thakur K, Rengasamy K R R, Khan M R, Busquets R and Wei Z-J 2023 Fortification of polysaccharide-based packaging films and coatings with essential oils: a review of their preparation and use in meat preservation *Int. J. Biol. Macromol.* **242** 124767
[144] Mu M, Lin Y T, Deflorio W, Arcot Y, Liu S, Zhou W, Wang X, Min Y, Cisneros-Zevallos L and Akbulut M 2023 Multifunctional antifouling coatings involving mesoporous nano-silica and essential oil with superhydrophobic, antibacterial, and bacterial antiadhesion characteristics *Appl. Surf. Sci.* **634** 157656
[145] Yasar S, Nizamlioğlu N M, Gücüş M O, Bildik Dal A E and Akgül K 2022 *Origanum majorana* L. essential oil-coated paper acts as an antimicrobial and antioxidant agent against meat spoilage *ACS Omega* **7** 9033–43
[146] Freitas E D, Bataglioli R A, Oshodi J and Beppu M M 2022 Antimicrobial peptides and their potential application in antiviral coating agents *Colloids Surf.* B **217** 112693
[147] Barbosa M, Costa F, Monteiro C, Duarte F, Martins M C L and Gomes P 2019 Antimicrobial coatings prepared from Dhvar-5-click-grafted chitosan powders *Acta Biomater.* **84** 242–56
[148] Wang S-H, Tang T W-H, Wu E, Wang D-W and Liao Y-D 2020 Anionic surfactant-facilitated coating of antimicrobial peptide and antibiotic reduces biomaterial-associated infection *ACS Biomater. Sci. Eng.* **6** 4561–72
[149] Lou M M, Zhu B, Muhammad I, Li B, Xie G L, Wang Y L, Li H Y and Sun G C 2011 Antibacterial activity and mechanism of action of chitosan solutions against apricot fruit rot pathogen burkholderia seminalis *Carbohydr. Res.* **346** 1294–301
[150] Huang Y-Y, Zhang L-P, Cao X, Tian X-Y and Ni Y-P 2024 Facile fabrication of highly efficient chitosan-based multifunctional coating for cotton fabrics with excellent flame-retardant and antibacterial properties *Polymers* **16** 1409

[151] Wang Q, Yao Q, Liu J, Sun J, Zhu Q and Chen H 2019 Processing nanocellulose to bulk materials: a review *Cellulose* **26** 7585–617

[152] Noremylia M B, Hassan M Z and Ismail Z 2022 Recent advancement in isolation, processing, characterization and applications of emerging nanocellulose: a review *Int. J. Biol. Macromol.* **206** 954–76

[153] Thomas B, Raj M C, B A K, H R M, Joy J, Moores A, Drisko G L and Sanchez C 2018 Nanocellulose, a versatile green platform: from biosources to materials and their applications *Chem. Rev.* **118** 11575–625

[154] Li M, Liu X, Liu N, Guo Z, Singh P K and Fu S 2018 Effect of surface wettability on the antibacterial activity of nanocellulose-based material with quaternary ammonium groups *Colloids Surf.* A **554** 122–8

[155] Liu Y, Li M, Qiao M, Ren X, Huang T S and Buschle-Diller G 2017 Antibacterial membranes based on chitosan and quaternary ammonium salts modified nanocrystalline cellulose *Polym. Adv. Technol.* **28** 1629–35

[156] Hongrattanavichit I and Aht-Ong D 2021 Antibacterial and water-repellent cotton fabric coated with organosilane-modified cellulose nanofibers *Ind. Crops Prod.* **171** 113858

[157] Jia R, Tian W, Bai H, Zhang J, Wang S and Zhang J 2019 Sunlight-driven wearable and robust antibacterial coatings with water-soluble cellulose-based photosensitizers *Adv. Healthcare Mater.* **8** 1801591

[158] Egghe T, Morent R, Hoogenboom R and De Geyter N 2023 Substrate-independent and widely applicable deposition of antibacterial coatings *Trends Biotechnol.* **41** 63–76

[159] Bai W B, Wang L T, Zheng X T, Dong Y Q, Lin Y C and Jian R K 2024 Construction of Ag-mediated organic-inorganic phosphorus-hybrid coating for flame-retardant and antibacterial cotton fabrics *Prog. Org. Coat.* **187** 108128

[160] Thuy L T, Kim S Y, Dongquoc V, Kim Y, Choi J S and Cho W K 2023 Coordination-driven robust antibacterial coatings using catechol-conjugated carboxymethyl chitosan *Int. J. Biol. Macromol.* **249** 126090

[161] Ruan H, Aulova A, Ghai V, Pandit S, Lovmar M, Mijakovic I and Kádár R 2023 Polysaccharide-based antibacterial coating technologies *Acta Biomater.* **168** 42–77

[162] Del Valle E M M 2004 Cyclodextrins and their uses: a review *Process Biochem.* **39** 1033–46

[163] Ha W, Hou G, Qin W, Fu X, Zhao X, Wei X, An Y and Shi Y 2021 Supramolecular hydrogel-infiltrated ceramics composite coating with combined antibacterial and self-lubricating performance *J. Mater. Chem.* B **9** 9852–62

[164] Fu X K, Cao H B, An Y L, Zhou H D, Shi Y P, Hou G L and Ha W 2022 Bioinspired hydroxyapatite coating infiltrated with a graphene oxide hybrid supramolecular hydrogel orchestrates antibacterial and self-lubricating performance *ACS Appl. Mater. Interfaces* **14** 31702–14

[165] Guo C, Cui W, Wang X, Lu X, Zhang L, Li X, Li W, Zhang W and Chen J 2020 Poly-L-lysine/sodium alginate coating loading nanosilver for improving the antibacterial effect and inducing mineralization of dental implants *ACS Omega* **5** 10562–71

[166] Liu Z, Liu T, Gu W, Zhang X, Li J, Shi S Q and Gao Q 2022 Hyperbranched catechol biomineralization for preparing super antibacterial and fire-resistant soybean protein adhesives with long-term adhesion *Chem. Eng. J.* **449** 137822

[167] Jo Y K, Seo J H, Choi B H, Kim B J, Shin H H, Hwang B H and Cha H J 2014 Surface-independent antibacterial coating using silver nanoparticle-generating engineered mussel glue *ACS Appl. Mater. Interfaces* **6** 20242–53

[168] Huang Z, Nazifi S, Hakimian A, Firuznia R and Ghasemi H 2022 'Built to last': plant-based eco-friendly durable antibacterial coatings *ACS Appl. Mater. Interfaces* **14** 43681–9
[169] Liu K, Zhang F, Wei Y, Hu Q, Luo Q, Chen C, Wang J, Yang L, Luo R and Wang Y 2021 Dressing blood-contacting materials by a stable hydrogel coating with embedded anti-microbial peptides for robust antibacterial and antithrombus properties *ACS Appl. Mater. Interfaces* **13** 38947–58
[170] Yan W, Tian T, Li M, Zong Y, Liao Y, Yang Y and Wu X 2023 Depleting conjugated C=C bonds in tung oil/eucommia rubber composite film: fundamentally improve the flexibility and crack resistance for high-performance protective coating *Ind. Crops Prod.* **191** 115943
[171] Zhou L, Li X, Kang Z, Liu X, LI Q, Ma L, Gao H and Nie Y 2022 Antibacterial cellulose fibers spun from ionic liquid and enriched with plant essential oils *ACS Appl. Polym. Mater.* **4** 6649–58
[172] Tominaga C, Shitomi K, Miyaji H and Kawasaki H 2018 Antibacterial photocurable acrylic resin coating using a conjugate between silver nanoclusters and alkyl quaternary ammonium *ACS Appl. Nano Mater.* **1** 4809–18
[173] Zhou Y, Liu R, Zhou C, Gao Z, Gu Y, Chen S, Yang Q and Yan B 2023 Dynamically crosslinked chitosan/cellulose nanofiber-based films integrated with γ-cyclodextrin/curcumin inclusion complex as multifunctional packaging materials for perishable fruit *Food Hydrocolloids* **144** 108996
[174] Farahani M, Shahidi F, Yazdi F T and Ghaderi A 2024 Antimicrobial and antioxidant effects of an edible coating of *Lepidium sativum* seed mucilage and *Satureja hortensis* L. essential oil in uncooked lamb meat *Food Control* **158** 110240
[175] Manzoor A, Yousuf B, Pandith J A and Ahmad S 2023 Plant-derived active substances incorporated as antioxidant, antibacterial or antifungal components in coatings/films for food packaging applications *Food Biosci.* **53** 102717
[176] Zhou Y, Zhong Y, Li L, Jiang K, Gao J, Zhong K, Pan M and Yan B 2022 A multifunctional chitosan-derived conformal coating for the preservation of passion fruit *LWT* **163** 113584
[177] Mondal K, Bhattacharjee S K, Mudenur C, Ghosh T, Goud V V and Katiyar V 2022 Development of antioxidant-rich edible active films and coatings incorporated with de-oiled ethanolic green algae extract: a candidate for prolonging the shelf life of fresh produce *RSC Adv.* **12** 13295–313
[178] Gagaoua M, Bhattacharya T, Lamri M, Oz F, Dib A L, Oz E, Uysal-Unalan I and Tomasevic I 2021 Green coating polymers in meat preservation *Coatings* **11** 1379
[179] Ali Ghoflgar Ghasemi M, Hamishehkar H, Javadi A, Homayouni-Rad A and Jafarizadeh-Malmiri H 2024 Natural-based edible nanocomposite coating for beef meat packaging *Food Chem.* **435** 137582
[180] Moriarty T F, Schlegel U, Perren S and Richards R G 2010 Infection in fracture fixation: can we influence infection rates through implant design? *J. Mater. Sci., Mater. Med.* **21** 1031–5
[181] Liu Y, Dong T, Chen Y, Sun N, Liu Q, Huang Z, Yang Y, Cheng H and Yue K 2023 Biodegradable and cytocompatible hydrogel coating with antibacterial activity for the prevention of implant-associated infection *ACS Appl. Mater. Interfaces* **15** 11507–19
[182] Lenguerrand E, Whitehouse M R, Beswick A D, Kunutsor S K, Foguet P, Porter M and Blom A W 2019 Risk factors associated with revision for prosthetic joint infection following

knee replacement: an observational cohort study from england and wales *Lancet Infect. Dis.* **19** 589–600

[183] Shirani A, Hu Q, Su Y, Joy T, Zhu D and Berman D 2019 Combined tribological and bactericidal effect of nanodiamonds as a potential lubricant for artificial joints *ACS Appl. Mater. Interfaces* **11** 43500–8

[184] Liang S, Wang M, Gao W, Diao H and Luo J 2022 Recyclable, UV-blocking, and radiative cooling multifunctional composite membranes *ACS Omega* **7** 25244–52

[185] Chen B, Liao M, Sun J and Shi S 2023 A novel biomass polyurethane-based composite coating with superior radiative cooling, anti-corrosion and recyclability for surface protection *Prog. Org. Coat.* **174** 107250

[186] Liu Z, Li Y and He Z 2023 Ice-phobic properties of MoS_2-loaded rice straw biogas residue biochar-based photothermal and anti-corrosion coating with low oxygen to carbon ratio *Biochar* **5** 74

[187] Yu J, Shang Q, Zhang M, Hu L, Jia P and Zhou Y 2024 Tung oil-based waterborne UV-curable coatings via cellulose nanofibril stabilized pickering emulsions for self-healing and anticorrosion application *Int. J. Biol. Macromol.* **256** 128114

Chapter 3

Bio-based gas barriers/films for sustainable packaging and preservation

Jiaxiu Wang and Dongmei Liu

3.1 Introduction

In our daily life, food is often spoiled due to improper preservation techniques. Similarly, pharmaceuticals and electronic equipment sensitive to moisture, etc, can sustain damage upon contact with water, reducing their lifespan and effectiveness. These issues can be prevented by the use of gas barrier films [1]. A gas barrier is defined by its ability to prevent or reduce the diffusion of permeable gases such as oxygen, water vapor, carbon dioxide, etc, from a region of higher concentration to a region of lower concentration, across a barrier film [2]. The gas barrier films on the market can be roughly divided into two types, namely, petroleum-based films and bio-based films. Petroleum-based films are the most widely used in the world today because of their highly mature manufacturing process and good gas shielding effect. Currently, materials such as polyvinyl alcohol (PVA), ethylene vinyl alcohol copolymer (EVOH), polyvinylidene chloride (PVDC), polyamide (PA), and polyethylene terephthalate (PET) are commonly used to prepare barrier films. EVOH is a semi-crystalline copolymer composed of ethylene and vinyl alcohol monomer units. It exhibits excellent barrier properties towards O_2, CO_2 and N_2, and the oxygen permeability of EVOH is one of the lowest among polymer materials [3]. PVDC films, made of vinylidene chloride homopolymer, usually possess a dense microstructure, good hydrophobicity and large cohesive energy density, resulting in excellent oxygen and water vapor barrier properties [4]. Despite their widespread use, petroleum-based materials themselves have some inherent drawbacks. For example, PVOH is sensitive to water because of its rich hydroxyl groups, and its degradation temperature is only 150 °C. Poly urethane (PU) materials need to other added organic or inorganic additives to achieve high barrier properties, which aggravate the greenhouse effect. Heavy use of petroleum-based gas barrier films causes serious environmental pollution and the exhaustion of non-renewable

doi:10.1088/978-0-7503-6184-2ch3

petroleum resources [5, 6]. Bio-based barrier films are made from biomass, which exists in nature and is widely available and renewable, as well as being low-cost. There are already some commercially available bio-based film materials, such as cellophane and polylactic acid (PLA) materials [7]. Cellulose derivatives, such as cellulose acetate and carboxymethyl cellulose, have proven to be excellent barrier materials in food packaging [7]. Compared to petroleum-based gas barrier membranes, bio-based alternatives offer several notable advantages. Bio-based gas barrier films utilize renewable biomass sources, providing a more environmentally friendly alternative to traditional petroleum-derived materials [8]. Second, bio-based gas barrier films often exhibit superior biodegradability. In the context of the circular economy, this characteristic is particularly important [9, 10]. Additionally, bio-based gas barrier films contribute to reducing greenhouse gas emissions. The production processes of petroleum-based materials typically release significant amounts of greenhouse gases into the atmosphere. In contrast, producing bio-based materials from renewable sources has the potential to be more carbon-neutral or even carbon-negative [11].

3.2 The gas permeation process and parameters for gas barriers

3.2.1 Introduction to the gas permeation process and related models

A fundamental understanding of the gas permeation process is important for the research and development of advanced gas barrier films. The gas diffusion process can typically be divided into three stages: adsorption, diffusion, and desorption. In the adsorption stage, gas molecules are initially adsorbed onto the surface of the adsorbent [12, 13]. This adsorption is usually accomplished through the attractive forces between active sites on the adsorbent surface and the gas molecules, which can be either chemical or physical in nature. During the adsorption process, gas molecules move from the gas phase to the solid phase, forming an adsorbed layer. Following the adsorption stage, the gas molecules adsorbed on the surface may diffuse into the pores of the adsorbent, which is known as the diffusion stage. The rate of diffusion is typically described by Fick's law, which states that the rate of diffusion is directly proportional to the concentration gradient and inversely proportional to the square root of the diffusion distance. Under certain conditions, gas molecules on the surface of the adsorbent may desorb back into the gas phase. This process is called desorption. Desorption can be physical, where gas molecules desorb from adsorption sites, or chemical, where gas molecules desorb after reacting with the adsorbent surface [14].

The gas permeation process is illustrated in figure 3.1.

Here, models corresponding to the three stages of gas diffusion process are presented in the following for a better understanding of the gas diffusion process and to offer theoretical guidance for the development of excellent gas barrier films. The adsorption process mainly includes the Langmuir model, Freundlich model, Brunauer–Emmett–Teller (BET) model, and Dubinin–Radushkevich (D–R) model. The Langmuir model is one of the most basic and commonly used adsorption models, primarily used to describe monolayer adsorption processes [15]. Langmuir's

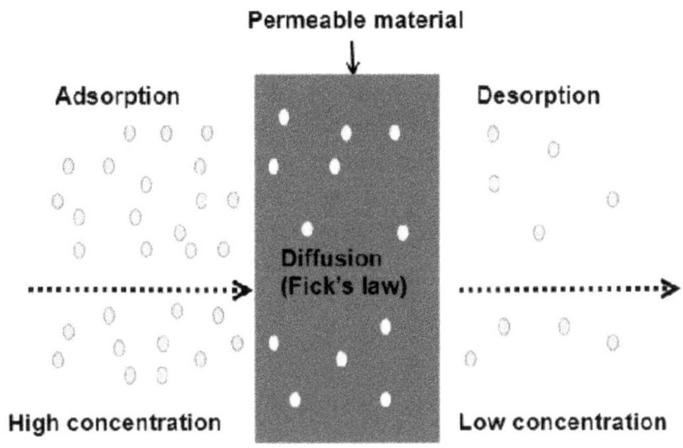

Figure 3.1. The three stages of gas permeation processes, including adsorption, diffusion and desorption.

law is based on three assumptions: (i) the rate of incidence of a large number of molecules in the gas phase on the surface of a unit area of adsorbent is proportional to the pressure at a constant temperature, (ii) the rate of adsorption depends not only on the rate of incidence of molecules on the surface, but also on the probability of adsorption, and (iii) the rate of desorption is equal to the rate of desorption at the maximum surface coverage multiplied by the fraction of adsorbed molecules occupying the surface sites [16]. Its nonlinear and linear expressions are

$$q_e = \frac{q_m K_L C_e}{1 + K_L C_e} \tag{3.1}$$

$$\frac{C_e}{q_e} = \frac{C_e}{q_m} + \frac{1}{K_L q_m}, \tag{3.2}$$

where K_L (l mg^{-1}) is the ratio of adsorption rate to desorption rate, q_m (mg g^{-1}) is the maximum adsorption capacity estimated by the Langmuir model, and q_e and C_e are the equilibrium adsorption capacity of the adsorbent and the equilibrium concentration of the adsorbate. The Freundlich and BET models are used to describe gas or solute adsorption on solid surfaces and can describe multi-layer adsorption processes. The Freundlich model is an empirical model used to describe adsorption processes on non-uniform surfaces, including multi-layer adsorption and non-ideal adsorption situations [17]. The linear and nonlinear forms of the Freundlich model are given by

$$\log q_e = \log K_F + \frac{1}{n} \log C_e \tag{3.3}$$

$$q_e = K_F {C_e}^{1/n}, \tag{3.4}$$

where K_F ($L^{1/n}$ mg$^{1-1/n}$ g^{-1}) and n are constants that simplify the Freundlich model to a linear model when $n = 1$.

The adsorption isotherm for gaseous substances based on the theory of multi-molecular adsorption was proposed by Brunauer, Emmet and Teller in 1938 [18]. The BET model is used to describe multi-layer adsorption processes, especially at relatively high gas pressures, based on specific molecular layer adsorption models. The BET model is based on the Langmuir model for monolayer adsorption but extends the adsorption capacity to multi-layer adsorption, assuming the presence of multiple adsorption layers [19]. The most general form of the BET model is

$$q_e = \frac{q_{mBET} C_{BET} C_e}{(C_{s,\,BET} - C_e)\left[1 + (C_{BET} - 1)\left(\frac{C_e}{C_{s,\,BET}}\right)\right]}. \tag{3.5}$$

The parameters q_{mBET} (mg g^{-1}), $C_{s,\,BET}$, and C_{BET} (l mg^{-1}) are defined as the maximum adsorption capacity of the BET isotherm, the monolayer saturation concentration of the isotherm, and the BET adsorption constant, respectively. In some cases, the Freundlich model can be used to describe adsorption situations that the BET model cannot handle well, especially on non-uniform surfaces or at low concentrations.

The D–R model was developed based on Polanyi's theory and the assumption that the distribution of pores in the adsorbent follows a Gaussian energy distribution [20]. The model takes into account the pore structure in the adsorbent and is applicable to heterogeneous surfaces [21]. The nonlinear D–R model is [22]

$$\ln q_e = \ln q_m - K\varepsilon^2 \tag{3.6}$$

$$\varepsilon = RT \ln\left(1 + \frac{1}{C_e}\right), \tag{3.7}$$

where q_e is the equilibrium adsorbent-phase concentration of the adsorbate (mg g^{-1}), q_m is theoretical saturation capacity (mg g^{-1}), K is the activity coefficient related to the mean free energy of adsorption (mol^2 kJ^{-2}), ε is the Polanyi potential (kJ mol^{-1}), R is the universal gas constant (8.314 J mol^{-1} K^{-1}), T is temperature in kelvins (K), and C_e is the equilibrium aqueous-phase concentration of adsorbate (mg l^{-1}).

The gas diffusion process is usually described by Fick's law. Fick's diffusion model is one of the basic diffusion models, applicable to relatively simple diffusion processes, such as gas diffusion in solids, liquids, or gases. The model assumes that the diffusion rate is proportional to the concentration gradient and inversely proportional to the square root of the diffusion distance [23]. Fick's first law describes the diffusion rate of solutes in a medium under the influence of a concentration gradient. The expression is

$$F = -D\frac{\partial C}{\partial x} = \frac{Q}{A_t} \tag{3.8}$$

where F is the flux or mass transfer rate, representing the quantity (Q) of diffused substance passing through unit area (A) per unit time (t), C is the concentration of diffused substance, and x is the moving direction of diffused substance.

Fick's second law describes the relationship between the concentration changes with time and position during the diffusion process. It is an extension of Fick's diffusion model, considering variations in diffusion rates at different positions and time points. If the diffusion coefficient is independent of concentration, distance (x) or time (t), equation (3.8) can be rewritten as the following equation, which represents Fick's second law:

$$\frac{\partial F}{\partial x} = \frac{\partial C}{\partial t} = D\frac{\partial^2 c}{\partial x^2}. \tag{3.9}$$

Various models can be used to describe gas desorption process. The Flory–Huggins model is typically used to describe the desorption process of solutes in a polymer solution. The basic assumption is that there is interaction between the solute and the solvent, and the desorption rate is influenced by the solute concentration in the solvent [24]. The above-mentioned models may be applicable to diverse situations, and the appropriate model should be selected for a particular situation.

In addition to understanding the models for the three stages of gas diffusion, there are some parameters and standards to evaluate gas permeability for judging the quality of gas barrier films.

The gas permeability can be described by the following formula [25]:

$$P = D \times S, \tag{3.10}$$

where P is permeability, D is the diffusion coefficient, which represents the dynamic movement of diffused molecular substances through the barrier film, and S is the solubility coefficient, which represents the degree of dissolution of a substance in the polymer. Ideally, permeability is primarily influenced by the nature of the penetration and matrix properties, such as chemical structure, polarity, crystallinity, density, crosslinking, molecular weight and degree of polymerization, as well as the presence of plasticizers [26]. The diffusion and solubility of gas largely depends on the interaction between the gas and polymer film. The stronger the interaction, the more molecules will diffuse and dissolve.

A lower transmittance rate indicates better barrier performance. Common gas transmittance rates include the water vapor transmission rate (WVTR), oxygen transmission rate (OTR), carbon dioxide transmittance, etc.

The American Society for Testing and Materials (ASTM) has developed a number of standard test methods for evaluating the barrier properties of films, such as ASTM D1434 (water vapor transmission rate), ASTM D3985 (gas transmission rate), etc [27]. The International Organization for Standardization (ISO) has also published a number of standard test methods for evaluating the barrier properties of films, such as ISO 15105 (oxygen transmission rate test) [28].

WVTR is commonly used to measure the barrier property of a film to water molecules. The WVTR expresses the mass of water that passes through a unit surface area of a barrier film at a constant temperature and relative humidity for a predetermined time interval and pressure difference. The expression for WVTR can be determined by Henry's law and Fick's first law:

$$\text{WVTR}\ (\text{g m}^{-2} \times 24\ \text{h}) = W/A, \tag{3.11}$$

where W is the weight increase a specific film area (A) experiences, within a measurement of 24 h [29].

Oxygen makes up about 21% of the atmosphere and is highly biologically active and reactive. Prevention of oxygen permeation is one of the important properties of barrier films. The traditional method of OTR determination is based on the ASTM D398517 (2017) standard [30]. Briefly, the barrier film needs to be placed between two dry chambers at an ambient atmospheric pressure of RH $< 1\%$ [29, 31]. One chamber contains nitrogen and the other oxygen, and the oxygen transmission rate is then measured by a coulometric detector.

The ASTM F2476-05 standard test method is used to evaluate the CO_2 transmission rate of packaging films [32]. The sample is mounted as a sealed semi-barrier between two chambers at ambient atmospheric pressure. One chamber is slowly purged with a stream of nitrogen gas and the other is slowly purged with carbon dioxide. As the carbon dioxide gas passes through the film into the nitrogen carrier gas, it is transferred to an infrared detector, where it produces an electrical output, the magnitude of which is proportional to the amount of carbon dioxide flowing into the detector per unit of time.

3.3 Strategies to construct bio-based barrier films

To create films with excellent barrier performance, it is necessary to comprehensively consider factors such as material selection, gas diffusion pathways, the addition of plasticizers, interactions between components, and crystal arrangement, etc. And the gas barrier performance can be enhanced through optimized material processing and manufacturing techniques. The selection of materials with excellent barrier properties is the primary consideration in preparing high-performance films. Table 3.1 provides an overview of various construction strategies and the specific materials and additives commonly employed to enhance gas barrier properties, offering insights into how different approaches and components contribute to film performance. Widely used materials include polymers such as starch, cellulose, and their composites.

As shown in figure 3.2, the design of gas diffusion pathways is crucial for the improvement of gas barrier films. Adding an appropriate amount of plasticizer can improve the flexibility and stretch ability of the film while potentially impacting its gas barrier properties. It is important to optimize the type and amount of plasticizer to balance the mechanical performance and gas barrier properties of the film. Molecular interactions and the crystal arrangements between materials have a significant impact on gas diffusion. Adding additional barrier layers, such as oxide layers, to the film structure, can effectively enhance gas barrier performance.

3.3.1 Tortuous path

Generally speaking, a longer gas diffusion path is less favorable for gas diffusion through a barrier material. Adding nanomaterials, barrier fillers, or coating to the films will increase the gas diffusion path and improve the gas barrier properties. In

Table 3.1. Construction strategies for bio-based barrier films and their gas barrier properties.

Polymers	Additive	WVTR	WVP	OTR	OP	Strategies	References
Gelatin/chitosan	Poly(l-lysine)	3.99×10^{-13} kg m^{-1} s^{-1} Pa^{-1}	—	1.89×10^{-13} kg m^{-1} s^{-1} Pa^{-1}	—	High crosslinking structures form complex gas diffusion pathways	[37]
Nanofibrillated cellulose	$Ti_3C_2T_x$, AgNPs	—	—	—	Less than 0.005 cm^3 m^{-2}·24 h·0.1 MPa	Tortuous gas diffusion pathways	[33]
Cellulose	Cinnamoyl chloride	—	$(0.94 \pm 0.03) \times 10^{-11}$ g m^{-1} s^{-1} Pa^{-1}	—	5.50×10^{-4} cm^3 μm m^{-2} day^{-1} Pa^{-1}	Dense film structure due to modification	[34]
Starch	Chitosan	—	2.8×10^{-12} g m^{-1} s^{-1} Pa^{-1} Less than 1.39×10^{-8} g m^{-1} s^{-1} Pa^{-1}	—	7.55×10^{-1} cm^3 μm m^{-2} day^{-1} Pa^{-1}	Intermolecular hydrogen bonds	[40]
CNC	Tannic acid, gelatin	—	Less than 0.05 g mm m^3 h^{-1} kPa^{-1}	—	—	Covalent bonds, hydrogen bonding, electrostatic interaction	[41]
Cellulose acetate	Layered double hydroxide	—	—	Lower than the instrument's detected minimum value	—	Hydrogen-bonding network structure	[35]
CNC	—	—	—	—	Less than 1×10^{-16} cm^3 cm $(cm^2\ s\ Pa)^{-1}$	Reducing free volume	[36]
CNC	—	—	—	—	4.21×10^{-16} cm^3 cm $(cm^2\ s\ Pa)^{-1}$	Reducing free volume	[42]

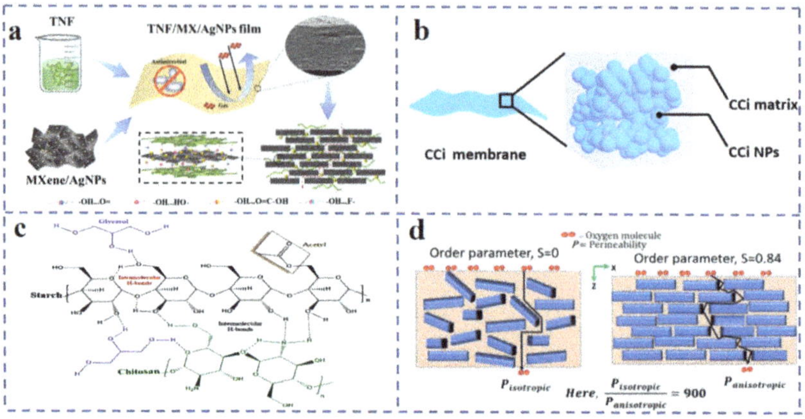

Figure 3.2. Representative works showing the factors influencing the barrier properties of bio-based films. (a) Hindered diffusion of gas molecules in the TNF/MX/AgNPS thin film. (Reproduced with permission from [33]. Copyright 2023 Elsevier.) (b) Dense film with two different morphologies of CCi film. (Reproduced with permission from [34]. Copyright 2021 American Chemical Society.) (c) Intramolecular and intermolecular interactions present between acetylated corn starch, chitosan and glycerol. (Reproduced from [35]. CC BY 4.0.) (d) Self-organized (chiral nematic) and shear-oriented CNC films. (Reproduced with permission from [36]. Copyright 2019 American Chemical Society.)

addition, films with a dense structure, crosslinked network, or more layers also hindered the gas from passing through the films, which also improves the gas barrier properties of the films. For example, Liu *et al* prepared highly crosslinked nanofibrous films by incorporating poly(l-lysine) (S-PL) into gelatin/chitosan-based polymers [37]. The highly crosslinked structure formed complex gas diffusion pathways, thereby improving the gas barrier properties. With an increasing dosage of ε-polylysine, the WVTR decreased from 8.57×10^{-13} kg m^{-1} s^{-1} Pa^{-1} to 3.99×10^{-13} kg m^{-1} s^{-1} Pa^{-1}, while the OTR decreased from 2.31×10^{-13} kg m^{-1} s^{-1} Pa^{-1} to 1.89×10^{-13} kg m^{-1} s^{-1} Pa^{-1}. Tang *et al* utilized high aspect ratio nanosheets of $Ti_3C_2T_x$ (MXene), silver nanoparticles (AgNPs), and TEMPO-oxidized cellulose nanofibers (TNFs) to self-assemble into a pearl layer-structured film [33]. The film demonstrated exceptional performance by obstructing the diffusion of a wide range of gases, including organic compounds such as ethyl acetate, diethyl ether, and ethanol, as well as inorganic gases such as oxygen and nitrogen, with the oxygen permeability even lower than the instrument's detected minimum value (0.005 cm^3 m^{-2}·24 h·0.1 MPa). The excellent barrier performance was explained on the one hand by the low solubility coefficient of oxygen, and on the other hand by the high aspect ratio of the nanosheets, increasing the tortuosity of gas diffusion pathways. Additionally, MXene and TNF can assemble into intertwined stacking structures through hydrogen bonding, making the TNF/MX film more compact and prolonging the gas diffusion pathways (figure 3.2(a)). However, the structure of the TNF/MX film still contains gaps, and the addition of AgNPs fills this gap. Taken together, these factors enable the TNF/MX/AgNP film to possess ultra-high gas barrier properties. Wang *et al* developed a self-compounding film using cellulose cinnamate, which consisted of a single component with two distinct morphologies: cellulose cinnamate nanoparticles and a cellulose cinnamate polymer matrix (figure 3.2(b)) [34]. The resulting film exhibited remarkable gas barrier properties, with a water vapor permeability (WVP) of $(0.94 \pm 0.03) \times 10^{-11}$ g m^{-1} s^{-1} Pa^{-1} and oxygen permeation (OP) of 5.50×10^{-4} cm^3 μm m^{-2} day^{-1} Pa^{-1}. This exceptional gas shielding performance can be attributed to the perfect coordination between the two morphologies. The nanoparticles were either embedded in the polymer matrix or fused with other CCi nanoparticles, leading to a dense internal structure within the film.

3.3.2 Intermolecular interactions

The interactions between the materials composing the film can lead to a denser film structure, reducing the mobility of the molecular chains and thereby enhancing gas barrier properties. For example, by introducing hydrogen-bonding crosslinking structures, the tightness and structural stability of the film can be enhanced, thereby improving the gas barrier performance [38, 39]. The following examples provide insight into the effect of intermolecular interactions on the gas barrier properties of thin films.

Dang *et al* incorporated chitosan into starch-based films and found that the addition of chitosan significantly improved the gas barrier properties of the starch-based films [40]. The presence of acetyl groups makes chitosan more hydrophobic than starch, allowing the enrichment of chitosan on the film surface to reduce the adsorption and diffusion of water molecules. The formation of intermolecular hydrogen bonds between chitosan and starch would reduce the chain mobility of the film and thus enhance the oxygen barrier properties. Furthermore, the surface of the films is enriched with chitosan of relatively high crystallinity, which make it difficult for oxygen to pass through the starch-based composite film.

Leite *et al* have prepared a ternary composite hybrid film consisting of cellulose nanocrystals (CNCs), tannic acid (TA), and gelatin [41]. The bonding between the components is a key factor in improving the barrier properties of this film, such as the noncovalent bonds formed between tannins and gelatin, including hydrogen bonding and the electrostatic interaction between the sulfate anion of the CNCs and the cationic amine of the gelatin. A variety of synergistic effects jointly enhanced the density of the composite film, thereby reducing the permeation of gas and water molecules.

Dou *et al* prepared flexible multi-layer films using alternating assemblies of cellulose acetate (CA) and layered double hydroxide (LDH) [35]. The films exhibited excellent oxygen barrier properties equal to or below the lowest value detectable by commercial instruments (<0.05 cm^3 m^{-2} d^{-1}). Molecular dynamics simulations revealed that the excellent oxygen barrier performance originated from the formation of a hydrogen-bonding network structure between LDH nanosheets and CA at the interface, which inhibited oxygen transport.

By adding chitosan and glycerol to acetylated corn starch, Jiménez-Regalado *et al* were able to form intramolecular and intermolecular interactions between the three components, resulting in a more compact composite film (figure 3.2(c)) [35]. Acetylated starch, chitosan and glycerol contain a large number of hydroxyl and amino groups, and water molecules can reduce the cohesive energy of the polymers by interacting with these polar groups. Moreover, the hydrophobicity of chitosan makes the interaction with water molecules weaker, thus improving the barrier to water vapor. The WVP of the composite membrane containing 75% chitosan was reduced from 26.20×10^{-11} g m^{-1} s^{-1} Pa^{-1} to 0.55×10^{-11} g m^{-1} s^{-1} Pa^{-1} compared to the pure starch film.

3.3.3 Crystal arrangements

It is widely accepted that the degree of crystallinity and the aspect ratio of crystalline domains play a crucial role in the barrier properties of bio-based nanomaterials. However, the degree of anisotropy (changing the packing density/free volume of the polymer system) would change the overall permeability.

Chowdhury *et al* prepared a range of pristine nanocellulose films with low to high barrier properties using a shear coating technique [36]. The induced anisotropy in nanocellulose films can control the free volume within the film, thus effectively controlling the gas diffusion path and providing excellent gas barrier properties. The more regular and ordered the arrangement of CNCs, the more favorable it is for constructing high energy barriers (figure 3.2(d)). Nuruddin *et al* changed the structural orientation of CNC films using shear forces [42]. Due to the alignment of individual CNCs in the shear direction, shear-aligned CNC films showed a lower free volume size and density compared to self-organized CNC films (chiral nematic), and the size and number of pores within the films were also affected. The reduction in free volume resulted in a 95% reduction in O_2 permeability and a 96% reduction in CO_2 permeability (table 3.1).

3.4 Types of bio-based barrier films

Bio-based gas barrier materials are garnering significant attention as a sustainable and environmentally friendly option. These materials stem from renewable resources, including cellulose, lignin, hemicellulose, chitosan, starch, and others. Crucially, many of these materials exhibit excellent gas barrier properties. In this section, biomass gas barrier films will be presented according to the types of biomass materials, which mainly include those made of bio-based polymers or bio-based nanomaterials. Special attention is given to the effect of bio-based materials on the gas barrier properties of the films.

3.4.1 Bio-based polymers and derivatives

In the following, we aim to provide an overview of the barrier properties of bio-based polymers and their derivatives for barrier films. Bio-based gas barrier films can be developed to help reduce reliance on finite fossil fuel resources, promote a circular economy, and foster sustainable development. For example, cellulose is often used as a matrix to enhance the barrier performance of composite films [43]. Lignin is commonly employed as a UV absorber in polymer films [44]. Hemicellulose, owing to its low migration rate and oxygen barrier properties, is frequently utilized to improve the oxygen barrier performance of composite films [45]. Chitosan, known for its excellent antimicrobial properties, is often compounded with other substances to enhance the properties of films [46]. Polylactic acid, with its high biodegradability, is widely used in environmentally friendly food packaging [47]. Additionally, starch and proteins are also employed in barrier films.

Cellulose possesses characteristics such as renewability, biodegradability, low cost, and lightweight. A large number of hydrogen bonds are formed between cellulose molecules. However, as gas barrier film material, it also has some drawbacks. For example, its abundance of hydroxyl groups makes it sensitive to moisture and prone to absorbing moisture, resulting in poor mechanical properties, and susceptible to deformation and tearing with high moisture. Cellulose exhibits relatively low gas barrier performance [43]. It is typically used as a substrate in combination with other polymers, or subjected to surface modification and the addition of reinforcing fillers to prepare high barrier films. Yang *et al* used a simple immersion method to surface modify regenerated cellulose films prepared by the LiOH/urea solvent system (AUC) in a cationic alkyl ketene dimer (AKD) dispersion [52]. The resulting films exhibited high water resistance and gas barrier properties. The modified films became hydrophobic, with the water contact angle increasing from an initial 50° to a maximum of 110°. The oxygen permeability of the modified films at 0% relative humidity (RH) was less than 0.0005 ml m^{-2} d^{-1} kPa^{-1}, which reached the lowest detection limit of the instrument. Even at 0.2% AKD content, the oxygen permeability of the modified films in a high humidity environment of 75% RH was only 2.1 ml m^{-2} d^{-1} kPa^{-1}, approximately one-third of the original value. Wu *et al* grafted chitosan (CS) onto an oxidized cellulose (OC) matrix. The chemical grafting process alters the microstructure of the matrix [53]. Upon loading chitosan onto the OC substrate, the pores in the OC matrix are occupied by the grafted chitosan, resulting in a denser structure. As a result, the OC/CS composite film exhibits excellent oxygen barrier properties and antimicrobial activity. The oxygen permeability of complex membranes of short time oxidized cellulose and longer oxidation time cellulose are 8.93×10^{-11} cm^3 mm^{-2} s^{-1} Pa^{-1} and 4.32×10^{-11} cm^3 mm^{-2} s^{-1} Pa^{-1}, respectively. Tedeschi *et al* grafted oleyl ester onto the repeating units of cellulose acetate (CA) to form covalent bonds, resulting in a more compact layered structure internally, thereby elongating the gas diffusion path [54]. The long-chain hydrophobic structure of the oleyl ester reduces the surface energy of the composite film, initially weakening the adsorption of liquid water on the film surface. As a result, the WVP decreased by 76%, while the OP decreased by 90%.

Lignin is a renewable biomass resource abundant in plants, characterized by its amphiphilic molecular structure, forming a natural aromatic biopolymer. Lignin possesses biodegradability, antioxidative properties, ultraviolet (UV) light-blocking properties, and structural diversity, allowing modulation of material properties by adjusting its chemical structure and polymerization degree. Additionally, lignin exhibits certain gas barrier properties, especially against oxygen and water vapor. It finds extensive applications in film preparation [55]. Sirvio *et al* prepared cationic wood nanofiber films (CWNF) containing lignin [56]. The presence of lignin resulted in high UV absorption and almost zero UV transmission below 400 nm. Moreover, at higher relative humidity, the oxygen barrier performance of CWNF films is comparable to or better than many bio-based films and synthetic packaging polymers. The OP of the CWNF film was about 2000–3000 $cm^3\ \mu m\ m^{-2}\ d^{-1}\ atm^{-1}$ at 92% RH. This is more than 20 times lower compared to polypropylene and polyethylene. Kim *et al* acetylated the lignin to solve the problem of degradation of film properties due to lignin aggregation and added the modified lignin to butylene terephthalate adipate (PBAT). An 80.3% addition of modified lignin had good dispersion and compatibility in PBAT [57]. The acetylated lignin blocked the UV photo-degradation of PBAT and impeded the movement of gas molecules within the composite, thus improving the gas barrier properties.

Starch is characterized by low cost, non-toxicity, and good biodegradability. However, the excellent water solubility, low mechanical strength, moisture sensitivity, and poor gas barrier performance of pure starch limit its application as a barrier material [58]. Starch is often functionalized or compounded to construct gas barrier films. In addition, starch has a wide range of applications in active and smart packaging due to its bioactivity, such as antimicrobial, antioxidant, and UV-blocking properties [59]. Zhang *et al* prepared pure starch films and a series of silica blended films with different grains using a casting method [60]. Compared with the pure starch film, the addition of silica enhanced the water vapor barrier performance, and the barrier effect increased with the increase of silica size. The oxygen atoms in silica formed hydrogen bonds with the hydroxyl groups in starch and, in addition, the good dispersibility of silica provided a zigzag path for water molecules. Ren *et al* added chitosan to starch-based films to measure the effect of chitosan concentration on the water vapor properties of starch films [61]. The WVP of the pure starch film was $7.89 \times 10^{-10}\ g\ m^{-1}\ s^{-1}\ Pa^{-1}$ at RH 75%, and the high WVP originated from the fact that starch contained more free hydroxyl groups, which enhanced the interaction with water molecules. The addition of chitosan can significantly decrease the WVP value of the composite film, and the amino group on chitosan forms hydrogen bonds with the free hydroxyl group of starch, which enhances the water vapor barrier. The WVP of the composite film increases with the concentration of chitosan. Zhao *et al* introduced cellulose nanofibers (CNFs) and lignin into starch-based films to develop multifunctional composite films [48] (figure 3.3(a)). Due to the high crystallinity of the CNFs, the hydrogen bond formed between the CNFs and starch, and the enhanced hydrophobicity by lignin, the oxygen permeability of the composite film is as low as 61.17 $cm^3\ \mu m\ m^{-2}\ d^{-1}\ atm^{-1}$ and the water contact angle reaches 107.50°.

Hemicellulose-based materials exhibit excellent biodegradability, low mobility, minimal oxygen permeability, and a dense macromolecular network [62]. However, its strong hydrophilicity and natural brittleness hinder the application of films consisting solely of hemicellulose. Its drawbacks can be improved by chemical modification since hemicellulose naturally comprises a heterogeneous polysaccharide with numerous functional groups, such as hydroxyl and carboxyl functional groups, which can be readily utilized for modification. Shao *et al* modified hemicellulose using epichlorohydrin as an alkylating agent to obtain nearly hydrophobic hemicellulose [63]. The OP value of the hemicellulose was 1053 $cm^3\ \mu m\ m^{-2}\ d^{-1}\ kPa^{-1}$, which was much lower than that of low-density polyethylene. With the increase of epoxy group content, the OP value of the modified hemicellulose film

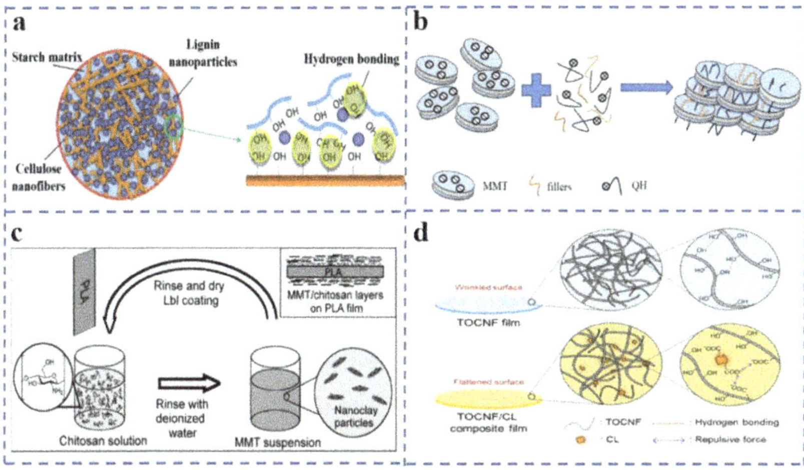

Figure 3.3. Representative works demonstrating bio-based polymers and their derivatives for gas barrier films. (a) Starch-based gas composite films reinforced by lignin nanoparticles and cellulose nanofibers. (Reproduced from [48]. CC BY 4.0.) (b) Hemicellulose-based barrier film reinforced with MTT filler. (Reproduced from [49]. CC BY 4.0.) (c) Tunable oxygen barrier properties by alternating MTT and chitosan assembled layer by layer with polylactic acid as the substrate. (Reproduced with permission from [50]. Copyright 2012 American Chemical Society.) (d) Enhancement of the gas and UV barrier of TEMPO-oxidized cellulose nanofibril (TOCNF) by carboxylated lignin. (Reproduced with permission from [51]. Copyright 2023 Elsevier.)

decreased significantly to 1.9 cm^3 μm m^{-2} d^{-1} kPa^{-1} when the feeding ratio of epichlorohydrin reached 2/3 ml g^{-1}. The hydrophobicity and oxygen barrier properties of hemicellulose can be improved significantly by a crosslinking reaction with hemicellulose using triformic acid (CA) as an esterifying agent. The water contact angle increased from 40.5° for the pure hemicellulose film to 87.5° for the CA-modified hemicellulose film. The OP decreased from 1053 cm^3 μm m^{-2} d^{-1} kPa^{-1} to 1.8 cm^3 μm m^{-2} d^{-1} kPa^{-1} [64]. Chen *et al* prepared hemicellulose-based films using quaternized hemicellulose (QH) by doping inorganic phase montmorillonite (MTT) and filler polyvinyl alcohol (PVA) (figure 3.3(b)) [49]. The presence of inorganic phase montmorillonite enlarged the gas diffusion path. The OTR values of the added PVA were lower compared to the unfilled PVA films, originating from the hydrogen bonding formed between QH-MMT-PVA.

Chitosan is a renewable material extracted from biological residues such as shells, shrimp, and crustaceans. The biomass of chitosan is surpassed only by that of cellulose in the biosphere. It possesses natural, sustainable, biocompatible, biodegradable, and environmentally friendly characteristics. Chitosan exhibits good gas barrier properties, capable of preventing the penetration of gases such as oxygen and carbon dioxide, and thus chitosan represents another highly interesting material as a biomaterial for gas barrier films. The structure of chitosan can be regulated through chemical modification and crosslinking to alter its physical and chemical properties, thereby achieving control and adjustment of gas transmission. However, the mechanical properties of pure chitosan are relatively weak, with lower strength and toughness. Additionally, chitosan has a certain degree of solubility in water, indicating poor water resistance, which can be improved by crosslinking or compounding with water-insoluble substances. Giannakas *et al* investigated the effect of low molecular weight PVOH and montmorillonite nanocomposites on the barrier properties of chitosan films [65]. The formation of intermolecular hydrogen bonds between PVOH and chitosan restricted the mobility of the intermolecular chains thereby blocking the transmission of water and oxygen. With the addition of sodium montmorillonite (NaMMT) and organically modified montmorillonite (OrgMMT) with dimethyldialkyl groups, the water and oxygen barrier properties of CS/PVOH-based nanocomposites were further improved. This is attributed to the intercalated structure of the nanocomposites, which forms a zigzag pathway for delaying the transition of water vapor and oxygen. Chitosan nanocomposite films with the addition of bacterial cellulose nanocrystals and silver nanoparticles were prepared by Salari [66] and the addition of bacterial cellulose nanocrystals decreased the sensitivity to water and the permeability. The WVP of the film containing 6% bacterial CNCs (BCNCs) decreased from 3.65×10^{-10} g m s^{-1} Pa^{-1} to 2.56×10^{-10} g m $s^{-1}Pa^{-1}$ compared to pure chitosan. Hydrogen bonds were found to form between the chitosan and BCNCs using FTIR, increasing the cohesion of the biopolymer matrix and decreasing the water sensitivity. FE-SEM and SEM showed that nanocomposite films containing BCNCs or AgNPs, or both, exhibited homogeneous and dense structures, with a filling effect that increased the interactions between the polymer chains. Gan *et al* [67] achieved the crosslinking of CNCs/chitosan using glutaraldehyde as the crosslinking agent, and solidified the material through microwave curing, resulting in a composite film with low water vapor permeability (WVP). The high crystallinity of the CNCs itself forms a dense barrier, while the strong hydrogen bonding between the hydroxyl groups of the CNCs and the amino groups in chitosan enhances the cohesion of the chitosan matrix. This hydrogen bonding prevents the disruption of hydrogen bonds by water molecules, thereby slowing down water vapor diffusion.

PLA is a biodegradable polymer, a type of α-hydroxyl acid containing carboxyl and hydroxyl groups. It is widely available and can be derived from various materials including corn and sugarcane [68]. PLA exhibits a certain level of gas barrier properties, being particularly effective against oxygen and carbon dioxide [69]. However, its mechanical performance is relatively poor, rendering it unsuitable for applications in high-strength and high-temperature environments. It also presents challenges in processing, requiring specialized techniques and equipment. Nevertheless, with the increasing demand for sustainable materials and technological advancements, PLA holds significant promise as a gas barrier [70, 71]. PLA is gaining traction as a growing option for packaging materials due to its true biodegradability, properties derived from renewable resources, and additional usage benefits for consumers. Svagan *et al* used a layer-by-layer method to assemble alternating layers of montmorillonite clay and chitosan on the surface of an extruded PLA film (figure 3.3(c)) [50]. When 70 bilayers were applied, the OP was reduced by 99% and 96% at 20% and 50% RH, respectively. Tang *et al* prepared PLA films in a ternary system to study the effect of filler ZnO nanoparticles on the gas barrier properties of PLA films [72]. As the content of ZnO nanoparticles increased, the films' oxygen and water vapor barrier improved. However, 15% ZnO nanoparticle content decreased the gas barrier of the films, which was speculated to be due to the aggregation of ZnO nanoparticles. The gas barrier properties of PLA can be improved by increasing its crystallinity [73]. Wang used lignin as a nucleating agent to improve the crystallinity of PLA, and the addition of 1 phr of lignin to the PLA matrix resulted in an approximate 50% reduction in oxygen permeability. Li *et al* used ethylene glycol diglycidyl ether (EGD) to modify organic montmorillonite (OMMT), and then utilized the modified montmorillonite as a nucleating agent to improve the crystallinity of PLA [74]. The OP of the PLA/EGDE4/OMMT-6 films was reduced by about 79% compared to pure PLA films. Table 3.2 highlights the improved barrier performance of bio-based films enhanced with specific nanoparticles and techniques, offering comparative insights into material effectiveness.

Proteins possess characteristics such as natural sourcing, biocompatibility, and flexibility. However, compared to some synthetic polymers, proteins generally exhibit poorer gas barrier properties, making them prone to gas permeation. Nevertheless, in specific applications such as food packaging or biomedical materials, their natural and biodegradable traits may serve as advantages. Protein films demonstrate a WVP typical of hydrophilic materials, thus necessitating their use in conjunction with other materials with barrier properties to enhance water vapor resistance [75]. Hosseini *et al* added 0.8% (w/v) *Origanum vulgare* L. essential oil to protein films and the maximum water vapor permeability of the films was reduced by 32% ($p < 0.05$) [76]. However, it is not the case that the more oil added, the lower the water vapor permeability, as this also depends on factors such as the type and content of oil. Ortiz *et al* developed active nanocomposite films of soybean isolate protein (SPI), microfibrillated cellulose (MFC) and clove essential oil (CEO) [77]. The effect of each component on the water vapor and oxygen barrier was investigated. The addition of oil did not give a significant improvement in the hydrophobicity of the films but caused a decrease in the WVP of the films. Moreover, the addition of oil led to a significant increase in the OP and ORT values of the films. The addition of MFC enhanced the water vapor and oxygen barrier of the films by making the gas diffusion path more tortuous, partly because of the hydrophobicity of oxygen, which is more affinitive to oil, and partly due to the potential effect of the oil acting as a plasticizer, facilitating gas diffusion (table 3.2).

3.4.2 Bio-based nanomaterials

Bio-based nanomaterials, with their natural origin, excellent mechanical properties and biocompatibility, bring new development opportunities for traditional barrier membrane technologies. These materials, such as nanocellulose, nanostarch, and chitosan nanoparticles, can not only effectively enhance the gas barrier performance of barrier membranes, but also endow them with new functional properties, such as antimicrobial and biodegradable properties [78, 79].

Nanocellulose is usually derived from natural cellulose, such as lignocellulose or plant cellulose, which is specially treated and processed (figure 3.4) [80, 81]. CNFs have a nanoscale fiber structure, which gives them excellent gas barrier properties

Table 3.2. Gas barrier properties of selected bio-based polymer films.

Polymer	WVTR	WVP	OTR	OP	References
Regenerated cellulose	—	—	Lower than the instrument's detected minimum value; 0% RH	—	[52]
Oxidized cellulose	—	$(3.67 \pm 0.39) \times 10^{-7}$ g·m m^{-2}·h·Pa; 30 °C, 97% RH	—	4.32×10^{-11} cm^3 mm^{-2} s^{-1} Pa^{-1}; 23 °C, 40% RH	[53].
Cellulose acetate	—	670 g m^{-2} d^{-1}; 25 °C, 100% RH	—	10 200 ml m^{-2} d^{-1}; 23 °C, 50% RH	[54]
Cationic wood nanofiber	—	—	(6.6 ± 0.0) cm^3 $(m^2\ d)^{-1}$; 23 °C, 50% RH	—	[56]
Starch	789.41 g m^{-2}·d; 25 °C, 90% RH	—	111.54 g m^{-2}·d; 25 °C, 90% RH	—	[60]
Starch	—	$1.22–3.04 \times 10^{-10}$ g m^{-1} s^{-1} Pa^{-1}; 25 °C, 75% RH	—	—	[61]
Starch	—	—	—	61.17 cm^3·μm m^{-2}·d·atm; 23 °C, 50% RH	[48]
Hemicellulose	—	—	—	1.9 cm^3 μm m^{-2} d^{-1} kPa^{-1}; 23 °C, 50% RH	[63].
Hemicellulose	—	—	1.37 cm^3 m^{-2}·24 h·0.1 MPa; 23 °C, 0% RH	—	[49]
Chitosan	—	2.02 ± 0.10 g m s^{-1} Pa^{-1}; 25 °C, 97% RH	—	—	[66]
Protein	—	0.487 ± 0.049 g mm kPa^{-1} h^{-1} m^{-2}; 20 °C, 0% RH	—	—	[76]
Soybean isolate protein	—	$(6.89 \pm 0.24) \times 10^{-11}$ g m s^{-1} Pa^{-1}; 20 °C, 75% RH	(115.00 ± 2.63) ml O_2 m^{-2} d; 23 °C, 65% RH	—	[77]

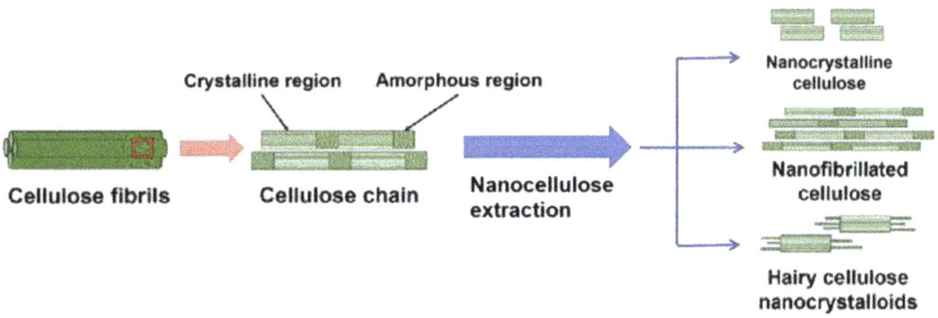

Figure 3.4. Schematic representation of the extraction of nanocrystalline cellulose, nanofibrillated cellulose and hairy cellulose nanocrystalloids from cellulose chains. (Reproduced with permission from [81]. Copyright 2018 Elsevier.)

that can effectively prevent the penetration of gases (e.g. oxygen, nitrogen, water vapor, etc) and prolong the shelf life of products [82, 83].

3.4.2.1 Cellulose nanocrystals

CNCs are relatively short rod-shaped nanostructured materials derived from cellulose, that are widely used due to their interesting intrinsic properties such as high crystallinity, high aspect ratio, high strength and stiffness, and optical properties [84]. Unlike the disorder of amorphous zones, CNCs are composed of well-arranged crystalline zones with relatively few free hydroxyl groups on the surface, a structure that theoretically contributes to the barrier against water vapor and oxygen [85]. Wang *et al* [86] compared the barrier properties of CNCs and CNFs and concluded that the crystal region of CNCs results in a barrier to the diffusion of external molecules, even at higher humidity. In addition, CNCs can be further used as an additive in film materials to enhance the overall properties of the material, as shown by Sun *et al*, who used different concentrations of CNCs to reinforce nanocomposite films prepared from *Eucommia ulmoides* gum (EUG) matrix [87]. The addition of 4% CNCs significantly improved the water vapor permeability of the nanocomposite films compared to the control with no CNC addition.

3.4.2.2 Cellulose nanofibers

Similar to CNCs, CNFs represent a cellulose nanomaterial with very high aspect ratio and high specific surface area [88]. These fibers form a dense network structure that lengthens the penetration path of water vapor and oxygen, as well as being a barrier to oils and fats, making them ideal for use as a barrier material [89]. However, even though CNFs shows a certain degree of impermeability, they are sensitive to moisture and cannot meet the requirements for high barrier properties. The plasticizing effect of water reduces the interaction between the CNFs, thereby decreasing the material's cohesion [90]. This necessitates the use of additional components or modifications to create a tailored structure to achieve high barrier performance [91]. Qin *et al* used a layer-by-layer deposition method to assemble CNFs containing quaternary ammonium functional groups with anionic vermiculite

(VMT) to form a nanobrick wall structure [92]. The OTR of the 20 layer CNF/VMT bilayer nanocoating with a thickness of 136 nm was only 0.013 cm^3 m^{-2} d^{-1} atm^{-1}. The excellent oxygen barrier capability was attributed to the highly aligned clay flakes, which created an extremely zigzagging diffusion pathway and a dense film structure. Tayeb *et al* [93] used CNFs (derived from mechanical milling) as the main component of a barrier film, cationic-modified montmorillonite as a filler, and polyamide epichlorohydrin (PAE)/thermosetting acrylic resin (ACR) as a cross-linking agent to produce a film material by heat treatment. The oxygen permeability of this barrier film is very low and even exceeds that of the PVA barrier film. Kim *et al* added carboxylated lignin (CL) to TEMPO-oxidized cellulose nanofibers (TOCNF) to improve the properties of the films (figure 3.4(d)) [51]. The addition of CL resulted in films with excellent UV-blocking ability, with the highest blocking ratios up to 97.1% and 99.9% for UVA and UVB, respectively. Furthermore, the addition of CL causes the surface of the film to be more hydrophilic and the WVTR to decrease, as water molecules were held more strongly resulting in delaying the transmission time.

3.4.2.3 Bacterial nanofibrillar cellulose

Bacterial nanofibrillar cellulose (BNC), a form of cellulose produced by certain bacteria, such as *Acetobacter xylinum*, during fermentation processes, has a dense 3D network structure which allows it to generate film with outstanding barrier properties. Wang *et al* prepared environmentally friendly antimicrobial films with bacterial nanocellulose, polyvinyl alcohol, and silver nanoparticles [94]. The oxygen barrier properties of these films were measured and showed that the peroxide value of BNC-based films was significantly lower than for PVA films, indicating that BNC acts as an oxygen barrier. This is mainly due to the dense 3D network structure of BNC. The inherent flexibility of BNC reduces the space between the original fibers and narrows the oxygen path.

In addition, some other bio-based nanomaterials that can be applied in gas barrier films, such as nanochitosan and nanostarch. Chitosan nanoparticles are applied in gas barrier membranes due to their unique physicochemical properties, such as small size, high surface area ratio, good biocompatibility and biodegradability [95]. By adding chitosan nanoparticles to gas barrier membranes, the mechanical properties and gas barrier properties of the films can be improved, in addition to providing antibacterial and antiviral effects. Jannatyha *et al* compared the effects of nanochitosan (NCH) and nanocellulose (NCL) added to carboxymethyl cellulose (CMC) on barrier properties [96]. Although both the NCH and NCL reduced the amount of water absorbed, the effect of CMC/NCH was significantly higher than that of CMC/NCL. Moreover, CMC/NCH also showed antimicrobial properties.

Natural starch-based films have limitations such as high water vapor transmission rate, poor solubility, and low tensile strength, which restrict their use. The development of nanostructured starch can improve its deficiencies [97, 98]. Nanostarch has higher oxygen barrier properties than pure starch polymers, and the addition of nanostarch can improve the gas barrier properties of composite films. Debranched starch nanoparticles (DSNPs) can significantly increase the barrier properties, mainly by filling the void space of the corn starch membrane matrix, resulting in denser nanocomposite films. Lin *et al* compared the OTR values of pure starch polymer films and nanocomposite membranes with 5% debranched starch nanoparticles (DSNPs). The OTR value of pure starch polymer films is 394.48 $cm^3\ m^{-2}\ d^{-1}\ atm^{-1}$, while the OTR value of the starch composite film containing 5% SNP is only 81.61 $cm^3\ m^{-2}\ d^{-1}\ atm^{-1}$ [99]. Jiang utilized potato nanostarch granules to enhance the water vapor barrier and tensile strength of pea starch-based films [100]. The addition of 6% potato nanostarch granules resulted in a decrease of WVP from $4.5 \times 10^{-10}\ g\ m^{-2}\ s^{-1}\ Pa^{-1}$ to $2.4 \times 10^{-10}\ g\ m^{-2}\ s^{-1}\ Pa^{-1}$ and an increase in tensile strength from 8.8 to 15 MPa. It is considered that the higher crystallinity of the potato nanostarch granules and hydrogen-bonding entanglement between the starch matrices enhances the water vapor barrier and tensile strength of the composite film.

3.5 Preparation methods of barrier films

In addition to the specific properties of the materials, the properties of gas barrier films are significantly determined by the preparation method. Therefore, developing efficient, sustainable, and low-cost methods for preparing gas barrier films has become a hot topic in current research. This section will give an overview of the pivotal role of preparation methods in the research and development of gas barrier films. Common technologies for preparing film material include casting [101], melt mixing (extrusion) [102], electrospinning [103], 3D printing [104] and polymerization. Among these methods, casting emerges as the simplest and most straightforward technique, while melt mixing is the method of choice for industrial applications as it offers a highly practical approach for combining two or more components into a new composite material through melt blending. Electrospinning can wrap and stabilize fiber films of bioactive molecules, while the polymerization technique for film preparation often involves toxic reactants and solvents making it a less green approach [105]. 3D printing technology, as a recent advancement, has already evolved into a prominent commercial manufacturing tool, finding extensive application in the field of biomedicine. By comparing and analysing the advantages and disadvantages of various preparation methods, theoretical and technical support for the efficient and qualified preparation of gas barrier films will be provided, thus promoting their widespread application in practical scenarios.

Solvent casting is simple and easy to operate. The general steps in casting include the preparation of the polymer solution (or casting solution), pouring of the solution into a mold, then drying, and finally post-processing. Depending on the solvent, it sometimes requires a long drying time and may consume a large amount of energy. Additionally, organic solvents may evaporate during the drying process, resulting in uneven film structure or the formation of pores, which can reduce gas barrier performance and also have environmental impacts.

Electrospinning is a simple and efficient method for preparing nanofibers, featuring a high specific surface area, which is conducive to enhancing gas barrier performance. The components of an electrospinning device include a syringe, metal needle, and collectors. By applying an electric field, the polymer solution is transferred from the syringe needle to the collector, thus generating nanofibers. Electrospinning includes three processes: spray initiation, elongation, and curing. The electrostatic force overcomes the surface tension through the electric field, and the charged jet of polymer solution is ejected from the tip of Taylor cone. When the jet passes through the collector, the solvent in the solution evaporates and the fibers are deposited on the collector [105, 106]. The factors that affect the quality of nanofibers are flow rate (Q), voltage (V) and the distance between the needle and the collector (D). In addition, the solution conductivity, solution viscosity, humidity and temperature are also important factors that need to be considered for the quality of the resulting nanofibers. By adjusting electrospinning process parameters, such as electric field intensity, solution concentration, and spinning distance, the structure and pore size distribution of the film can be controlled, thereby regulating gas barrier performance. In a comprehensive study, Feng *et al* compared the two methods with one another. They prepared PLA/TiO_2 composite films by solvent casting and electrospinning, and found that the water vapor transmission rate of films prepared via solvent casting significantly surpasses that of films crafted through electrostatic spinning [107]. The main reason for these results can be found in the fine structure of the fibers. While electrospinning adjusted the fine structure of the fibers, the solvent casting method simply mixed the polymers and TiO_2 leading to the agglomeration of TiO_2 nanoparticles, which causes the difference in the water vapor transmission rate.

Additive manufacturing (AM), commonly known as 3D printing, is a manufacturing technology that emerged in the 1990s. 3D printing does not require molds to complete the manufacture of various shapes but only requires the use of computer technology. 3D printing technology can customize the shape, structure, and size of films according to specific needs, enabling personalized design to meet the requirements of particular applications. Currently, there are many 3D printing methods, such as selective laser sintering (SLS), stereolithography equipment (SLA), fused deposition modeling (FDM), selective laser melting (SLM) and electron beam melting (EBM) [108]. However, the range of gas barrier materials available for 3D printing is relatively limited at present, and the layer-by-layer stacking during the 3D printing process may increase surface roughness, thereby affecting gas barrier performance. Kim *et al* compared the water vapor shielding properties of Pec/CMC/ZnO prepared by a 3D printing technique and a solvent casting method, and observed that the surface of the samples prepared by the 3D printing technique was denser than that of solvent casting, and in addition the water vapor shielding properties of the films prepared by the 3D printing technique were improved by 38% compared to the solvent casting method [109].

To sum up, choosing the appropriate film preparation method is crucial for achieving the desired gas barrier performance. When designing gas barrier films, it is necessary to consider the impact of the preparation method on the gas barrier properties to meet the optimal performance requirements.

3.6 Applications

As demand for sustainable and environmentally friendly solutions is increasingly pressing, bio-based gas barrier films have emerged as a noteworthy technological innovation. Their potential applications in fields such as food packaging, flexible electronic devices, renewable energy, and the circular economy have garnered extensive attention (figure 3.5). Furthermore, with the continuous progress and innovation of bio-based material technology, bio-based gas barrier films will be applied in a wider range of fields, providing important support for the development of environmentally friendly, high-performance packaging materials and energy technologies, and promoting the realization of sustainable development goals. This section will provide an overview of the applications of bio-based gas barrier films.

3.6.1 Food packaging

Packaging plays a crucial role in the maintenance of food quality. It helps prevent undesirable biological and chemical changes, blocks unwanted gases and microorganisms, and maintains the flavor and quality of food, thereby preventing spoilage [110, 111]. Gas barrier films play a critical role in food packaging. They are designed to protect food from the effects of gases, humidity, light, and odors in the external environment [112]. However, aside from preservation, there is an increasing concern about potential toxic effects stemming from the packaging material, making safety awareness among consumers a more and more important consideration for

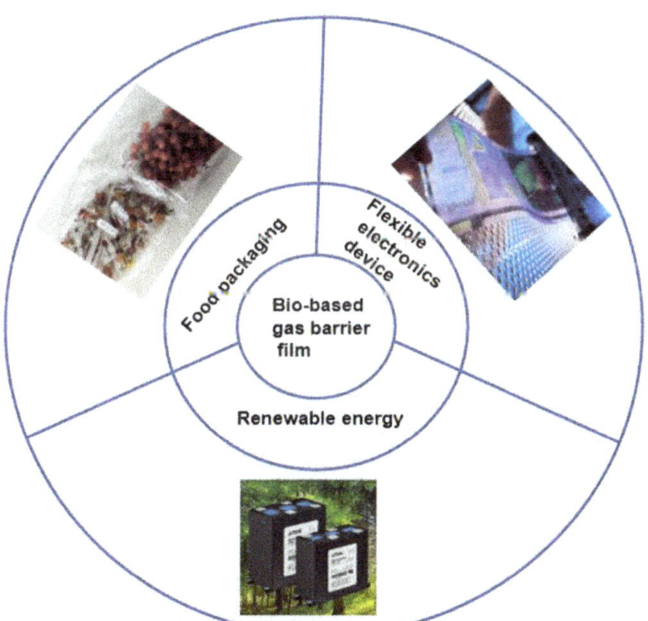

Figure 3.5. Overview of the applications of bio-based barrier films.

improvement [113]. For example, traditional plastic packaging may release a certain number of micro-plastics into food, which then enter the human body, while metal packaging may release traces of metals, potentially causing health issues. This leads to a wide range of requirements that the packaging film must meet, including gas barrier properties, non-toxicity, biocompatibility, antibacterial properties, mechanical strength, and hydrophobic properties. In essence, it ensures that the food remains intact, and the packaging film maintains its integrity. Bio-based packaging films with excellent gas barrier performances meet the above-mentioned requirements for food packaging, considering their biodegradability, abundance of sources, and their suitability to be complexed with other functional ingredients [114]. Cellophane is one of the most interesting products produced from recycled cellulose and has good barrier properties against water and oil, which makes this material suitable for the food packaging industry. It is currently commercialized in food packaging [115].

Adding bio-based materials to food packaging films not only enhances the gas barrier properties of the packaging film but also imbues it with new functionalities, thereby extending the preservation of food flavor and quality for a longer period. Grapeseed lignin has high oxidation resistance, and the influence of free radicals can be eliminated by adding grapeseed lignin to packaging barrier film instead of traditional antioxidants such as hindered phenols. Vostrejs *et al* prepared films of poly(3-hydroxybutyrate) and polyhydroxy fatty acid ester blends with grapeseed lignin and found that the gas transport rate of the blended films with grapeseed lignin was significantly decreased. In addition, scanning electron microscopy results showed that the presence of grapeseed lignin would lead to smaller nanopores on the surface of the blended films, limiting the gas transport [116]. Wen *et al* incorporated two color indicators, thymol and purple sweet potato anthocyanin, into TEMPO-oxidized bacterial cellulose to develop a novel smart active food packaging film. This composite film serves the dual purpose of freshness preservation and real-time monitoring of food quality [117]. Additionally, the inclusion of these two color indicators enhances the UV protection and water vapor barrier properties of the film.

3.6.2 Flexible electronics device

With the rapid development and application of organic electronics, there is growing concern regarding their protection and preservation. Water vapor can cause degradation in organic electronic devices, thereby impacting their performance and lifespan. The WVTR of polymers is typically in the range of 0.1–100 g m^{-2} d^{-1}. This range usually meets the water vapor barrier requirements for food packaging but falls short of the standards for organic electronics. To provide sufficient lifespan for organic electronic products, it is necessary to deposit an additional barrier film on the surface of encapsulation films or directly coat electronic devices with a permeation barrier. This can significantly reduce the permeation of water vapor and oxygen, effectively protecting organic electronic devices from moisture and oxidation [118]. Usually, organic electronic gas barrier films require WVTR below

10^{-4} g m^{-2} d^{-1}, while at the same time being flexible, stretchable, deformable, and provides long-term stability under environmental conditions [119–121]. In addition, the organic electronic barrier films composed of two or more components should also increase the interfacial adhesion between the organic/inorganic components. Otherwise, with the repeated use of the product, the mechanical strength of the barrier film will be weakened, and the performance of the product will be diminished [122]. Zhou *et al* prepared a novel nanocomposite film consisting of three substances, cellulose acetate (CA)/polyethyleneimine (PEI)/reduced graphene oxide (rGO)-$NiCoFeO_x$ [123]. The presence of PEI enhances the CA/rGO interfacial interactions, and the oxygen-containing groups in PEI and CA form hydrogen bonding with rGO, which results in stronger interactions with the nanocomposite film. The film has a low water vapor permeability and OTR. The film structure remains nearly unchanged even after being bent at least one hundred times, with stable mechanical properties, rendering it suitable for electronic instrument packaging. Lang *et al* [124] used cellulose and chitin to prepare renewable barrier films for the encapsulation of polymer-based electrochromic devices. The OTR of the film was 29 cm^3 m^{-2} d^{-1} atm^{-1}, which is comparable to commercially available PET films, and the devices with encapsulation film showed photo-degradation about ten times slower than that without encapsulation.

3.6.3 Renewable energy

Renewable energy refers to those energy sources that can be continuously replenished, including but not limited to solar energy, wind energy, and biomass energy. These energy sources, compared to traditional fossil fuels such as coal, oil, and natural gas, are more environmentally friendly during their usage and also more cost-effective from a technological standpoint [125, 126].

Capacitors play a crucial role in renewable energy systems for storing and managing electricity generated by renewable energy sources such as solar and wind [127, 128]. By establishing close connections between capacitors and renewable energy sources, they offer more sustainable and efficient solutions for the transition to renewable energy. Arantes *et al* used cellulose and chitosan to prepare bio-based films and added magnetite and glycerol [129]. The films containing glycerol had a higher dielectric constant than those without glycerol, as tested by capacitance tests, and the magnetite could alter the charge movement, resulting in an increase in the dielectric constant. The addition of both components increased the charge storage capacity, giving the cellulose/chitosan bio-based films potential as capacitors. Commercial examples for the applications of bio-based film materials are rare. However, the general drive towards a circular economy has led TDK Corporation to launch a 100% bio-circular polypropylene film capacitor, the first step towards a fully sustainable film capacitor that has already been certified by the International Sustainability and Carbon Certification (ISCC) and is already in mass production [130]. Traditional polypropylene is obtained from fossil feedstocks, while bio-based polypropylene is derived from agricultural renewable raw material or wastes and residues and is a 100% bio-circular film.

3.7 Circular economy integration

A circular economy requires maximizing the value of resources by utilizing virgin polymers made from renewable or recyclable materials and ensuring that products can be reused or recycled at the end of their life cycle. A circular economy tries to reduce the carbon footprint associated with newly produced materials from natural resources and aims to maximize the recycling of materials for further life cycles, thereby minimizing the amount of waste [131]. Bio-based films are part of a circular economy where virgin polymers derive from renewable or recycled resources and the final products can be reused or recycled at the end of their lifetime [132]. Bio-based plastics have a low carbon footprint compared to petroleum-based plastics. The use of renewable biomass not only reduces the burden of the fossil resource crisis, but also spurs the development of a bio-based economy, which contributes to the United Nations' goal of sustainable development. In 2021, global plastics production reached 390.7 million tons, with recycled plastics accounting for about 9.8% of global plastics production, of which bio-based plastics accounted for only 1.5% [133]. With the continuous innovation and improvement of technology, the integration of bio-based films in the circular economy is becoming more crucial. Figure 3.6 is a schematic diagram of a circular economy, using PLA as an example [134]. Wang *et al* [135] developed hydroplastics that can be molded with water using renewable cellulosic biomass. The hydroplastics demonstrated excellent mechanical properties and recyclability. Hydroplastics can be molded in water into a variety of shapes without heat treatment, which is an eco-friendly and sustainable process, decreasing energy consumption significantly, and avoiding expensive and complex shaping devices, as well as harsh conditions (figure 3.7). Moreover, the same plastic can be molded more than 15 times with this hydrosetting method, while retaining their good mechanical properties. Bio-based hydroplastics have opened a new window for the circular economy.

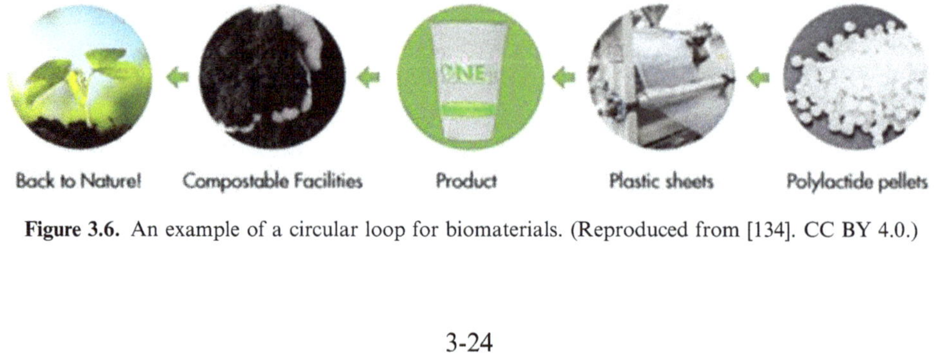

Figure 3.6. An example of a circular loop for biomaterials. (Reproduced from [134]. CC BY 4.0.)

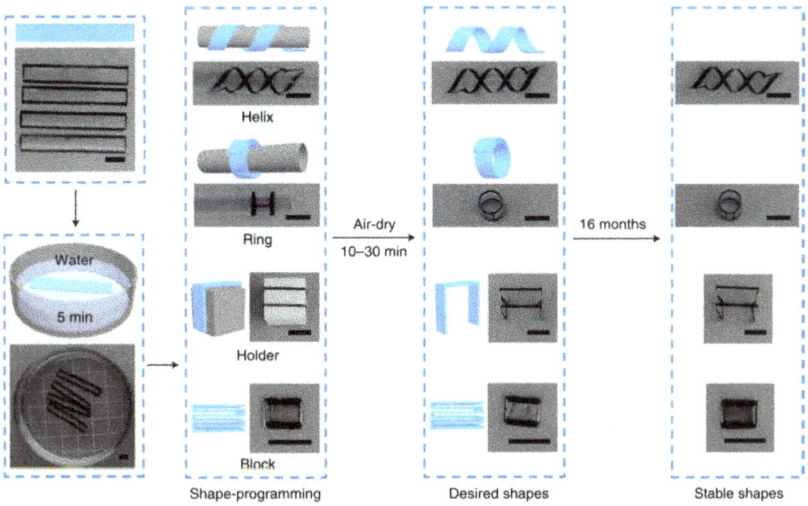

Figure 3.7. Sustainable and highly facile hydrosetting shape-programable CCi films. (Reproduced with permission from [135]. Copyright 2021 Springer Nature.)

3.8 Challenges

Barrier films play a crucial role in preventing gas hazards. Barrier films made from biomass materials are more in line with today's green philosophy and the pursuit of sustainable living than traditional petroleum-based barrier films. Materials for bio-based barrier films mainly include bio-based polymers, bio-based nanomaterials. The design of high-performance barrier films requires attention to the crystallinity of the film, the compatibility and structure of the blended substances, the free volume, and the interaction of the permeable molecules with the film interface, all of which will affect the barrier properties of the film. With regards to barrier films prepared from polymers, in particular polysaccharides, attention should be paid to the fact that the polymers themselves are sensitive to moisture due to the large number of hydroxyl groups, and are not yet suitable for some highly moisture-sensitive products. The addition of nanocomponents, including CNCs, CNFs, nanochitosan, montmorillonite and so on, usually endow the bio-based film with a more compact structure, resulting in a more complex gas diffusion path and therefore better gas barrier properties.

However, there are some challenges for bio-based barrier films. (i) In the context of green environmental protection, efforts are still needed to realize the sustainable development of barrier films from use to reuse. (ii) The development of high barrier films is still a direction of continued effort, as for barrier films prepared from biomass rich in hydrophilic groups, water sensitivity is still a challenging problem to be solved; for flexible packaging, how to maintain a certain degree of flexibility after repeated use while maintain the barrier performance is still quite difficult at present. (iii) It is necessary to develop more biomass-based barrier films to meet the needs of diverse industries. In the field of flexible device packaging, the types of biomass-based barrier films are relatively rare, and most of them are still petroleum-based products. Compared with petroleum-based barrier films, most biomass barrier films are still under laboratory scales, and the realization of mass production still needs much effort. (iv) Despite their potential environmental friendliness and sustainability, bio-based barrier films may have relatively low recognition in the market. Therefore, there is a need to increase market awareness and the acceptance of bio-based barrier films through promotion and publicity as well as the development of relevant standards.

Bibliography

[1] Tseng M H, Su D Y, Chen G L and Tsai F Y 2021 Nano-laminated metal oxides/polyamide stretchable moisture- and gas-barrier films by integrated atomic/molecular layer deposition *ACS Appl. Mater. Interfaces* **13** 27392–9

[2] Feldman D 2001 Polymer barrier films *J. Polym. Environ.* **9** 49–55

[3] Maes C, Luyten W, Herremans G, Peeters R, Carleer R and Buntinx M 2018 Recent updates on the barrier properties of ethylene vinyl alcohol copolymer (EVOH): a review *Polym. Rev.* **58** 209–46

[4] Liu Y R *et al* 2023 Crystalline polyvinylidene chloride embedded in epoxy composite coating for oxygen gas barrier and anti-corrosion *Chem. Eng. J.* **474** 145848

[5] Zubair M and Ullah A 2020 Recent advances in protein derived bionanocomposites for food packaging applications *Crit. Rev. Food Sci. Nutr.* **60** 406–34

[6] Lamb J B *et al* 2018 Plastic waste associated with disease on coral reefs *Science* **359** 460–2

[7] Paunonen S 2013 Strength and barrier enhancements of cellophane and cellulose derivative films: a review *Bioresources* **8** 3098–121

[8] Sid S, Mor R S, Kishore A and Sharanagat V S 2021 Bio-sourced polymers as alternatives to conventional food packaging materials: a review *Trends Food Sci. Technol.* **115** 87–104

[9] Sikorska W, Musioł M, Zawidlak-Węgrzyńska B and Rydz J 2021 End-of-life options for (bio) degradable polymers in the circular economy *Adv. Polym. Tech.* **2021** 1–18

[10] Tarazona N A, Machatschek R, Balcucho J, Castro-Mayorga J L, Saldarriaga J F and Lendlein A 2022 Opportunities and challenges for integrating the development of sustainable polymer materials within an international circular (bio) economy concept *MRS Energy. Sustain.* **9** 28–34

[11] Gironi F and Piemonte V 2011 Bioplastics and petroleum-based plastics: strengths and weaknesses *Energy Sources* A **33** 1949–59

[12] Choi J-G, Do D D and Do H D 2001 Surface diffusion of adsorbed molecules in porous media: monolayer, multilayer, and capillary condensation regimes *Ind. Eng. Chem. Res.* **40** 4005–31

[13] Dąbrowski 2001 Adsorption—from theory to practice *Adv. Colloid Interface Sci.* **93** 135–224

[14] Joshi M, Adak B and Butola B S 2018 Polyurethane nanocomposite based gas barrier films, membranes and coatings: a review on synthesis, characterization and potential applications *Prog. Mater Sci.* **97** 230–82

[15] Saadi R, Saadi Z, Fazaeli R and Fard N E 2015 Monolayer and multilayer adsorption isotherm models for sorption from aqueous media *Korean J. Chem. Eng.* **32** 787–99

[16] Swenson H and Stadie N P 2019 Langmuir's theory of adsorption: a centennial review *Langmuir* **35** 5409–26

[17] Freundlich H 1907 Über die Adsorption in Lösungen *Z. Phys. Chem.* **57U** 385–470

[18] Brunauer S, Emmett P H and Teller E 1938 Adsorption of gases in multimolecular layers *J. Am. Chem. Soc.* **60** 309–19
[19] Králik M J C P 2014 Adsorption, chemisorption, and catalysis *Chem. Pap.* **68** 1625–38
[20] Polanyi M L 1932 Section III—theories of the adsorption of gases. A general survey and some additional remarks. introductory paper to section III *Trans. Faraday Soc.* **28** 316–33
[21] Dubinin M M 1960 The potential theory of adsorption of gases and vapors for adsorbents with energetically nonuniform surfaces *Chem. Rev.* **60** 235–41
[22] Liu L, Luo X-B, Ding L and Luo S-L 2019 Application of nanotechnology in the removal of heavy metal from water *Nanomaterials for the Removal of Pollutants and Resource Reutilization* (Amsterdam: Elsevier) pp 83–147
[23] Wu F, Misra M and Mohanty A K 2021 Challenges and new opportunities on barrier performance of biodegradable polymers for sustainable packaging *Prog. Polym. Sci.* **117** 101395
[24] Jr W W J, LeBoeuf E J, Young T M and Huang W 2001 Contaminant interactions with geosorbent organic matter: insights drawn from polymer sciences *Water Res.* **35** 853–68
[25] Choudalakis G and Gotsis A D 2009 Permeability of polymer/clay nanocomposites: a review *Eur. Polym. J.* **45** 967–84
[26] Delgado J F, Peltzer M A, Wagner J R and Salvay A G 2018 Hydration and water vapour transport properties in yeast biomass based films: a study of plasticizer content and thickness effects *Eur. Polym. J.* **99** 9–17
[27] Hu Y, Topolkaraev V, Hiltner A and Baer 2001 Measurement of water vapor transmission rate in highly permeable films *J. Appl. Polym. Sci.* **81** 1624–33
[28] Firpo G, Setina J, Angeli E, Repetto L and Valbusa U 2021 High-vacuum setup for permeability and diffusivity measurements by membrane techniques *Vacuum* **191** 110368
[29] Siracusa V, Rocculi P, Romani S and Dalla Rosa M 2008 Biodegradable polymers for food packaging: a review *Trends Food Sci. Technol.* **19** 634–43
[30] Velásquez E *et al* 2023 Feasibility of valorization of post-consumer recycled flexible polypropylene by adding fumed nanosilica for its potential use in food packaging toward sustainability *Polymers* **15** 1081
[31] Helanto K, Matikainen L, Talja R and Rojas O J 2019 Bio-based polymers for sustainable packaging and biobarriers: a critical review *Bioresources* **14** 4902–51
[32] Curtzwiler G, Vorst K, Palmer S and Brown J W 2008 Characterization of current environmentally-friendly films *J. Plast. Film Sheeting* **24** 213–26
[33] Tang S *et al* 2023 Nacre-inspired biodegradable nanocellulose/MXene/AgNPs films with high strength and superior gas barrier properties *Carbohydr. Polym.* **299** 120204
[34] Wang J *et al* 2021 Self-compounded nanocomposites: toward multifunctional membranes with superior mechanical, gas/oil barrier, UV-shielding, and photothermal conversion properties *ACS Appl. Mater. Interfaces* **13** 28668–78
[35] Jiménez-Regalado E J, Caicedo C, Fonseca-García A, Rivera-Vallejo C C and Aguirre-Loredo R Y 2021 Preparation and physicochemical properties of modified corn starch–chitosan biodegradable films *Polymers* **13** 4431
[36] Chowdhury R A, Nuruddin M, Clarkson C, Montes F, Howarter J and Youngblood J P 2019 Cellulose nanocrystal (CNC) coatings with controlled anisotropy as high-performance gas barrier films *ACS Appl. Mater. Interfaces* **11** 1376–83
[37] Liu F *et al* 2020 Preparation and antibacterial properties of epsilon-polylysine-containing gelatin/chitosan nanofiber films *Int. J. Biol. Macromol.* **164** 3376–87

[38] Song P and Wang H J A M 2020 High-performance polymeric materials through hydrogen-bond cross-linking *Adv. Mater.* **32** 1901244

[39] Qin S, Song Y, Floto M E and Grunlan J C 2017 Combined high stretchability and gas barrier in hydrogen-bonded multilayer nanobrick wall thin films *ACS Appl. Mater. Interfaces* **9** 7903–7

[40] Dang K M and Yoksan R 2016 Morphological characteristics and barrier properties of thermoplastic starch/chitosan blown film *Carbohydr. Polym.* **150** 40–7

[41] Leite L S F *et al* 2021 Effect of tannic acid and cellulose nanocrystals on antioxidant and antimicrobial properties of gelatin films *ACS Sustain. Chem. Eng.* **9** 8539–49

[42] Nuruddin M, Chowdhury R A, Lopez-Perez N, Montes F J, Youngblood J P and Howarter J A 2020 Influence of free volume determined by positron annihilation lifetime spectroscopy (PALS) on gas permeability of cellulose nanocrystal films *ACS Appl. Mater. Interfaces* **12** 24380–9

[43] Yu Z, Ji Y, Bourg V, Bilgen M and Meredith J C 2020 Chitin- and cellulose-based sustainable barrier materials: a review *Emerg. Mater.* **3** 919–36

[44] Zhang Y and Naebe M 2021 Lignin: a review on structure, properties, and applications as a light-colored UV absorber *ACS Sustain. Chem. Eng.* **9** 1427–42

[45] Gröndahl M, Eriksson L and Gatenholm P 2004 Material properties of plasticized hardwood xylans for potential application as oxygen barrier films *Biomacromolecules* **5** 1528–35

[46] Cazón P and Vázquez M 2020 Mechanical and barrier properties of chitosan combined with other components as food packaging film *Environ. Chem. Lett.* **18** 257–67

[47] Mangaraj S, Yadav A, Bal L M, Dash S and Mahanti N K 2019 Application of biodegradable polymers in food packaging industry: a comprehensive review *J. Packag. Technol. Res.* **3** 77–96

[48] Zhao Y, Troedsson C, Bouquet J M, Thompson E M, Zheng B and Wang M 2021 Mechanically reinforced, flexible, hydrophobic and UV impermeable starch-cellulose nanofibers (CNF)-lignin composites with good barrier and thermal properties *Polymers* **13** 4346

[49] Chen G G *et al* 2015 Hemicelluloses/montmorillonite hybrid films with improved mechanical and barrier properties *Sci. Rep.* **5** 16405

[50] Svagan A J *et al* 2012 Transparent films based on PLA and montmorillonite with tunable oxygen barrier properties *Biomacromolecules* **13** 397–405

[51] Kim J C *et al* 2023 Fabrication of transparent cellulose nanofibril composite film with smooth surface and ultraviolet blocking ability using hydrophilic lignin *Int. J. Biol. Macromol.* **245** 125545

[52] Yang Q L, Saito T and Isogai A 2012 Facile fabrication of transparent cellulose films with high water repellency and gas barrier properties *Cellulose* **19** 1913–21

[53] Wu Y *et al* 2016 Green and biodegradable composite films with novel antimicrobial performance based on cellulose *Food Chem.* **197** 250–6

[54] Tedeschi G *et al* 2018 Thermoplastic cellulose acetate oleate films with high barrier properties and ductile behaviour *Chem. Eng. J.* **348** 840–9

[55] Gillet S *et al* 2017 Lignin transformations for high value applications: towards targeted modifications using green chemistry *Green Chem.* **19** 4200–33

[56] Sirviö J A, Ismail M Y, Zhang K, Tejesvi M V and Ämmälä A 2020 Transparent lignin-containing wood nanofiber films with UV-blocking, oxygen barrier, and anti-microbial properties *J. Mater. Chem.* A **8** 7935–46

[57] Kim J *et al* 2023 Enhanced barrier properties of biodegradable pbat/acetylated lignin films *Sustain. Mater. Technol.* **37** e00686
[58] Xie F W, Pollet E, Halley P J and Avérous L 2013 Starch-based nano-biocomposites *Prog. Polym. Sci.* **38** 1590–628
[59] Cui C L, Ji N, Wang Y F, Xiong L and Sun Q J 2021 Bioactive and intelligent starch-based films: a review *Trends Food Sci. Technol.* **116** 854–69
[60] Zhang R, Wang X and Cheng M 2018 Preparation and characterization of potato starch film with various size of nano-SiO_2 *Polymers* **10** 1172
[61] Ren L, Yan X, Zhou J, Tong J and Su X 2017 Influence of chitosan concentration on mechanical and barrier properties of corn starch/chitosan films *Int. J. Biol. Macromol.* **105** 1636–43
[62] Huang B B, Tang Y J, Pei Q Q, Zhang K J, Liu D D and Zhang X M 2018 Hemicellulose-based films reinforced with unmodified and cationically modified nanocrystalline cellulose *J. Polym. Environ.* **26** 1625–34
[63] Shao H *et al* 2020 Barrier film of etherified hemicellulose from single-step synthesis *Polymers* **12** 2199
[64] Shao H, Sun H, Yang B, Zhang H and Hu Y 2019 Facile and green preparation of hemicellulose-based film with elevated hydrophobicity via cross-linking with citric acid *RSC Adv.* **9** 2395–401
[65] Giannakas A *et al* 2016 Preparation, characterization, mechanical, barrier and antimicrobial properties of chitosan/PVOH/clay nanocomposites *Carbohydr. Polym.* **140** 408–15
[66] Salari M, Khiabani M S, Mokarram R R, Ghanbarzadeh B and Kafil H S 2018 Development and evaluation of chitosan based active nanocomposite films containing bacterial cellulose nanocrystals and silver nanoparticles *Food Hydrocoll.* **84** 414–23
[67] Gan P G, Sam S T, Abdullah M F, Omar M F and Tan W K 2021 Water resistance and biodegradation properties of conventionally-heated and microwave-cured cross-linked cellulose nanocrystal/chitosan composite films *Polym. Degrad. Stab.* **188** 109563
[68] Auras R A, Lim L-T, Selke S E and Tsuji H 2022 Poly (lactic acid): synthesis, structures, properties, processing *Applications, and End of Life* (New York: Wiley)
[69] Marano S, Laudadio E, Minnelli C and Stipa P 2022 Tailoring the barrier properties of PLA: a state-of-the-art review for food packaging applications *Polymers* **14** 1626
[70] Singha S and Hedenqvist M S 2020 A review on barrier properties of poly(lactic acid)/clay nanocomposites *Polymers* **12** 1095
[71] Mohan S and Panneerselvam K 2022 A short review on mechanical and barrier properties of polylactic acid-based films *Mater. Today-Proc.* **56** 3241–6
[72] Tang Z, Fan F, Chu Z, Fan C and Qin Y 2020 Barrier properties and characterizations of poly(lactic acid)/ZnO nanocomposites *Molecules* **25** 1310
[73] Wang N, Zhang C and Weng 2021 Enhancing gas barrier performance of polylactic acid/lignin composite films through cooperative effect of compatibilization and nucleation *J. Appl. Polym. Sci.* **138** 50199
[74] Li F, Zhang C and Weng Y 2020 Improvement of the gas barrier properties of PLA/OMMT films by regulating the interlayer spacing of OMMT and the crystallinity of PLA *ACS Omega* **5** 18675–84
[75] Calva-Estrada S J, Jiménez-Fernández M and Lugo-Cervantes E 2019 Protein-based films: advances in the development of biomaterials applicable to food packaging *Food Eng. Rev.* **11** 78–92

[76] Hosseini S F, Rezaei M, Zandi M and Farahmandghavi F 2016 Development of bioactive fish gelatin/chitosan nanoparticles composite films with antimicrobial properties *Food Chem.* **194** 1266–74

[77] Ortiz C M, Salgado P R, Dufresne A and Mauri A N 2018 Microfibrillated cellulose addition improved the physicochemical and bioactive properties of biodegradable films based on soy protein and clove essential oil *Food Hydrocoll.* **79** 416–27

[78] Li J *et al* 2018 Nanocellulose-based antibacterial materials *Adv. Healthc. Mater.* **7** 1800334

[79] Dufresne A 2013 Nanocellulose: a new ageless bionanomaterial *Mater. Today* **16** 220–7

[80] Moon R J, Martini A, Nairn J, Simonsen J and Youngblood J 2011 Cellulose nanomaterials review: structure, properties and nanocomposites *Chem. Soc. Rev.* **40** 3941–94

[81] Phanthong P, Reubroycharoen P, Hao X, Xu G, Abudula A and Guan G 2018 Nanocellulose: extraction and application *Carbon Resour. Convers.* **1** 32–43

[82] Nazrin A, Sapuan S M, Zuhri M Y M, Ilyas R A, Syafiq R and Sherwani S F K 2020 Nanocellulose reinforced thermoplastic starch (TPS), polylactic acid (PLA), and polybutylene succinate (PBS) for food packaging applications *Front. Chem.* **8** 213

[83] Gao Q, Lei M, Zhou K, Liu X and Wang S 2020 Nanocellulose in food packaging—a short review *Carbohydr. Polym.* **14** 431–43

[84] Kwon G, Lee K, Kim D, Jeon Y, Kim U J and You J 2020 Cellulose nanocrystal-coated TEMPO-oxidized cellulose nanofiber films for high performance all-cellulose nanocomposites *J. Hazard. Mater.* **398** 123100

[85] He Y, Boluk Y, Pan J, Ahniyaz A, Deltin T and Claesson P M 2020 Comparative study of CNC and CNF as additives in waterborne acrylate-based anti-corrosion coatings *J. Dispers. Sci. Technol.* **41** 2037–47

[86] Wang J W, Gardner D J, Stark N M, Bousfield D W, Tajvidi M and Cai Z Y 2018 Moisture and oxygen barrier properties of cellulose nanomaterial-based films *ACS Sustain. Chem. Eng.* **6** 49–70

[87] Sun Q, Zhao X, Wang D, Dong J, She D and Peng P 2018 Preparation and characterization of nanocrystalline cellulose/*Eucommia ulmoides* gum nanocomposite film *Carbohydr. Polym.* **181** 825–32

[88] Benítez A and Walther 2017 Cellulose nanofibril nanopapers and bioinspired nanocomposites: a review to understand the mechanical property space *J. Mater. Chem.* A **5** 16003–24

[89] Bardet R *et al* 2015 Substitution of nanoclay in high gas barrier films of cellulose nanofibrils with cellulose nanocrystals and thermal treatment *Cellulose* **22** 1227–41

[90] Aulin C, Gällstedt M and Lindström T 2010 Oxygen and oil barrier properties of microfibrillated cellulose films and coatings *Cellulose* **17** 559–74

[91] Tyagi P, Lucia L A, Hubbe M A and Pal L 2019 Nanocellulose-based multilayer barrier coatings for gas, oil, and grease resistance *Carbohydr. Polym.* **206** 281–8

[92] Qin S *et al* 2019 Super gas barrier and fire resistance of nanoplatelet/nanofibril multilayer thin films *Adv. Mater. Interfaces* **6** 1801424

[93] HT A and Tajvidi M 2019 Sustainable barrier system via self-assembly of colloidal montmorillonite and cross-linking resins on nanocellulose interfaces *ACS Appl. Mater. Interfaces* **11** 1604–15

[94] Wang W, Yu Z L, Alsammarraie F K, Kong F B, Lin M S and Mustapha A 2020 Properties and antimicrobial activity of polyvinyl alcohol-modified bacterial nanocellulose packaging films incorporated with silver nanoparticles *Food Hydrocoll.* **100** 105411

[95] Jha R and Mayanovic R A 2023 A review of the preparation, characterization, and applications of chitosan nanoparticles in nanomedicine *Nanomaterials* **13** 1302

[96] Jannatyha N, Shojaee-Aliabadi S, Moslehishad M and Moradi E 2020 Comparing mechanical, barrier and antimicrobial properties of nanocellulose/CMC and nanochitosan/CMC composite films *Int. J. Biol. Macromol.* **164** 2323–8

[97] Chavan P *et al* 2022 Nanocomposite starch films: a new approach for biodegradable packaging materials *Starch* **74** 2100302

[98] Hubbe M A, Tyagi P and Pal L 2019 Nanopolysaccharides in barrier composites *Advanced Functional Materials from Nanopolysaccharides* (Singapore: Springer) pp 321–66

[99] Lin Q *et al* 2020 Fabrication of debranched starch nanoparticles via reverse emulsification for improvement of functional properties of corn starch films *Food Hydrocoll.* **104** 105760

[100] Jiang S, Liu C, Wang X, Xiong L and Sun Q 2016 Physicochemical properties of starch nanocomposite films enhanced by self-assembled potato starch nanoparticles *LWT—Food Sci. Technol.* **69** 251–7

[101] Silagy D, Demay Y and Agassant J F J P E 1996 Study of the stability of the film casting process *Polym. Eng. Sci.* **36** 2614–25

[102] Chokshi R and HJIjopr Z 2004 Hot-melt extrusion technique: a review *Iran. J. Pharm. Res.* **3** 3–16

[103] Chronakis 2005 Novel nanocomposites and nanoceramics based on polymer nanofibers using electrospinning process—a review *J. Mater. Process. Technol.* **167** 283–93

[104] Xu W H *et al* 2021 3D printing for polymer/particle-based processing: a review *Composites* B **223** 109102

[105] Ahari H *et al* 2022 Bio-nanocomposites as food packaging materials; the main production techniques and analytical parameters *Adv. Colloid Interface Sci.* **310** 102806

[106] Aydogdu A, Sumnu G and Sahin S 2019 Fabrication of gallic acid loaded hydroxypropyl methylcellulose nanofibers by electrospinning technique as active packaging material *Carbohydr. Polym.* **208** 241–50

[107] Feng S Y, Zhang F, Ahmed S and Liu Y W 2019 Physico-mechanical and antibacterial properties of Pla/TiO_2 composite materials synthesized via electrospinning and solution casting processes *Coatings* **9** 525

[108] Ligon S C, Liska R, Stampfl J, Gurr M and Mulhaupt R 2017 Polymers for 3D printing and customized additive manufacturing *Chem. Rev.* **117** 10212–90

[109] Kim Y H, Priyadarshi R, Kim J W, Kim J, Alekseev D G and Rhim J W 2022 3D-printed pectin/carboxymethyl cellulose/ZnO bio-inks: comparative analysis with the solution casting method *Polymers* **14** 4711

[110] Gaikwad K K, Singh S and Ajji A 2019 Moisture absorbers for food packaging applications *Environ. Chem. Lett.* **17** 609–28

[111] Wang J, Euring M, Ostendorf K and Zhang K 2022 Biobased materials for food packaging *J. Bioresour. Bioprod.* **7** 1–13

[112] Moeini A, Pedram P, Fattahi E, Cerruti P and Santagata G 2022 Edible polymers and secondary bioactive compounds for food packaging applications: antimicrobial, mechanical, and gas barrier properties *Polymers* **14** 2395

[113] Alamri M S *et al* 2021 Food packaging's materials: a food safety perspective *Saudi J. Biol. Sci.* **28** 4490–9

[114] Rhim J W 2007 Potential use of biopolymer-based nanocomposite films in food packaging applications *Food Sci. Biotechnol.* **16** 691–709

[115] Nešić A, Cabrera-Barjas G, Dimitrijević-Branković S, Davidović S, Radovanović N and Delattre C 2020 Prospect of polysaccharide-based materials as advanced food packaging *Molecules* **25** 135

[116] Vostrejs P *et al* 2020 Active biodegradable packaging films modified with grape seeds lignin *RSC Adv.* **10** 29202–13
[117] Wen Y Y *et al* 2021 Development of intelligent/active food packaging film based on TEMPO-oxidized bacterial cellulose containing thymol and anthocyanin-rich purple potato extract for shelf life extension of shrimp *Food Packag. Shelf Life* **29** 100709
[118] Jarvis K L *et al* 2017 Comparing three techniques to determine the water vapour transmission rates of polymers and barrier films *Surf. Interfaces* **9** 182–8
[119] Nehm F, Dollinger F, Fahlteich J, Klumbies H, Leo K and Muller-Meskamp L 2016 Importance of interface diffusion and climate in defect dominated moisture ultrabarrier applications *ACS Appl. Mater. Interfaces* **8** 19807–12
[120] Cai C and Dauskardt R H 2015 Nanoscale interfacial engineering for flexible barrier films *Nano Lett.* **15** 6751–5
[121] Ochirkhuyag N *et al* 2022 Stretchable gas barrier films using liquid metal toward a highly deformable battery *ACS Appl. Mater. Interfaces* **14** 48123–32
[122] Kim Y, Bulusu A, Giordano A J, Marder S R, Dauskardt R and Graham S 2012 Experimental study of interfacial fracture toughness in a $SiN_{(x)}$/PMMA barrier film *ACS Appl. Mater. Interfaces* **4** 6711–9
[123] Zhou H *et al* 2022 High-performance cellulose acetate-based gas barrier films via tailoring reduced graphene oxide nanosheets *Int. J. Biol. Macromol.* **209** 1450–6
[124] Lang A W, Ji Y, Dillon A C, Satam C C, Meredith J C and Reynolds J 2021 Photostability of ambient-processed, conjugated polymer electrochromic devices encapsulated by bioderived barrier films *ACS Sustain. Chem. Eng.* **9** 2937–45
[125] Bilgen S, Kaygusuz K and Sari A 2004 Renewable energy for a clean and sustainable future *Energy Sources* **26** 1119–29
[126] Herzog A V, Lipman T E, Edwards J L and Kammen D M 2001 Renewable energy: a viable choice *Environ. Sci. Policy Sustain. Devel.* **43** 8–20
[127] Kinjo T, Senjyu T, Urasaki N and Fujita H 2006 Output levelling of renewable energy by electric double-layer capacitor applied for energy storage system *IEEE Trans. Energy Convers.* **21** 221–7
[128] Amrouche S O, Rekioua D, Rekioua T and Bacha S 2016 Overview of energy storage in renewable energy systems *Int. J. Hydrogen Energy* **41** 20914–27
[129] Arantes A C C *et al* 2019 Bio-based thin films of cellulose nanofibrils and magnetite for potential application in green electronics *Carbohydr. Polym.* **207** 100–7
[130] Sustainable film capacitors. TDK Electronics AG. (2023, July 5). https://www.tdk-electronics.tdk.com/en/374108/tech-library/articles/products-technologies/products-technologies/more-sustainable-film-capacitors/3168412
[131] AliAkbari R *et al* 2021 High value add bio-based low-carbon materials: conversion processes and circular economy *J. Clean. Prod.* **293** 126101
[132] Rosenboom J G, Langer R and Traverso G 2022 Bioplastics for a circular economy *Nat. Rev. Mater.* **7** 117–37
[133] Plastics – The Facts 2022 • Plastics Europe. (2023, March 14). https://plasticseurope.org/knowledge-hub/plastics-the-facts-2022/
[134] Zhu Z, Liu W, Ye S and Batista L 2022 Packaging design for the circular economy: a systematic review *Sustain. Product. Consump.* **32** 817–32
[135] Wang J X, Emmerich L, Wu J F, Vana P and Zhang K 2021 Hydroplastic polymers as eco-friendly hydrosetting plastics *Nat. Sustain.* **4** 877–83

Chapter 4

Bio-based hydrogels in biomedical applications

Haodong Zhang and Jinping Zhou

4.1 Biopolymers and their properties for biomedical applications

Biopolymers, obtained from renewable resources such as plants, microbes, and animals, play key roles across all the major kingdoms of organisms. The biopolymers fall into three major types: polysaccharides, proteins, and polyesters. Table 4.1 provides an overview of the most typical bio-based materials discussed in this chapter. Among the materials that are made from biopolymers, hydrogels, which are three-dimensional (3D) networks of polymer chains containing significant amounts of water, display similarity to the tissue extracellular matrix (ECM), which is beneficial for biomedical applications [1]. Hydrogels have attractive physicochemical properties, including flexibility, porosity, and injectability. Therefore, bio-based hydrogels have gained attention in the biomedical field [2]. Although many developments have been made with the use of synthetic polymers or bio-based chemicals (e.g. sugar monomer and oligomers) to prepare hydrogels, the specific functionality of hydrogels often requires precise control over chemical structures, which requires high processing costs. In contrast, biopolymers consist of plenty of functional groups onto repeat units that allow for diverse methods of cross-linking regarding hydrogel formation. Accordingly, the formed hydrogels could be driven by the extent of biopolymer modification, the precursor concentration, the type of cross-linking strategies, and the crosslinking density. Notably, some unique properties such as self-healing ability and strong adhesion could also be derived through the elaborate design of physical/chemical structures. Moreover, these macromolecules also possess desired properties for biomedical applications, including abundant bioactive functional groups, high biocompatibility/biodegradability, and non-toxic degradation products [3].

The main goal of this chapter is to offer readers a concise overview of the utilization of biopolymers and their derivatives in fabricating hydrogels, focusing on design and preparation strategies, as well as surface properties in biomedical applications.

doi:10.1088/978-0-7503-6184-2ch4

Table 4.1. Sources, properties, and products of typical biopolymers.

Type	Polymer component	Natural resource/synthetic route	Properties in biomedical applications[a]	Commercial products[b]
Polysaccharides	Cellulose	Structural component of plant cell walls	No desired bioactivity, but with abundant hydroxyl groups for derivatization	SURGICEL® Hydrofiber®
	Chitin	Component of fungal cell walls and arthropod exoskeletons	N-acetyl groups for degradability *in vivo*	None
	Chitosan	Derived from chitin deacetylation	Positive amino groups for antibacterial property	HemCon™
	Alginate	Component of brown algae cell walls and biofilms	Easy and mild crosslinking method (i.e. with multivalent cations)	VIVAPHARM®
	Xylan	Hemicellulose extracted from plants and algae	Targeted degradation by enzymes produced by colon bacteria	Novozymes®
	Hyaluronic acid	Component of connective, epithelial, and neural tissues	Lubricity and moisture retention ability	Deflux™
	Chondroitin sulfate	Component of cartilage and connective tissues	High sulfation degrees for anti-coagulant and antithrombotic activities	Droi-Kon®
	Heparin sulfate	Component of mast cells and basophil secretory granules		Clotclear™
Proteins	Silk fibroin	Structural protein of silk fibers	The formation of β-sheets for tunable mechanical properties	SERI®
	Collagen	Structural protein in animal connective tissues	Unique triple helical structure and the close parallel alignment for structural stability	Integra® AlloDerm®
	Gelatin	Derived from collagen hydrolysis	RGD sequence (Arg-Gly-Asp) for cell adhesion	Quali-Pure™
	Albumin	Plasma protein produced in the liver	Endogenous protein with minimal immunogenicity	Recombumin®
	Elastin	Structural protein in connective tissues	Multiple hydrophobic domains for unique resilient behavior	BioCoat™
	Keratin	Structural protein in hair, nails, horns, etc	Abundant cell adhesion sites and cysteine content for formation of disulfide bonds	Keragel®
Others	Lignin	Components of lignocellulosic biomass	Abundant polyphenolic structure for antioxidant activity	LignoBase™
	Cyclodextrins	Metabolic product from starch	Ring structure with a hydrophobic interior for drug loading	KLEPTOSE®

[a] All bio-based polymers have biocompatibility and biodegradability, not to be repeated here.
[b] Approval from the Food and Drug Administration (FDA).

4.1.1 Polysaccharides

Polysaccharides are carbohydrate polymers formed of repeating monosaccharide units bound together by glycosidic bonds. They possess excellent characteristics, such as biocompatibility, biodegradability, hydrophilicity, and easy processing, for biomedical applications [4]. Currently, as presented in figure 4.1, lots of polysaccharides and their derivatives (e.g. cellulose, chitin, chitosan (CS), alginate (Alg), xylan, hyaluronic acid (HA), chondroitin sulfate (ChS), heparin sulfate, and so on) have been utilized in wound healing, tissue engineering, and drug delivery [5].

Cellulose, the most abundant natural polysaccharide and the main structural component of plant cell walls, is composed of D-glucose units linked by β-1,4-glycosidic bonds. The hydroxyl (–OH) groups on cellulose are responsible for its excellent hydrophilicity while forming intra- and intermolecular hydrogen bonding, thereby it cannot dissolve in common solvents. Thus, suitable solvent systems for cellulose have been developed, including NaOH/carbon disulfide (CS_2), cuprammonium, N-methyl-morphine-N-oxide (NMMO), lithium chloride/dimethylacetamide (LiCl/DMAc), ionic liquids (ILs), and alkali/urea aqueous solutions [6]. Subsequently, with the reconstitution of hydrogen bonding under specified conditions (e.g. acid or salt solution), cellulose can form into a hydrogel with a 3D network structure, which could be used in wound dressings, tissue scaffolds, and drug delivery systems [7]. The high hydrophilicity and mechanical properties of cellulosic hydrogels enable the absorption of exudates and integration into tissues [8]. However, cellulose does not have any desired bioactivity for humans, and occasionally causes unnecessary complement and platelet activation [9]. With abundant reactive –OH groups, cellulose can be chemically modified into various functional derivatives, such as carboxymethyl cellulose (CMC), quaternized cellulose (QC), and sulfated cellulose (SC) [10]. Because the most of –OH groups are substituted (no distinct hydrogen bonding), these derivatives can form into hydrogels assisted by chemical crosslinker or copolymerization [11]. By interacting with proteins and

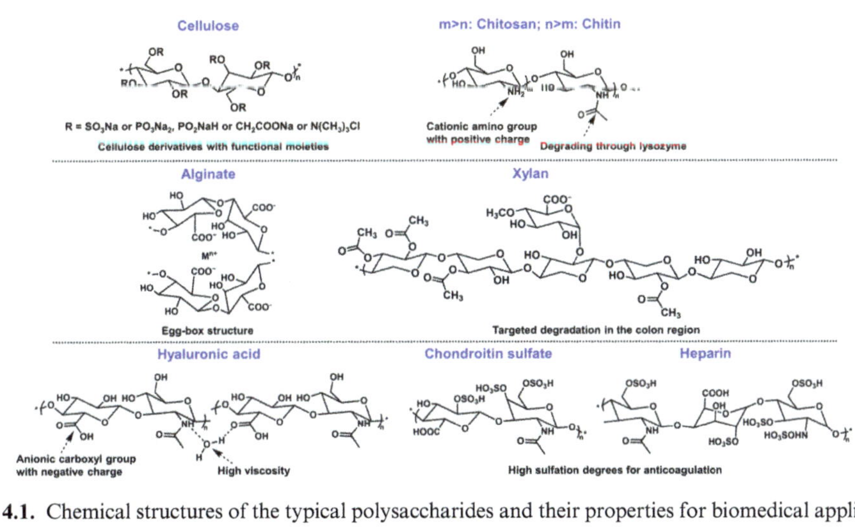

Figure 4.1. Chemical structures of the typical polysaccharides and their properties for biomedical applications.

regulating cell behavior, ionic cellulose derivatives show immense potential for biomedical applications. For example, the SC could accelerate the inhibition of antithrombin III (AT-III) on coagulation factors FIIa and FXa in plasma, resulting in anticoagulation activity [12]. Moreover, cellulose nanocrystals (CNCs) and cellulose nanofibers (CNFs) can be prepared by 'top-down' approaches such as mechanochemical methods, TEMPO-oxidization, and acid treatment. Benefiting from their high aspect ratio and crystallinity, CNCs/CNFs are often used as nanofiller to improve the mechanical properties of bio-based hydrogels and enhance the desirable properties for target applications [13].

Chitin, the second most abundant natural polysaccharide and the main structural component of fungal cell walls and crustacean shells, consists of β-1,4-linked N-acetyl-D-glucosamine units. Similar to cellulose, chitin is insoluble in common solvents due to the strong intermolecular hydrogen bonding. However, it can be converted to the more deacetylated derivative CS through alkaline hydrolysis, which is soluble in dilute acidic solutions. CS is composed of D-glucosamine and N-acetyl-D-glucosamine units linked by β-1,4-glycosidic bonds. The primary amino ($-NH_2$) groups on CS enable pH-dependent solubility and electrostatic interactions with negatively charged molecules, which are well situated to form hydrogel networks [14]. Additionally, the cationic nature of CS enables binding to negatively charged bacteria/cell surfaces. These properties make CS bacteria resistant, biocompatible, and useful for biomedical applications such as tissue engineering scaffolds, wound dressings, drug delivery carriers, and gene therapy vectors [15]. Furthermore, with the emergence of dissolution systems (i.e. LiCl/DMAc, ILs, $CaCl_2$/methanol, and alkali/urea aqueous solutions), chitin and its derivatives have been developed [16]. The hydrogel can be prepared by hydrogen bonding or crosslinker similar to cellulose. Unlike CS, chitin will degrade *in vivo* because lysozyme can interact with the N-acetyl groups. Therefore, chitin can be better applied in bone regeneration without requiring secondary surgery [17].

Alg as an anionic polysaccharide derived from brown algae, consists of (1,4)-linked β-D-mannuronic acid (M) and α-L-guluronic acid (G) monomers. Divalent cations such as Ca^{2+} ions can interact with G blocks in Alg to form hydrogels. This biocompatible gelation method can be used under mild, aqueous conditions, which is suitable for cell culture and drug delivery [17]. Furthermore, the mechanical properties and degradation rates of Alg-based materials can be controlled by monomer sequence distribution and molecular weight [18]. Nowadays, Alg hydrogels and microspheres have been applied in wound dressings and for delivery of various pharmaceutically active agents [19]. Additionally, injectable Alg hydrogels can be used for cell transplantation and 3D printing for tissue regeneration [20].

Xylan as the main hemicellulose, consists of xylose residue onto backbone as characteristic feature [21]. It is mainly found in the plant cell walls (7%–30% of the dry weight of wood and annual plants). The structure of xylan can be varied with the degree of polymerization/substitution and the type of substitution molecule. In particular, xylan is a biopolymer that cannot be digested in the stomach and intestine, while being degradable by bacteria in the human colonic microflora. Therefore, xylan can be used for colon-targeted drug delivery systems [22].

The xylan-based hydrogel as a robust carrier, can be prepared by physical cross-linking (using sodium trimetaphosphate), achieving controlled drug release in the colon region [23].

In addition to naturally sourced polysaccharide mentioned above, glycosaminoglycans (GAGs) as native components of the ECM, involved in many essential biological cell processes [24]. One common characteristic of GAGs is the presence of distinct disaccharide units, which consist of either N-acetylglucosamine (GlcNAc) or N-acetylgalactosamine (GalNAc) combined with a uronic ring such as glucuronic acid (GlcA) or iduronic acid (IdoA). HA has a high molecular weight GAG, is composed of repeating disaccharide units of D-glucuronic acid and N-acetyl-D-glucosamine linked via alternating β-1,3 and β-1,4 glycosidic bonds. As a major component of the connective, epithelial, and neural tissues, HA can interact with cell surface receptors such as clusters of differentiation 44 (CD44) receptors to facilitate cell proliferation, migration, and adhesion [25]. The carboxyl (–COOH) and acetyl groups of HA provide lots of hydrogen bonds with H_2O, conferring high viscosity and elasticity [26]. HA can be degraded by oxidative species or enzymes *in vivo* due to the presence of N-acetyl groups in its polymeric structure. It has been used for many biomedical applications including dermal fillers for soft-tissue augmentation, wound dressings, and intra-articular injections to manage symptoms of osteoarthritis [27]. Moreover, chemical modifications such as methacryloyl-grafting and oxidation to dialdehyde have been introduced to provide crosslinking points for hydrogel formation [28].

ChS as a sulfated GAG, is composed of 40–100 repeating disaccharide units of β-1,3-linked N-acetylgalactosamine and glucuronic acid connected by β-1,4 glycosidic bonds. ChS has been used as a nutritional supplement or nonprescription drug for osteoarthritis. ChS-based materials play a vital role in articular cartilage injury by modulating tissue hydration, swelling, and lubrication, benefiting from its anti-inflammatory properties and negative surface charges [29]. Moreover, ChS shows high affinity for CD44 receptors and glycosylation enzymes, and it degrades under the condition of physiological stimuli, hyaluronidase enzyme or reactive oxygen species. This makes it an ideal biomaterial for targeted drug delivery and tissue engineering [30].

Heparin, also a sulfated GAG, is composed of repeating disaccharide units of uronic acid (D-glucuronic or L-iduronic acid) and D-glucosamine linked by β-1,4 glycosidic bonds. Found on cell surfaces and in the ECM, heparin plays critical roles in blood-contacting medical devices [31]. Benefiting from the high negative charges, heparin and its sulfated derivative can interact with thrombin and factor X for anticoagulation or inflammatory mediators for angiogenesis [32]. The quality and application of heparin depend on its molecular weight. Low or ultra-low molecular weight heparin is known as an anticoagulant to prevent blood clot formation. Conversely, higher molecular weight heparin finds used in tissue engineering scaffolds due to its high affinity for endothelial cell adhesion and low degradation rate [33].

In summary, polysaccharides possess numerous desired properties such as low-cost, biodegradability, biocompatibility, and hydrophilicity, which make them essential for biomedical applications. Particularly, some polysaccharides and their derivatives can interact with different components such as growth factors, glycoprotein receptor, active proteins, and endogenous mediators. Therefore, they will indirectly participate in cell behavior regulation, inflammatory response, and coagulation cascade *in vivo*. Of note, owing to the interactions derived from their unique physicochemical structures, various types of polysaccharide-based platforms, especially hydrogels, have been developed using a range of crosslinking methods. These platforms have been utilized for drug delivery, tissue engineering, wound healing, and regenerative medicine.

4.1.2 Proteins

Proteins are biopolymers comprised of repeating amino acid units that serve as fundamental building blocks of life, playing essential roles in the structural integrity and functional processes of organisms. For example, they can act as enzymes to facilitate chemical reactions, hemoglobin to transports oxygen in the blood, antibodies to combat infections, and hormones such as insulin to regulate blood sugar levels. Benefiting from their unique cell interaction sites and similar functions to the ECM, a variety of proteins has been processed into materials for biomedical applications. Key protein-based biopolymers have gained attention including silk fibroin, collagen, gelatin, albumin, elastin, keratin, and so on (figure 4.2).

Silk fibroin (SF), a protein extracted from silkworm cocoons, consists of repeating amino acid sequences including glycine, alanine, and serine. It can form antiparallel

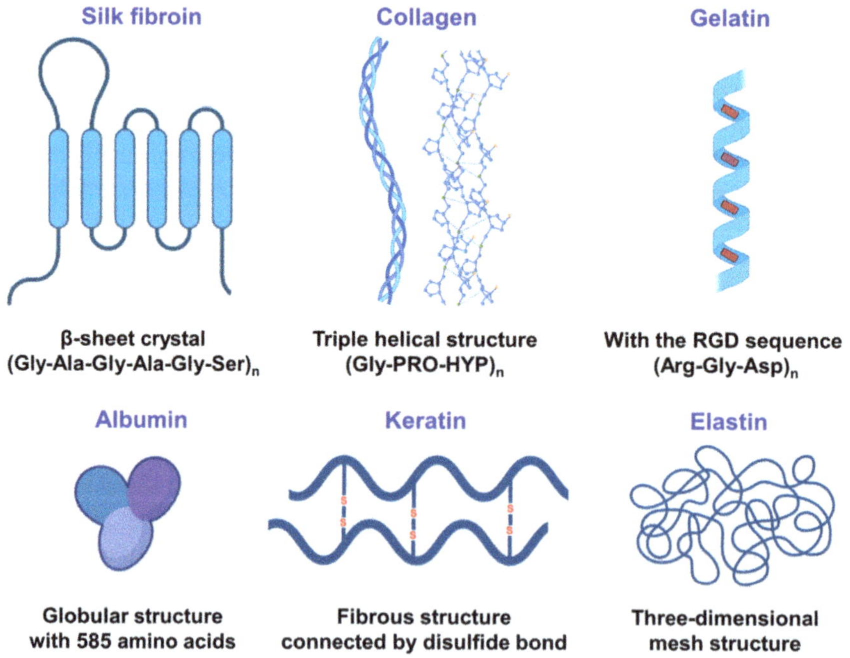

Figure 4.2. Illustration of the chemical structure of typical proteins.

β-sheet crystals connected by amorphous regions (around 65% crystalline region and 35% amorphous), resulting in superior mechanical properties. In addition, SF supports cell attachment/growth and can be degraded by proteases without immune rejection [34]. Moreover, in order to retain the original fibrillar structure of natural silk, various techniques (bottom-up self-assembly and top-down disintegration) have been developed to generate silk nanofibrils (SNFs) [35]. On this basis, SF and SNFs have been processed into high toughness hydrogels for biomedical applications. However, native SF hydrogels are still undesirable for cartilage tissue engineering as well as having a high enzymatic degradation rate and low cell adhesion. Silk proteins usually need to be combined with natural or synthetic polymers to optimize hydrogels via physical and chemical cross-linking methods. Some of SF-based products including surgical scaffolds and injectable implants are already commercialized [36].

Collagen, a structural protein that provides biological integrity in connective tissues, consists of three polypeptide chains and a characteristic triple helical structure stabilized by hydrogen bonding. There are at least 29 types of collagens with varying chain compositions, biological roles, and distributions in the body. Collagens show high biocompatibility (after rigorous isolation and purification), biodegradability, and excellent binding capacity with cell transmembrane receptors [37]. Collagen-based hydrogels are often used as wound dressings and skin substitutes. However, due to the absence of covalent crosslinking, the actual mechanical properties of biomaterials from the extracted collagen are undesirable. At present, combined with physical treatment, structural modification, and chemical crosslinking, decellularized ECM and regenerated collagen biomaterials are used extensively for tissue regeneration [38].

Gelatin, a product of the alkaline hydrolysis of collagen, is composed of polypeptide chains in a single-stranded, non-helical conformation. The chemical structure of gelatin remains similar to collagen, but the RGD sequence (Arg-Gly-Asp), i.e. a well-known cell adhesion site and integrin ligand, is exposed. Thus gelatin-based biopolymers have been introduced to improve cell adhesion in many applications [39]. However, they showed poor mechanical properties and an uncontrollable degradation rate. Gelatin methacryloyl (GelMA) has been demonstrated as an effective strategy to overcome these obstacles, and can be tailored via chemical crosslinking or compositing with other polymers. To date, GelMA-based biomaterials have been widely studied in hemostatic agents and cartilage/bone defect regeneration [40].

Albumin, an endogenous protein, is predominantly produced in the liver and secreted into the blood plasma. Within the category of albumin, human serum albumin (HSA), a globular protein composed of 585 amino acids, is the most common protein in blood plasma, making up approximately 50%–60% of the total plasma protein content. It is not only the main transporter of molecules in the circulatory system, but also can regulate fluid distribution through colloid osmotic pressure [41]. Albumin-based hydrogels can bind with bioactive compounds, recruit stem cells, assimilate nutrients, and facilitate cell growth and proliferation [42]. Therefore, they are frequently utilized in drug delivery and tissue engineering.

HSA has been utilized as a plasma volume expander to restore and maintain circulating blood volume after trauma, surgery, and hemorrhage [43]. In addition, bovine serum albumin has also garnered interest as a more abundant and cost-effective alternative protein source.

Keratin, the structural component of skin, hair, and nails, is a fibrous, cysteine-rich protein. Keratins are classified as α-keratins (α-helical) and β-keratins (β-sheet). The high cysteine content enables disulfide bond formation, conferring excellent mechanical properties to keratinous tissues. With abundant cell adhesion sites such as the RGD sequence within the structure, keratin is gaining interest as a sustainable, cost-effective raw material [44]. Breaking disulfide bonds by chemical reduction is widely used to generate free thiol (–SH) groups on keratin, which provide antioxidant properties and oxidation responsiveness, as well as opportunities for further crosslinking or functionalization [45]. Moreover, keratin could serve as a ligand (or perhaps a pseudo-ligand) for cell adhesion receptors, which can promote cell proliferation and adhesion. Particularly, keratin has lots of cell motifs similar to the proteins in the ECM. Therefore, keratin-based hydrogels have been used in wound healing through highly simulating the physiological structure and function of skin tissues [46].

Elastin, a vital ECM structural component, provides resilience and elasticity in connective tissues. It is extremely stable, with an approximately 70 year half-life period [47]. Elastin exhibits characteristic α-helical and β-turn conformations that contribute to its mechanical properties, while also providing sites for cells adhesion through receptors and integrins [48]. Up to now, various elastin derivatives have been developed such as α-elastin (a water-soluble derivative solubilized with oxalic acid), tropoelastin (water-soluble at low temperature), and elastin-like polypeptides (with hydrophobic motif Val-Pro-Gly-X-Gly) [49]. Elastin-based biopolymers can be crosslinked by chemical crosslinking (such as disuccinimidyl suberate and disuccinimidyl glutarate) to form hydrogel [50]. Due to the unique resilient behavior of polypeptides, elastin-based hydrogels showed great potential as scaffolds for tissue engineering applications.

In comparison to polysaccharides, proteins typically contain hydrophobic domains, which contribute to the formation of stable crystalline structures but also result in their insolubility in water. Moreover, the high risk of transmitting animal diseases and expensive processes of isolation and purification significantly limit their commercial applications. Nonetheless, proteins exhibit abundant cell adhesion sites and excellent degradability, and can be fabricated as nanoparticles, nanofibers, hydrogels, and sponges for further biomedical applications.

4.1.3 Others

In addition to polysaccharides and proteins discussed above, some of natural polymers such as cyclodextrins (CDs) and lignin also can be as fundamental building blocks for hydrogel fabrication.

CDs as cyclic oligosaccharides, consist of α-1,4-linked glucopyranose units. In the CDs family, they are differentiated by the number of glucoses repeat units. The most common species are α-, β-, and γ-CDs, which contain 6, 7, and 8 glucose repeat units, respectively [51]. The ring structure of CDs gives them a hydrophilic outer edge and hydrophobic interior cavity. This unique structure allows CDs to form non-covalent inclusion complexes through host–guest interaction (including van der Waals force, Coulomb force, hydrophobic interaction, and hydrogen bonding) [52]. Guest molecules with appropriate polarity and size can embed within the CD cavity, imparting protection from degradation, controlled release, increased solubility, and stimuli-responsive capabilities. Due to these molecular encapsulation and delivery capabilities, CDs have been applied for drug delivery in a large-scale [53]. Moreover, CDs can be threaded onto polymer chains with diverse guest polymers, leading to hydrogels with a necklace-like supramolecular structure [54]. For instance, Li *et al* developed supramolecular hydrogels by combining CD with high molecular weight polyethylene glycol (PEG) in an aqueous solution [55]. In this system, the terminals of the PEG chain were capable of threading through the inner cavities of CD molecules. Consequently, CD rings could link together via hydrogen bonding, forming stoichiometric CD/PEG polypseudorotaxanes. As the hydroxyl groups on CD were consumed, they transitioned to a hydrophobic state. This induced aggregation of the α-CD/PEG complexes, acting as physical cross-linkers and facilitating the formation of supramolecular hydrogels. Remarkably, these hydrogels exhibited thixotropic behavior and reversibility: under shear force, their viscosity decreased significantly, but upon cessation of the force, they reverted to their original viscosity within hours. This unique property paved the way for the development of an injectable drug delivery system [56].

Lignin as a phenolic polymer derived from plant cell wall, consists of three monomeric units including syringyl (S), guiaiacyl (G) and p-hydroxy-phenyl (H), which are linked by C–O–C/–C(O)–O–C, and C–C bonds [57]. Lignin exhibits prominent antioxidant activity due to the presence of a polyphenolic structure, which can eliminate excess radicals by breaking O–H bonds [58]. In addition, lignin is also an effective microbicide that facilitates the fracture of bacterial membrane and inhibits viral replication [59]. The abundant aromatic groups in the lignin structure also endow it with anti-ultraviolet activity [60]. Moreover, the abundant functional hydrophilic and active groups, such as phenolic hydroxyls, carboxyls and methoxyls, have great potential to be applied in the preparation of hydrogels [61]. Interestingly, lignin has been also employed as an effective catalyst in hydrogel formation. It can participate in a dynamic redox reaction, and rapidly activate ammonium per sulfate (oxidant), which can trigger the polymerization of acrylic acid (AA), forming multifunctional polyacrylic acid (PAA) hydrogels [62].

4.2 Crosslinking mechanisms for bio-based hydrogels

There are two crosslinking mechanisms for the design and preparation of bio-based hydrogels. (i) *Physical crosslinking*: Physically crosslinked hydrogels are formed through non-covalent interactions such as ionic crosslinking, coordinative crosslinking, hydrogen bonding, hydrophobic association, cation–π or π–π interaction, electrostatic interaction, host–guest interactions, and chain entanglement. The physical crosslinks are reversible and can be broken and re-formed dynamically.

Table 4.2. The crosslinking mechanisms and components for bio-based hydrogels.

Mechanism	Mode	Components	References
Physical crosslinking	Ionic crosslinking	Alg/Ca^{2+}	[64]
	Coordinative crosslinking	QC/β-GP	[66]
	Hydrogen bonding	Adenine-modified CS	[68]
	Chain entanglement	Cellulose/ethanol	[70]
	Host–guest interactions	Adamantane-modified cellulose	[77]
	Cation–π interaction	Nucleosides/CS	[72]
	Hydrophobic association	Lipid-grafted peptide/sulfated polysaccharide	[73]
		Gelatin/$(NH_4)_2SO_4$	[75]
Chemical crosslinking	Introducing chemical agents	GA	[77]
		ECH	[78]
	Click chemistry	Amino-yne	[81]
		Thiol–ene	[82]
	Free radical polymerization	GelMA/methacryloyl-grafted peptides	[83]
	Enzymatic reaction	Tyrosinase induced EGCG/CS	[85]
	Dynamic crosslinking	Boric acid ester	[86]
		Schiff base crosslinking	[87]

(ii) *Chemical crosslinking*: Chemical crosslinking is achieved by introducing chemical components to create covalent bonds between polymer chains, such as click chemistry, free radical polymerization, enzymatic crosslinking, and dynamic crosslinking (e.g. imine bonds, acylhydrazone bonds, disulfide bonds, and borate ester bonds). Chemically crosslinked bio-based hydrogels exhibit improved mechanical stability. However, the potential presence of unreacted crosslinking components may pose toxicity risks to cells and tissues. Specifically, to improve the mechanical strength or introduce some unique properties of bio-based hydrogels, the two crosslinking mechanisms always come in pairs in practical application, i.e. hybrid crosslinking. Several examples for bio-based hydrogels based on these mechanisms are listed in table 4.2.

4.2.1 Physical crosslinking

Physically crosslinked bio-based hydrogels utilize non-covalent interactions between polymer chains to form network structures, as shown in figure 4.3. The reversible physical interactions endow them with unique properties and do not require the use of potentially harmful reagents, and they can thus provide cytocompatible microenvironment similar to ECM.

Direct ionic crosslinking between oppositely charged biopolymers represent the simplest method for preparing hydrogels. For example, Jing *et al* [63] prepared a

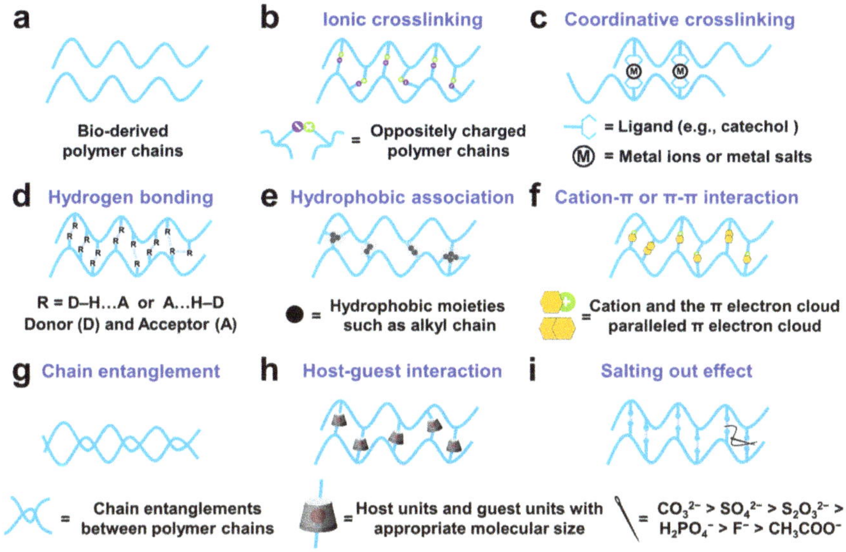

Figure 4.3. The typical physical crosslinking of (a) bio-derived polymer chains for the preparation of bio-based hydrogels according to different types of interaction: (b) ionic crosslinking, (c) coordinative crosslinking, (d) hydrogen bonding, (e) hydrophobic association, (f) cation–π or π–π interaction, (g) chain entanglement, (h) host–guest interactions, and (i) the salting out effect.

pH-responsive carboxymethyl CS (CMCS)/Alg composite hydrogel as protein carrier for oral delivery. In addition, Ma *et al* [64] used hydrogel microspheres based on Alg/Ca^{2+} as a protein drug carrier to realize oral administration. It has been proved that Alg/Ca^{2+} encapsulation can not only protect proteins from enzyme degradation, but also promote gastrointestinal tract penetration and absorption. Moreover, bio-based hydrogels can be prepared through coordinative crosslinking between polymer chains consisting of chelating ligands (e.g. bisphosphonate, catechol, and histidine) and low molecular components (e.g. metal ions or metal salts) [65]. You *et al* [66] prepared a thermosensitive hydrogel through coordinative crosslinking between –NH_4^+ on QC and the PO_4^{3-} on β-glycerophosphate (β-GP), with the sol–gel transition temperature/time depending on various factors such as QC and β-GP concentrations, molecular weight, degree of substitution (DS) of QC, and solvent medium. This bio-hydrogel remained stable under physiological conditions and could be utilized as a drug delivery carrier or *in situ* gel-forming material for tissue engineering. Xie *et al* [67] prepared a self-healing hydrogel dressing through the crosslinking between precoordinated europium-ethylenediaminetetra-acetic acid complexes and CMC. As the Eu^{3+} ions had pH-sensitive fluorometric properties, the hydrogel dressing showed the ability to promote angiogenesis and real-time pH monitoring for chronic diabetic wounds.

Hydrogen bonding as another interaction can be used to form hydrogels by combining biopolymers containing hydrogen acceptors (such as –NH_4^+ groups) and polymers containing hydrogen donors (such as –OH, –NH_2, and –COOH groups).

Ding *et al* [68] used the aqueous alkali/urea system as solvent to fabricate regenerated CS hydrogels. The hydrogen bonding between the $-NH_2$ and –OH groups of CS chains facilitated adhesion and proliferation of bone mesenchymal stem cells (BMSCs). Deng *et al* [69] synthesized adenine-modified CS using the same alkali/urea aqueous system as mentioned. This derivative could form a hydrogel through intra- and intermolecular hydrogen bonding in the presence of adenine (hydrogen bond donor and acceptor) and CS (containing numerous –OH and $-NH_2$ groups), which accelerated wound healing. Except for intrinsic hydrogen bonding, Zhao *et al* [70] discovered that neutralization in an ethanol aqueous solution induced the formation of the crystallite hydrate structure of cellulose II, thereby providing high mechanical strength of hydrogels for tissue scaffolds.

In addition to conventional methods, special forms of crosslinking can be utilized. Host–guest interactions between adamantane-grafted cellulose and CD-grafted glycerol ethoxylate have been employed as a drug delivery carrier [71]. π–π stacking and cation–π interactions between the aromatic rings of nucleosides and protonated $-NH_3^+$ groups of CS have been used as a cell growth platform [72].

Furthermore, amphiphilic biopolymers can form hydrogels through hydrophobic association. Liu *et al* [73] developed a non-covalent hydrogel by using a lipid-grafted peptide and a sulfated polysaccharide. The hydrophobic associations among lipid chains contributed to structural stability, interaction with hydrophobic drugs, improved drug loading efficiency, and sustained release. Remarkably, it was shown that the addition of anions enhanced hydrophobic interactions by removing the hydration water of proteins and causing their folding and precipitation. This phenomenon aligns with the Hofmeister sequence of ions ($CO_3^{2-} > SO_4^{2-} > S_2O_3^{2-} > H_2PO_4^- > F^- > CH_3COO^- > Cl^- > Br^- > NO_3^- > I^- > ClO_4^- > SCN^-$) [74]. He *et al* [75] prepared a gelatin hydrogel through simple soaking in an ammonium sulfate solution, utilizing the salting-out effect to enhance hydrophobic interactions and chain bundling, thereby improving mechanical properties and scaffold capabilities for tissue engineering.

Significantly, the double physical crosslinked method is a common strategy to improve the physicochemical properties of bio-based hydrogel. Guo *et al* [76] employed both the ureido–pyrimidinone hydrogen bonding and coordinative cross-linking between Fe^{3+} ions and catechol groups. The obtained double physical crosslinked hydrogels showed fast self-healing properties and high mechanical strength, making them ideal for wound dressing requiring complex motion.

4.2.2 Chemical crosslinking

Most of the previously discussed physical crosslinked hydrogels exhibit poor mechanical strength (as tissue scaffold) and structure stability (in terms of external ion concentration and pH). Therefore, various approaches have been investigated to prepare crosslinked bio-based hydrogels through strong covalent bonds (figure 4.4).

The most commonly used strategy for covalent crosslinking involves the introduction of chemical crosslinking agents such as glutaraldehyde (GA), poly (ethylene glycol diglyceryl ether) (PEGDE), genipin, epichlorohydrin (ECH), and N,

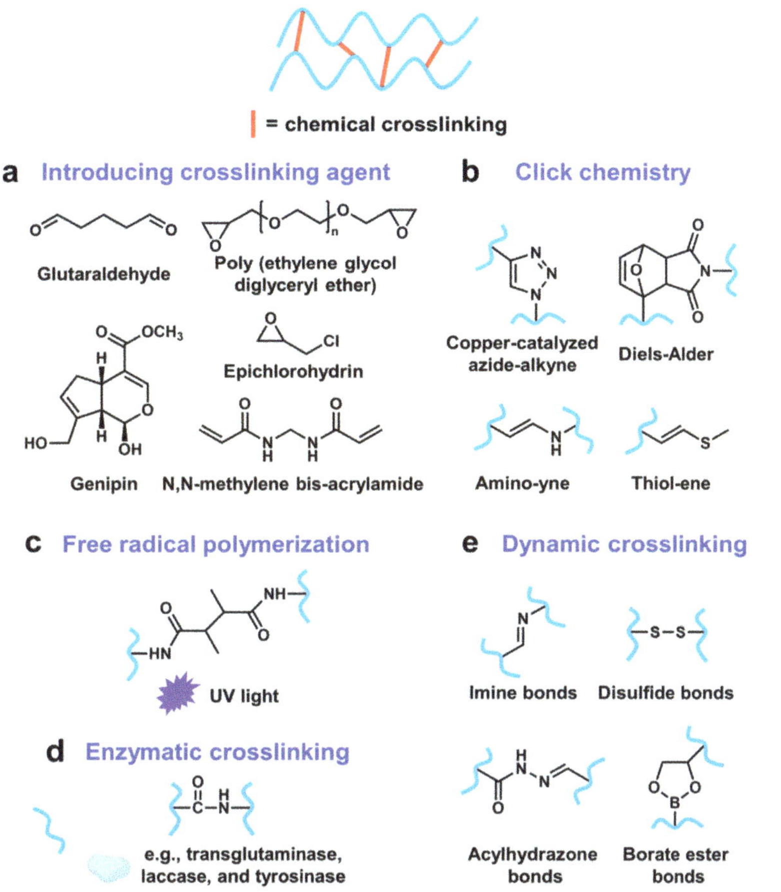

Figure 4.4. The typical chemical crosslinking of bio-based hydrogels: (a) introducing crosslinking agent, (b) click chemistry, (c) free radical polymerization, (d) enzymatic crosslinking, and (e) dynamic crosslinking.

N-methylene bis-acrylamide (MBAA). Ding *et al* [77] incorporated GA-crosslinked CS microspheres into GelMA hydrogel. By adjusting the dosage of GA, the hydrogels were able to exhibit injectability and softness, rendering them suitable for irregular and non-compressible wounds. Moreover, Chang *et al* [78] used ECH as chemical crosslinker in NaOH/urea aqueous solution to fabricate cellulose hydrogels. Owing to the basicity of the solvent, the chlorine and epoxy groups on ECH can react with –OH groups on cellulose chains via Williamson etherification and the alkali-catalyzed oxalkylation. It was expected that the covalent bonds formed from ECH would improve the mechanical properties by efficiently dispersing the stress upon cracking into pieces, However, the increase in the chemical cross-link density of the hydrogel network will lead to more brittle properties. Moreover, the molecular weight and chain length of chemical cross-linkers also have certain influence on mechanical properties. The network containing short chains is brittle due to small extensibility, while those containing long chains have good extensibility, but low ultimate strength. Therefore, keeping a balance will make the material suitable for tissue engineering applications that require high strength. Ye *et al* [79] proposed a new type of chemically crosslinked hydrogels by combining long-chain and short-chain linking. The breaking of the short-chain crosslinking within the networks efficiently dissipated the mechanical energy, and the extensibility of the relatively long-chain crosslinking retained the elasticity of hydrogel, ultimately leading to the excellent strength and toughness.

However, the toxicity of maintaining unreacted crosslinkers in the hydrogel is a strong argument against this strategy regarding biocompatibility, limiting the application of this method in the biomedical field. As a result, various biocompatible chemical crosslinking methods, including click chemistry, free radical crosslinking, enzymatic crosslinking, and dynamic crosslinking, have been developed. Click chemistry, known for its biorthogonal reactions with high yields, stereospecificity, and benign by-products, can crosslink and functionalize biopolymers under physiological conditions. Copper-catalyzed azide–alkyne, Diels–Alder, thiol–ene, and amino-yne reactions have been extensively investigated for bio-based hydrogels. Nonetheless, because most biopolymers lack specific reactive groups for click reactions, this method often requires prefunctionalization of the bio-polymer. Hu *et al* [80] synthesized allyl cellulose using an aqueous NaOH/urea solution and then utilized a thiol–ene reaction to prepare novel fluorescent cellulose derivatives, which are expected to be used as biological monitoring sensors. Huang *et al* [81] developed an injectable and degradable pH-responsive hydrogel by using an amino-yne click reaction between alkyne group-modified polyethylene glycol (PEG) and CMCS. Due to the pH sensitivity of the enamine bonds formed during the crosslinking reaction, the hydrogel exhibited the ability to undergo pH-induced sol–gel transition, enabling pH-controlled drug release *in vivo*. In the study by Lee *et al* [82], to achieve sustained-release delivery, epidermal growth factor was modified with an acrylate group and subsequently grafted onto the thiolated HA hydrogel matrix via thiol–ene reaction. The release behavior of EGF depended on the degradation rate of the HA hydrogel, breaking the diffusion-based limitations.

Moreover, the introduction of polymerizable C=C groups onto the molecular chain can also be used to induce free radical polymerization for hydrogel formation. Qiao *et al* [83] reported the preparation of a photo crosslinked polypeptide hydrogel by co-crosslinking GelMA with methacryloyl-grafted peptides under UV light. Thanks to the negligible activity loss during the polypeptide modification process, the hydrogel retained its ability to promote cell adhesion, enhance gene expression, and accelerate bone regeneration.

In order to achieve milder and safer reaction conditions, an enzymatic reaction is developed for the preparation of bio-based hydrogel [84]. Studies have demonstrated that polyphenol oxidases (such as laccase and tyrosinase) and peroxidases (such as horseradish peroxidase) exhibit crosslinking effects, forming peptide bonds within and between protein molecules, thereby resulting in the crosslinking of proteins and other materials. Kim *et al* [85] used tyrosinase to induce an oxidative reaction between the phenolic rings of epigallocatechin gallate and the $-NH_2$ groups on CS.

The resulting hydrogels could maintain intrinsic antibacterial and antioxidant effects, which greatly facilitated skin regeneration.

Furthermore, benefiting from the inherent cis-diol groups on Alg, $-NH_2$ groups on CS or gelatin, and –SH groups on proteins, constructing boric acid ester, acylhydrazone, imide, and disulfide bonds have also been identified as an effective strategy for preparing bio-based hydrogels. These dynamic covalent bonds can enable the preparation of a self-healing bio-based hydrogel. Deng *et al* [86] utilized the catechol structure of lignin and combined it with phenylboric acid modified hydroxypropyl cellulose to form a dynamic borate ester bond. The hydrogels exhibited self-healing and shape-adaptive properties, which could significantly enhance wound healing. A Schiff base crosslinked injectable hydrogel was prepared using gelatin, CMCS, and oxidized sodium Alg [87]. Due to its fast gelation rate, adhesion ability, and drug loading capacity, the hydrogel could be used as *in situ* tissue adhesives and hemostatic material. Jiang *et al* [88] fabricated injectable cellulose-based hydrogels using dynamic acylhydrazone bonds between carboxyethyl cellulose-grafted adipic dihydrazide and ethyl-1-adamantane 4-formylbenzoate, demonstrating pH responsiveness and hydrolysis resistance.

4.2.3 Hybrid crosslinking

As discussed above, the two types of crosslinking mechanisms have advantages and disadvantages. It is proposed that they be combined in practical applications. This approach can not only improve the physicochemical properties, especially mechanical strength, but also introduce some unique properties (e.g. self-healing, non-swelling, shape memory abilities) of bio-based hydrogels.

On this basis, Cai *et al* [89] developed a sequential chemical and physical crosslinking method to fabricate double-cross-linked hydrogels based on crystalline polysaccharides. ECH as the chemical crosslinker could form chemical crosslinking. Subsequent neutralization in an aqueous ethanol solution led to physically cross-linked domains through hydrogen bonding, hydrophobic interactions, and crystalline hydrates formation. The incorporation of chemically crosslinked domains and physically crosslinked domains could largely improve the mechanical properties of the polysaccharide-based hydrogels, which exceed those cellulose hydrogels with single chemical crosslinking and physical crosslinking. Deng *et al* [90] unitized the electrostatic interaction between the $-NH_2$ groups of CS (alkaline polysaccharide) and phenylboric acid groups (weak acid molecule) of phenylboric acid-modified hydroxypropyl chitosan, as well as the dynamic borate bond between phenylboric acid and the catechol structure of PDA-decorated CNTs. As a result, this hydrogel exhibited ultrafast self-healing, high shape adaptability and strong adhesive properties. A similar synthetic route for biopolymers (poly(γ-glutamic acid)-crosslinked amino-functionalized PEGylated poly(glycerol sebacate)) and gallic acid-modified CS was also reported [91]. The introduced coupling reaction and electrostatic interaction lead to high mechanical resilience and effective energy dissipation, which could assist in healing wounds under a moist and dynamic physiological environment. Likewise, Liang *et al* [92] used coordinate bonds (catechol–Fe) and dynamic

Schiff base bonds based on protocatechualdehyde@Fe tricomplex molecule and quaternized CS (QCS). The dual-dynamic crosslinking with reversible breakage and re-formation caused the hydrogel to have excellent autonomous healing and on-demand dissolution or removal properties.

4.3 Design and preparation of bio-based hydrogels and their biomedical applications

According to the aforementioned mechanisms, a series of design strategies and preparation methods are developed: (a) *Molecular self-assembly*: Under the influence of non-covalent interactions, the biopolymer chains can spontaneously turn into ordered network structures by self-assembly process. These processes take place in absence of toxic chemical crosslinkers and result in dynamic and reversible network structures. (b) *3D printing (or 4D printing)*: In this process a solution of the biopolymer is put into a suitably shaped mold either with light irradiation, through extrusion, inkjet, or stereolithographic printing techniques, which enable precise layer-by-layer fabrication of complex bio-based hydrogel structures. (c) *Double network (DN) strategy*: DN hydrogels that consist of an interpenetrating polymer network (IPN) or semi-IPN have been developed to achieve high strength and toughness compared to individual networks.

Some examples of each type of bio-based hydrogel made using these methods are listed in table 4.3.

Table 4.3. The typical preparation methods and their biomedical applications of bio-based hydrogels.

Methods	Mode	Components	Biomedical applications	References
Molecular self-assembly	Peptide	Jigsaw-shaped peptide	Cell culture	[98]
	Protein	SF	Drug delivery	[100]
	Polysaccharide	Chitin/polyphenols	Tissue engineering	[103]
	DNA	DNA/silver nanoclusters	Wound healing	[106]
3D printing	Extrusion	Alg/cellulose/gelatin	Tissue engineering	[110]
	Light/photo polymerization	GelMA	Cell culture	[113]
		Methacrylated HA	Tissue engineering	[114]
	Inkjet	Alg	Tissue engineering	[117]
	DIW	CS	Tissue engineering	[118]
	4D printing	Alg/MC	Drug delivery	[121]
DN strategy	Two-step	Chitin nanofiber/PEGDE	Tissue engineering	[124]
	Molecular stent	ChS/PDMAAm	Tissue engineering	[128]
	One-pot	QCS/PAA	Skin patch	[131]

4.3.1 Molecular self-assembly

Molecular self-assembly can fabricate bio-based hydrogels without the need of external crosslinkers. In self-assembly, polypeptides, protein, DNA, and polysaccharide will undergo the spontaneous organization of molecular components into ordered structures, and ultimately form hydrogels (figure 4.5) [93]. Molecular self-assembly utilize non-covalent interactions that previously described physical cross-linking in section 4.2.1, including hydrogen bonding, ionic interaction, hydrophobic interaction, π–π stacking, and van der Waals forces to drive ordered arrangement of the biopolymer [94].

Peptides first undergo self-assembly into secondary structures such as β-sheet, β-hairpin, and α-helix, before aggregating into nanofibers [95]. These nanofibers then elongate, associate, and entangle to form hydrogels [96]. The self-assembly process is induced through various methods, typically involving the control of pH, ionic strength, and temperature to inhibit charge repulsion between amino acid side chains. On this basis, Mei *et al* [97] developed a series of RGD-derived peptides with a unique self-assembling motif. By designing a hydrophobic region to regulate hydrophobicity and hydrophilicity balance, sensitivity to mild acidic conditions, and electrostatic interaction with drugs, this peptide hydrogel enabled controlled drug release in a typical tumor microenvironment. Yaguchi *et al* [98] designed a cell-adhesive jigsaw-shaped fiber-forming peptide. Under physiological conditions, this self-assembling peptide underwent a helix-to-strand transition, forming supramolecular nanofibers several micrometers long, ultimately resulting in the formation of a hydrogel. It can be applied to sustain drug release and cell transplantation scaffolds.

Similarly, self-assembled protein hydrogels can be obtained through reversible or irreversible aggregation of protein segments, driven by chemical reactions and/or non-covalent interactions [99]. Meleties *et al* [100] prepared a thermo-responsive hydrogel using a coiled-coil protein and investigated the correlation between gelation ability and pH value. The results indicated that electrostatic repulsions play a dominant role at lower pH levels, especially when the pH is close the isoelectric point. This facilitated the stacking and assembly of coiled-coils into fibers, which subsequently entangled to form hydrogels. Feng *et al* [101] combined a small

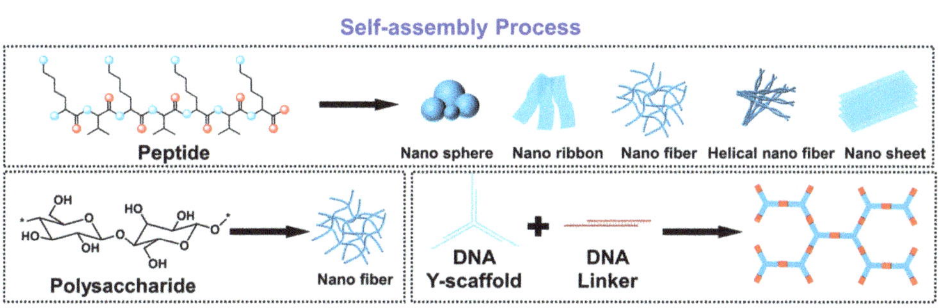

Figure 4.5. The representative molecular self-assembly of three types of biopolymers.

peptide with SF to fabricate a hybrid hydrogel, enabling continuous drug delivery for neural regeneration. The diffusion of SF micelles into the peptide solution, driven by the dynamic synergy between osmotic pressure and mutual electrostatic interactions, resulted in the rearrangement of SF micelles and the subsequent formation of a hydrogel. Shi *et al* [102] developed a dynamic metal–ligand coordination strategy to assemble an SF-based hydrogel between SF microfibers and a polysaccharide binder. These reversible interactions endowed the hydrogel with shear-thinning and self-healing properties, rendering it suitable for regenerating irregularly shaped bone defects.

Furthermore, polysaccharides also exhibit a hierarchical structure and unique self-assembly behavior. In particular, the polysaccharides including cellulose, chitin, and chitosan exhibited a stiff chain conformation in alkali/urea solvent system. The polysaccharide chains were surrounded by an alkali–urea complex and then formed a water-soluble sheath-like structure. However, when the solvent system suffered from disruption, the hydrogen bonding would drive the extended chains to self-assemble in parallel into nanofibers. Lots of conditions such as acid/salt solutions, organic solvent, and elevated temperature, were used to break the solution thermodynamic stability and *in situ* induce nanofiber formation. In this, Duan *et al* [103] reported a polyphenol-mediated chitin self-assembly strategy. The functional groups on chitin chains interact with polyphenols through noncovalent interactions (hydrogen bonding, ionic and hydrophobic interactions), resulting in the self-assembly of chains into α-chitin crystalline hydrate structures. The obtained hydrogels exhibited excellent mechanical properties, considerable antibacterial properties, and solubility in stomach acid, making them promising for tissue engineering. Further, Duan *et al* [104] developed a solvent consumption strategy to disrupt the thermodynamic stability of a polysaccharide solution, inducing the parallel self-assembly of molecular chains into nanofibers without gelation. The weak Lewis acid fumed silica was partially dissolved in the solvent and slowly consumed part of the alkali solvent. As a result, it was possible to fabricate a polysaccharide-based ink, which exhibited excellent rheological properties for 3D printing and could further be used for cell-loaded bone tissue engineering.

DNA stands as an outstanding example of precise biological self-assembly, demonstrating remarkable macromolecular properties. Leveraging these properties, DNA can serve as a blueprint for the creation of self-assembled hydrogels as well. A variety of self-assemble methods, including nuclease amplification techniques, covalent and physical interactions between DNAs, and enzymatic reactions, can be used [105]. Geng *et al* [106] utilized enzymatic polymerization to elongate DNA chains, forming a distinctive hydrogel network through physical entanglement. By incorporating silver nanoclusters onto DNA scaffolds, this hydrogel exhibited fluorescent and antibacterial properties, making it suitable for wound dressing. Jiang *et al* [107] employed electrostatic interaction and hydrogen bonding between DNA and tetrakis (hydroxymethyl) phosphonium sulfate (THPS) to drive hydrogel formation. This hydrogel possessed broad-spectrum antibacterial ability by releasing THPS, significantly accelerating wound healing. In Nayak's work [108], DNA derived from onion biomass was converted into DNA-dots through hydrothermal pyrolysis. These DNA-dots were then used as crosslinkers to prepare hydrogel via hybridization-mediated self-assembly with untransformed genomic DNA.

4.3.2 3D printing

Given the high complexity of human tissues, the above-mentioned approaches fail to meet the immediate demands of tissue engineering and regenerative medicine. 3D printing (or 4D printing with the dimension of time) as a successive layer-by-layer deposition technology that is digitally controlled with pre-defined 3D models through computer-aided design/manufacturing (figure 4.6), has revolutionized the biomedical field and driven numerous key advancements [109].

Currently, there are several approaches to 3D printing including extrusion, inkjet, light/photo polymerization, and direct ink writing (DIW). Erkoc *et al* [110] used biopolymers such as Alg, cellulose, and gelatin to develop a hydrogel ink for extrusion-based 3D printing. The printability of the hydrogel ink could be adjusted by introducing crosslinking agents such as GA and $CaCl_2$. This scaffold was suitable for tissue engineering as it supported cell growth and proliferation. However, hydrogel inks for extrusion-based 3D printing require appropriate rheological properties (such as viscosity, shear-thinning, and yield stress), which limits the material options. A gellan fluid support bath for fabricating freeform 3D hydrogel was developed by Huang *et al* [111]. The ink filaments are deposited along designed paths due to the solid–fluid–solid transition of the support bath. The deposited ink structure is then crosslinked to form an intact structure, after which the support bath material can be washed away. Gellan fluid gel allows for enzymatic, thermal, and ionic crosslinking, as well as easy recovery of printed structures while maintaining cell viability and structural fidelity.

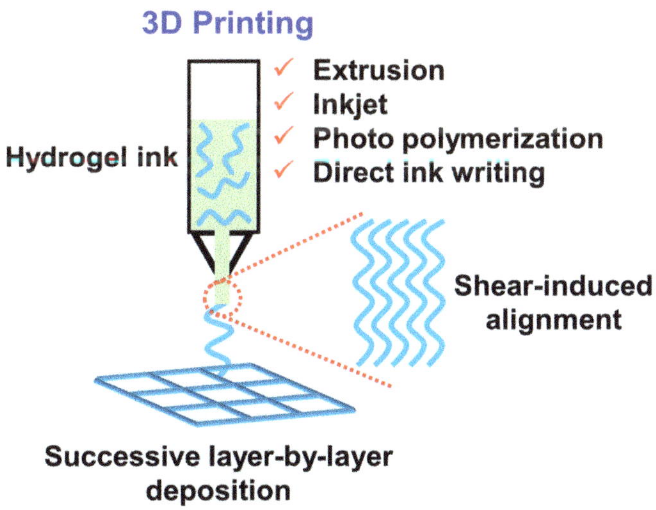

Figure 4.6. Schematic fabrication of 3D printing for bio-based hydrogels.

However, the resolution of hydrogels based on extrusion is poor. Light/photo polymerization is a possible printing technique to overcome this issue and manufacture high-resolution hydrogels [112]. Photoinitiators convert photolytic energy into reactive species, which drive a chain growth polymerization in the hydrogel ink and leads to a crosslinked 3D network. GelMA, which contains photo-crosslinkable methacrylate groups, is commonly used in light-based 3D printing systems. Weber *et al* [113] combined digital light processing (DLP) printing with a microfluidic chip system to fabricate GelMA hydrogel matrices. By creating size-tunable uniform gas bubbles within a hydrogel matrix, they prepared a porous scaffold with narrow size distributions that supported cell growth, spreading, proliferation, and migration. Methacrylated HA was also synthesized for extrusion-based and DLP-based 3D printing [114]. The resulting hydrogel provided an ideal environment for cell growth and can be used in tissue engineering scaffolds.

Photo-induced click chemistry has also been applied in 3D printing. Pereira *et al* [115] synthesized norbornene-modified pectin, which was then crosslinked with divalent ions and thiolated cell-adhesive peptide ligand. The rheology of the hydrogel ink was controlled by ionic crosslinking, while the mechanical and biochemical properties of the hydrogels were enhanced via a post-printing thiol–ene reaction. This unique feature allowed for the modulation of cell proliferation and ECM deposition, enabling the creation of functional cell/tissue-specific scaffolds with controlled cellular behavior.

Inkjet printing is an emerging technology that allows for the patterning and construction of unprintable and incompatible materials in a non-contact manner [116]. This technology can directly deposit the ink onto a receiving tray through the inkjet printer, reproducing the original image from the computer. Compared with traditional printing involving printing operations on pre-made templates, inkjet printing shows the advantages of non-contact with the printed pattern, fast printing speed, and high printing accuracy. Alg has been investigated as a common system in inkjet printing due to its easy processing and rapid gelation rate [117]. By laminating layers of the inkjet precursor (Alg) in the bath (calcium ion solution) on the target substrate, structures such as microshells, tubes, pyramids, and 'pie-shaped' hydrogels are reported.

Furthermore, DIW technology can be applied to most biopolymer-based hydrogel inks, allowing the fabrication of complex 3D structures without the need for supplementary sacrificial materials or UV irradiation curing. The DIW method equipped with a temperature controller was developed to achieve *in situ* gelation printing of high-strength CS hydrogels, utilizing the unique temperature-induced gelation property of CS in an alkali/urea aqueous solution [118].

Specifically, 4D printing is an emerging technology that allows for the creation of dynamic devices capable of changing their shape, function, or properties over time when exposed to external stimuli after fabrication [119]. Smart hydrogels sensitive to environmental stimuli have been demonstrated for use in actuators, cellular scaffolds, and drug release devices [120]. Lai *et al* [121] fabricated a shape memory Alg/methyl cellulose hydrogel by strategically controlling the network density gradients during the 3D printing process. The patterned 2D architectures were encoded with anisotropic stiffness and swelling behavior, allowing them to transform into various 3D morphologies.

4.3.3 DN strategy

Bio-based hydrogels formed via a single chemically or physically crosslinked network usually show poor mechanical properties and weak energy dissipation during deformation, which makes them soft, weak, and brittle, severely limit their applications. In the last decades, many researchers have strengthened the mechanical properties of hydrogels by designing novel structure or introducing different energy dissipation mechanisms. DN strategy that consist of IPN or semi-IPN have been proved to achieve high strength and toughness compared with either individual network [122] (figure 4.7).

There are three types of methods of DN strategy for bio-based hydrogels, including conventional two-step sequential methods, molecular stent methods, and one-pot methods. For the two-step methods, the initial step involves employing a polysaccharide, such as methacrylated ChS, to establish a covalently cross-linked, rigid, and brittle primary network through UV photo-polymerization. Subsequently, the first network is immersed and swelled in a precursor solution comprising neutral monomers, photoinitiators, and crosslinkers for the formation of the second network. In the following polymerization step, the second network is formed within the swollen first network, causing a combination of brittle and ductile networks [123]. In this, a long-chain polymer such as polyacrylamide (PAAm)/PAA/polyvinyl alcohol (PVA)/polyethylene glycol diacrylate (PEGDA) is often used as the second network. Huang *et al* [124] utilized the regenerated chitin nanofiber/PEGDE as the first network and then immersed it into acrylamide solutions to build the second network of PAAm. The DN network enhanced the mechanical properties of the hydrogel, making it more suitable for use as a rapid soft tissue repairing material.

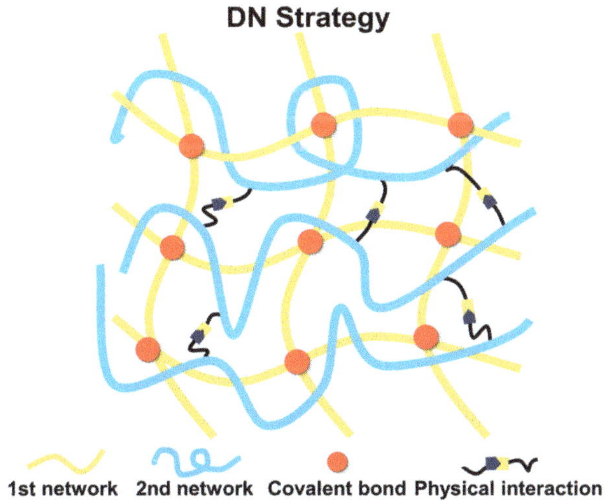

Figure 4.7. Schematic fabrication of DN network consisting of covalent bonds and physical interactions.

In addition, Choi *et al* [125] used the Alg/Ca^{2+} as the first network in PAAm, embedding with mesoporous silica microrods. Benefiting from the multiple interactions, such as ionically crosslinked of Alg, covalently cross-linked of PAAm, and van der Waals interactions between silica particles and polymers, the hydrogel exhibited excellent *in vivo* mechanical stabilities, which could be applied in controlled drug delivery systems such as insulin delivery. An Alg derivative also can be used to construct DN hydrogels. For example, oxidized Alg was used to form a dynamic first network by the Schiff base reaction, and PEGDA acted as the second covalent network [126]. The synergetic effect of both networks provides the hydrogel with excellent self-healing ability and injectability, which could withstand external damage as a wound dressing.

Moreover, molecular stent methods were developed by Nakajima *et al* [127]. The swollen neutral gel containing cationic or anionic polyelectrolytes is referred as 'St gel' or 'Stent gel'. Then, following the conventional two-step sequential methods, the highly swollen St gel is immersed in the precursor solution, followed by the second network polymerization to create resilient St-DN gels. The molecular stent method can extend the DN hydrogel concept to lots of functional polymers without chemical modifications. Zhao *et al* [128] utilized polysaccharides (e.g. ChS and HA) as molecular stents into the precursor monomer solution of the first neutral network. After the polymerization, the overall osmotic pressure in the hydrogel is increasing to stretch the neutral first-network chains to the extended conformation, resulting in a highly swollen degree like the strong polyelectrolyte hydrogels. Then, a large amount of second neutral network (poly(N,N-dimethyl acrylamide) (PDMAAm)) was introduced to form the contrasting topological structure required for a tough DN hydrogel. Yang *et al* [129] infiltrated the BC network into a PVA–poly-2-acrylamide-2-methylpropanesulfonic acid (PAMPS) network to create a St-DN hydrogel. Here, BC imparts the hydrogel with tensile strength similar to that of collagen nanofibers in cartilage. PVA contributed elastic restoring force, viscoelastic energy dissipation, and mitigates stress concentration on individual BC fibers. PAMPS can offer a negative charge akin to ChS, potentially inducing osmotic pressure to swell cartilage and enhance its compressive strength. Therefore, this hydrogel presents itself as an outstanding candidate material for repairing cartilage lesions.

However, both the two-step and molecular stent methods require extended periods of swelling and diffusion processes, and it is difficult to control the molar ratio of the two networks. Thus, a one-pot method was developed as an efficient and controllable approach to fabricate hybrid chemically/physically crosslinked DN hydrogels. In this, all the reagents react in just one reactor with successive physical (i.e. heating–cooling and freezing–thawing) and chemical reactions (photo/thermo-polymerization). Additionally, the reaction can be changed in a fast and controllable manner, resulting in consistent and reproducible gels. For instance, Chen *et al* [130] took advantage of the thermoreversible sol–gel transition of agar to fabricate highly mechanical and recoverable agar/PAAm DN hydrogels. The properties of DN hydrogels could be optimized just by changing the number of monomers. Moreover, Shi *et al* [131] developed a multi-stimuli-responsive DN hydrogel by *in situ* polymerization of hydrophilic anion monomers AA in a cationic polysaccharide of QCS. Notably, owing to the reversible physical cross-linked network and its thermosensitivity, the mechanical properties, adhesion, and visual effect of the hydrogel can be adjusted by altering the density of hydrogen bonding through enthalpy-driven UCST phase transition behavior, rendering it an outstanding biocompatible skin patch to recognize temperature change signals. Furthermore, one-pot methods have been reported for synthesizing biopolymer-based IPN hydrogels, facilitated by the simultaneous crosslinking of two independent networks using light [132]. These methods involve: (i) free-radical crosslinking of methacrylate-modified HA to establish the primary network, and (ii) thiol–ene crosslinking of norbornene-modified HA with thiolated guest–host assemblies comprising adamantane and β-CD to construct the second network. The hydrogel exhibited good work of fracture, tensile strength, and low hysteresis. Moreover, the capacity to adjust mechanical properties through minor alterations in the concentrations of individual networks establishes these IPN hydrogels as adaptable and versatile contenders for a broad array of biomedical applications.

In summary, hydrogels maintain solid-like structures with large amounts of water, providing a wet and fluid-like environment within their networks. Various preparation methods, including molecular self-assembly, and 3D (or 4D) printing, and DN strategy have been developed for biomedical applications to adapt to changes in the human microenvironment. However, these methods require pre-functionalization of polymer chains, adequate viscosity for post-processing, and additional synthetic polymers, which significantly increases operating and production costs. Therefore, it is important to find the most suitable and simple methods for various types of bio-based hydrogels.

4.4 Properties at the interface of bio-based hydrogels and their surroundings

As discussed above, many methods and techniques have been developed for the design and preparation of bio-based hydrogels. Depending on the resulting internal interactions with bio-based hydrogels, they show unique properties such as flexibility, injectability, adaptability, recoverability, programmability, and self-healing ability. These properties enable bio-based hydrogels be applied in various biomedical applications. Previous reviews have thoroughly discussed the intrinsic properties of bio-based hydrogels [15, 133]. Here we want to focus on the interfacial properties of bio-based hydrogels instead and take a closer look at the interaction of hydrogels with their surroundings. Beyond that, the surface properties of hydrogels and their systematic investigation play an important role in their applications. It is necessary to systematically investigate surface properties such as antibacterial properties, hemostatic ability, anticoagulation activity, adhesive capacity, immobilization of biomolecules and cells, inhibition of protein/cell adhesion, and hydrophobization. Understanding these properties can greatly expand the applicability of these materials in the biomedical field.

4.4.1 Antibacterial property

Antibacterial properties play a crucial role in the biomedical field, particularly in wound healing, where the prevention of severe bacterial infections is essential. However, the overuse or improper use of antibiotics has led to an increase in bacterial resistance, necessitating the development of antibacterial biomaterials as alternatives to antibiotics [134]. This section will discuss several representative antibacterial hydrogels based on cationic polymers and antibacterial peptides, as well as their mechanisms and bacteria-killing effects (figure 4.8).

Cationic biopolymers such as CS, chitin, and QC possess inherent antibacterial properties due to their positive surface charge. For example, CS contains abundant $-NH_2$ groups that can bind with the negative charges on the surface of bacteria through electrostatic interaction, thereby altering cell permeability and causing cell membrane lysis [135]. Currently, several CS-based dressings have been commercialized in wound healing such as ChitoFlex® and ChitoClear®. However, CS with >90% deacetylation (DA) may not be the most effective therapeutic agent for wound healing. Therefore, the impact of the degree of acetylation on wound healing efficiency should be investigated. To explore this, a series of chitin samples with varying DA were prepared. The results showed that the regenerated chitin with 71% DA significantly improved wound healing, via promoting re-epithelialization and collagen deposition [136]. Lin *et al* [137] developed a non-crosslinked CS/HA hybrid hydrogel via a novel method to enhance wound healing. By adding acetic acid and phosphates to the system to weaken the interaction between CS and HA, the positive charge and intrinsic antibacterial activity of CS could be maintained. In order to further improve the antibacterial properties of CS, quaternary ammonium groups are introduced onto the polymeric backbone. QCS can be synthesized via a one-pot reaction between epoxy groups of glycidyltrimethylammonium chloride and $-OH/-NH_2$ groups of CS in the presence of acetic acid [138] or alkali/urea aqueous solutions [139]. The antibacterial activity of QCS is enhanced compared with CS due to the increased positive charges. Qu *et al* [140] developed an antibacterial hydrogel using dynamic Schiff base crosslinking between QCS and benzaldehyde-terminated Pluronic®F127. The hydrogel exhibited an excellent killing ratio (>90%) for both *Staphylococcus aureus* and *Escherichia coli* due to the positively charged $-NH_2$ groups and quaternary ammonium groups. Similarly, other biopolymers can also be

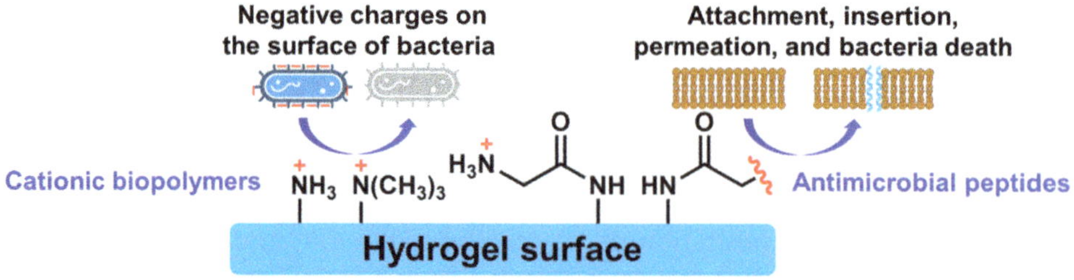

Figure 4.8. The antibacterial mechanism of bio-based hydrogels based on cationic biopolymers or antimicrobial peptides.

modified to achieve antibacterial properties. QC was homogeneously synthesized via an alkali-catalyzed oxalkylation reaction between the –OH groups of cellulose and 3-chloro-2-hydroxypropyltrimethylammonium chloride in alkali/urea aqueous solutions. A polycationic hydrogel was prepared through chemical crosslinking between QC and cellulose for killing *Saccharomyces cerevisiae* [141]. Meanwhile, Xie *et al* [142] explored the relationship between chemical structure and biological activity, taking quaternized chitin as an example. Optimizing the degree of quaternization (DQ) and degree of alkylation (DA) proportions is essential to achieve maximal antimicrobial efficacy while minimizing cytotoxicity. The results demonstrated quaternized chitin with a DQ of 0.46 and a DD of 82%, exhibiting optimized broad-spectrum antimicrobial properties for *S. aureus*, *Pseudomonas aeruginosa*, *E. coli*, and MRSA.

Furthermore, antimicrobial peptides (AMPs) such as polylysine, indolicidin, and tritrpticin also possess excellent antibacterial properties. Due to their unique amino acid composition, amphipathicity, cationic charge, and size, the antibacterial mechanism of AMPs is believed to proceed through hydrophobic and electrostatic interactions with bacterial membranes, leading to AMP attachment, insertion, permeation, and bacterial death. Gram-positive bacteria, which have negative charges on their surface, are particularly sensitive to electrostatic interactions with AMPs, while gram-negative bacteria, with a higher lipid content on their surface, are more sensitive to hydrophobic interactions. Thus, several gram-selective AMP hydrogels were synthesized through the rational design of the hydrophilicity, hydrophobicity, and charge properties of the peptide molecules [143]. As discussed in section 4.3.1, peptides can self-assemble into self-supportive hydrogels via π–π interactions. These hydrogels show strong gram-selective antibacterial and wound healing abilities, which could treat drug-resistant infections without harming non-selected strains. Gao *et al* [144] developed a high-efficiency screening method for AMPs using bacterial membrane extraction of *E. coli* and *S. aureus*. Subsequently, these were immobilized on silica microspheres to create a bacterial membrane chromatography stationary phase. An antibacterial hydrogel consisting of optimal AMP and oxidized dextran (ODex) was also prepared through Schiff base cross-linking. However, AMPs degrade rapidly upon contact with human serum. A novel strategy was reported to sustain activity and prolong the half-life by incorporating AMP conjugates into CS derivatives [145]. The synergistic antibacterial effects of the conjugates enabled the hydrogel to exhibit excellent *in vitro* activity against *P. aeruginosa*. In addition to surface modification strategies, various incorporation methods including antibiotics, metal nanoparticle, and photothermal agent-loaded have been proposed to provide antibacterial properties for bio-based hydrogels. For further details, comprehensive summaries of these works can be found elsewhere [146–148].

4.4.2 Hemostatic ability

Excessive bleeding and hemorrhaging present major challenges to and burdens on the healthcare system. Traditional methods of stopping bleeding have limitations

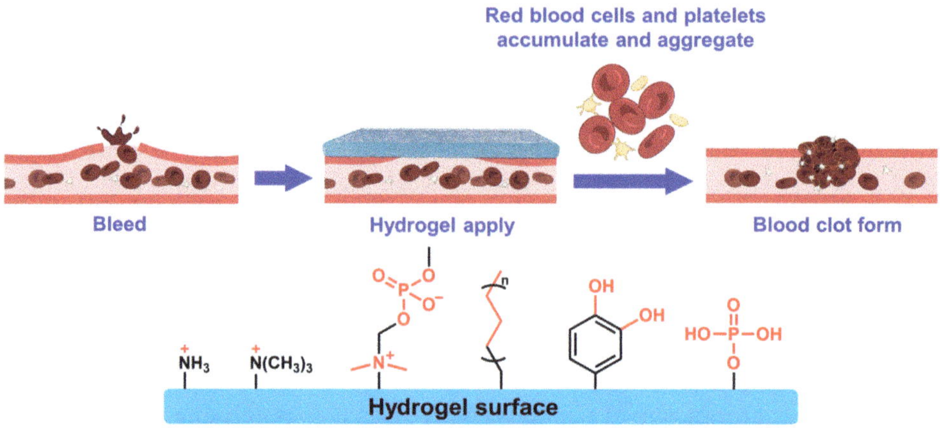

Figure 4.9. The hemostatic mechanism of bio-based hydrogels.

that may result in ineffective treatment and potential secondary injuries. Hemostasis is the natural process triggered by vascular endothelial impairment, which initiates a series of events to form a clot and prevent excessive blood loss. There are two main stages of hemostasis: primary and secondary. Primary hemostasis involves vasoconstriction and platelet activation, leading to the formation of a soft platelet plug. This is followed by secondary hemostasis, where the soft plug is transformed into a hard, insoluble fibrin clot through the conversion of fibrinogen to fibrin (figure 4.9) [149]. On this basis, hemostasis can be achieved through three types of mechanisms including: (i) physical adhesion to the tissue, (ii) chemical interactions and bonding with coagulation factors, and (iii) absorption of plasma [150]. Notably, hydrogel can seal the wound by interacting with the bleeding site to form a protective adhesion barrier quickly and safely. After stopping the bleeding, they form an antibacterial surface over the bleeding site, which protects it from bacterial infection and further facilitate wound healing. Moreover, introducing hemostatic components within the hydrogel to trigger the blood coagulation cascade and accelerate blood clot formation. For instance, cationic bio-based hydrogel can interact with red blood cells (RBCs) and platelets membrane containing negative charges (sialic acid), resulting in an excellent hemostatic ability [151].

CS, a classical hemostatic material, has been a commercial success such as Celox™ and HemCon™. Numerous attempts have been made to enhance the hemostatic properties of CS through chemical modification. A bio-based hydrogel with rapid hemostatic capability was created by mixing QCS and tannic acid (TA) [152]. Due to the presence of amine groups (–NH_2) and quaternary ammonium groups in QCS, as well as phenolic groups in TA, this hydrogel can effectively stop bleeding in arterial and deep incompressible wounds in various animal models, including mouse tail amputation, femoral artery hemorrhage, and liver incision. Additionally, it has been reported that zwitterionic choline phosphoryl (CP) can create electrostatic interactions with the headgroups of phospholipids in the cell membrane, specifically with the phosphatidyl choline (PC) hydrophilic head group

[153]. The hemostatic hydrogel was produced by crosslinking choline phosphoryl-functionalized CS (CS-g-CP) and ODex through Schiff base reactions. Due to the inverse orientation, CP groups can bind to PC headgroups, promoting adhesion and aggregation of RBCs and resulting in rapid hemostasis in models of rat tail amputation, rat liver, and spleen injury. Likewise, hydrophobic alkyl chains can insert themselves into and firmly attach to phospholipids in the cell membrane, thus conferring the hydrogel with exceptional coagulation and hemostasis capabilities. Importantly, the hydrophobic groups also contribute to blood gelation, leading to the transformation of liquid blood into a self-supporting hydrogel that effectively seals the bleeding site [154]. Furthermore, researchers have developed mussel-inspired catechol-conjugated CS derivatives to enhance hemostatic performance through the creation of blood protein barriers via interactions with serum proteins. By grafting pyrocatechol derivatives onto the $-NH_2$ groups of the CS backbone, the resulting hydrogel can capture and immobilize blood cells and platelets, forming a resilient thrombus at the bleeding sites [155]. Furthermore, polyphosphate can bind to a range of blood proteins, platelets, and coagulation factors, thereby facilitating the blood coagulation cascade. Hence, Wang *et al* [156] conjugated phosphate groups onto a CS backbone to enhance hemostatic ability. The enhanced zwitterionic nature of phosphorylated CS expands its interaction spectrum with coagulation factors.

4.4.3 Anticoagulation activity

In addition to advancing hemostatic abilities for wound healing, there is equal interest in achieving hemocompatibility for biomaterials *in vivo*, with one approach focusing on enhancing interfacial interactions, while the other aims to avoid them (figure 4.10). For instance, in critical care medicine, cardiopulmonary by-pass is commonly employed during open-heart surgery to sustain circulation. However, this procedure can induce the activation of the hemostatic response in clinical scenarios. Hence, full-dose heparin is necessary in such cases, which may increase the risk of bleeding. Moreover, extracorporeal membrane oxygenation, commonly utilized in patients experiencing cardiac or respiratory failure, necessitates a relatively moderate dose of heparin for anticoagulation, even so, hematologic issues remain as

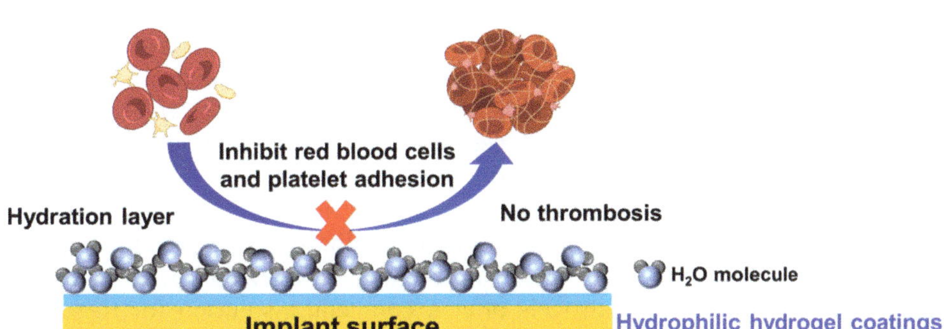

Figure 4.10. Anticoagulation mechanism of hydrophilic hydrogel coatings.

possible side effects [157]. Therefore, all biomaterials in contact with whole blood or blood plasma should ideally prevent thrombus formation and coagulation initiation, achieved through strategies that mitigate interfacial interactions

Grafting bio-functional hydrogel coatings onto the surface of biomaterials has been identified as an effective approach [158, 159]. Glycosaminoglycans, such as heparin, sulfated fucans, and ChS, contain sulfate, sulfamide, and carboxylate groups, enabling them to bind with antithrombin (ATIII) and induce a conformational change in the serpin. This interaction enhances surface blood compatibility and exerts anticoagulant effects [160]. In Jiang's work, heparin and ChS were incorporated into a HA and gelatin hydrogel coating [161]. Heparin was found to significantly reduce platelet deposition and interact with ATIII, thus preventing thrombus formation. However, heparin-based coatings tend to lose stability when exposed to other ionic substances in the blood. As a result, zwitterion coatings have been developed for anticoagulant coating purposes [162]. By incorporating both positively and negatively charged groups within the repetitive unit of a polymer, zwitterion-coated hydrogel surfaces exhibit enhanced solvation capacity through electrostatic interactions with water molecules. This property effectively inhibits the adhesion of RBCs [163]. Notably, polysaccharide-based zwitterion hydrogels can combine the benefits of zwitterionic polymers (preventing thrombosis) and polysaccharide chains (facilitating cellular adherence). Yao *et al* [164] synthesized sulfobetaine-derived starch using Williamson etherification between a sulfobetaine-based zwitterionic etherifying agent and the –OH groups of starch. This novel compound exhibited an excellent anti-clotting property attributed to the electrostatically induced high hydration. Subsequently, a hydrogel formed by disulfide bridge formation was coated onto the polydopamine-modified PET surface [165]. The results demonstrated the hydrogel's ability to resist non-specific protein adsorption and promote competitive adhesion between endothelial cells and smooth muscle cells, making it a promising candidate for long-term blood-contacting devices. Moreover, the presence of zwitterionic units mimics the biological surface of cells, which can give an advantage in cell-to-cell communication through changes in charge and composition, further enhancing the compatibility [166].

4.4.4 Adhesive capacity

In addition to its antibacterial and hemostatic properties, the adhesive capacity of bio-based hydrogels is an important surface property for biomedical applications [167]. Commercialized fibrin glue exhibits poor adhesion to tissues and can be easily washed away by blood flow. Moreover, cyanoacrylates, commonly utilized as components of superglue for strong adhesion purposes in the biomedical sector, possess significant toxicity and exhibit poor biodegradability. Hence, there is an urgent need for the development of new bio-based adhesives that are biocompatible, biodegradable, and capable of strong adhesion and sealing of injured bleeding tissues. Bio-based adhesive hydrogels offer a compelling solution for these goals. They can effectively seal wounds to stop bleeding initially and subsequently promote tissue remodeling while preserving the original function of the tissue throughout the

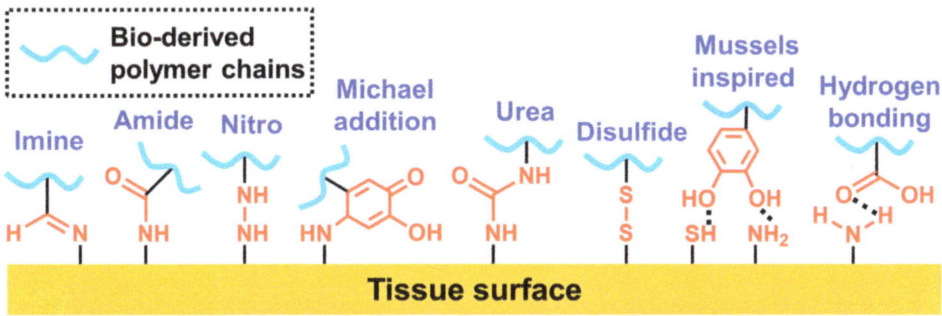

Figure 4.11. Adhesive mechanism of the bio-based hydrogels at the tissue interface. (Reproduced with permission from [168]. Copyright 2015 Royal Society of Chemistry.)

remodeling process. The surface adhesiveness of hydrogels is primarily deriving from chemical bonding and/or non-covalent interactions between active functional groups in hydrogels and biological tissues [168]. As depicted in figure 4.11, numerous functional groups present in hydrogels, including $-NH_2$, –COOH, –SH, –OH, and catechol groups, can form covalent bonds with nucleophilic groups (–OH, –SH, $-NH_2$) at the interface with biological tissues through condensation and addition reactions. Examples include Schiff base crosslinking and Michael addition, which consequently enhance adhesion strength [169]. Additionally, some non-covalent interactions such as H electrostatic interaction, hydrogen bonding, and cation–π interactions have been used for interfacial adsorption. Covalent bonding is typically robust yet irreversible, and non-covalent bonding is relatively weak but reversible, potentially enabling repeated adhesion and detachment [170]. Furthermore, it is a great challenge to achieve adhesion under aqueous conditions for bio-based adhesives. In the presence of liquids, a hydration layer forms on the surface of the adhered material, hindering the formation of molecular bridges or leading to hydrogel swelling. Thus conventional gluing techniques have low adhesive strength to tissues with blood or fat. Much effort is focused on improving the underwater adhesion of hydrogel. Currently, introducing reactive functional groups that can covalently stick to the tissue surface or hydrophobic groups that can assist water discharge at the interface to promote underwater adhesion are good solutions.

In Lei's work, thioether bonds, imide bonds, and disulfide bonds were integrated to strengthen tissue adhesion [171]. This hydrogel could be used as a rapid wound sealant to facilitate postoperative wound healing. However, the tissue surfaces are always humid and dynamic and the existence of surrounding water at the contact interfaces will weaken the molecular interactions between hydrogels and tissues [172], thus limiting their reliable functions *in vivo*. Marine mussels and barnacle can attach rapidly and securely to various surfaces even in turbulent seawater. Mussel foot proteins and cement proteins are known for their universal adhesiveness, and their analogs have been widely used in the development of adhesive hydrogels. Pan *et al* [173] proposed a dual biomimetic hydrogel with excellent repeatable and durable adhesiveness. In addition to hydrogen bonding, cation–π, and electrostatic interactions between the hydrogel and tissue surface, the numerous hydrophilic groups can absorb water molecules at the interface, effectively removing surface-bound water layers. Therefore, the wet adhesion enabled excellent hemostatic performance for rabbit/pig models of cardiac penetration holes and femoral artery injuries.

Unfortunately, numerous adhesive hydrogels are formulated for dual-sided adhesion, potentially causing postsurgical tissue adhesion and scar tissue formation. This is attributed to their indiscriminate adhesion to surrounding tissues and organs, alongside an adhesive-induced inflammatory response. To address this issue, Wu *et al* [174] developed a smart adhesive Janus hydrogel. They synthesized a photocurable precursor by grafting 3,4-dihydroxyphenylalanine and 2-aminoethyl methacrylate onto the HA chain to achieve an asymmetric adhesive capability. The hydrogel acts as a wet adhesive on the injured cecum, while its outward-facing side is nonadherent after photo-crosslinking, which can prevent the formation of adhesions after minimally invasive surgeries. Another approach is the design of on-demand removable hydrogels through reversible non-covalent interactions at the interface. Trivalent metal ion-crosslinked CMCS hydrogel was designed to accelerate skin tissue regeneration [175]. Moreover, after soaking in Na_2SO_4 aqueous solution, thanks to the interaction between the $-NH_2$ groups on CMCS and SO_4^{2-}, the hydrogel underwent phase separation and lost its interfacial adhesion strength, making it easy to detach from the skin with little residue.

4.4.5 Immobilization of biomolecules and cells

The favorable surface properties of bio-based hydrogels, combined with the presence of reactive functional groups, render this material class promising for biomedical applications such as drug delivery, wound adhesion, tissue scaffolds, and biosensing materials. However, undesirable bioactivity has hindered its implementation as a bio-material scaffold. To overcome this obstacle, biomolecules (e.g. peptides and growth factors) have been immobilized within hydrogel networks, resulting in functional materials with catalytic activity, fluorescence, and the ability to regulate cell behaviors (figure 4.12).

To preserve biomacromolecule activity during aqueous-based reactions at physiological pH and temperature, bioorthogonal chemistry has garnered attention for

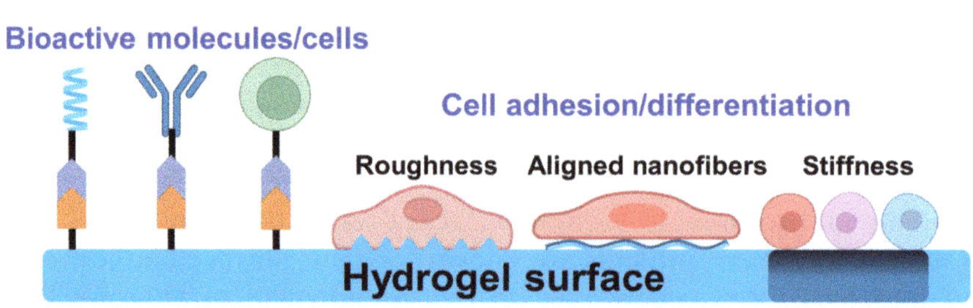

Figure 4.12. Enhancing properties and functionalities for biomolecules and cell immobilization on bio-based hydrogel surfaces: a schematic overview.

its high efficiency and low toxicity [176]. Arslan *et al* [177] designed a hydrogel with grafted disulfide groups on CS, serving as a reversible immobilization platform for biomolecules containing –SH groups. The tripeptide glutathione and dye methyl red-SH could rapidly conjugate and cleave onto the hydrogel surface through a mild thiol-disulfide exchange reaction, simplifying the hydrogel functionalization for biomedical applications. Thiol-methacrylate Michael addition has also been used to conjugate the Jagged-1 mimetic peptide ligand onto HA hydrogels [178], enhancing gene expression, mechanotransduction, and osteogenesis of stem cells, ultimately improving calvarial defect regeneration. A Zn^{2+}/Fe_3O_4-loading CS hydrogel microsphere was developed for promoting wound healing [179]. Zn^{2+} ions have a high affinity for binding to histidine-tagged proteins, and combined with the magnetic feature of Fe_3O_4 for desorption, it provides an efficient method for capturing and delivering proteins.

Moreover, developing new strategies to enhance cell adhesion on hydrogel surfaces is crucial for improving cell viability and function in biomedical applications. Cells initially adhere to hydrogels through electrostatic interactions and then proliferate due to its similarity to the ECM [180]. The strength of cell adhesion can be regulated by electric density, chemical composition, surface topography, etc, on the hydrogel surface. The adhesive peptide sequence RGD interacts with integrin adhesion receptors on the cell surface [181]. On this basis, an RGD-conjugated polypeptide-based hydrogel was reported as a promising platform mimicking cell-interactive ECM [182]. It also takes advantage of the reversibility of disulfide bonds, enabling dynamic changes through treatment with glutathione. Consequently, cell behavior, such as adhesion and migration, can be dynamically regulated, which is beneficial for cell culture. Additionally, soluble fibronectin was coated onto an amyloid-based albumin hydrogel surface to support prolonged cell adhesion and growth [183]. Huang *et al* [184] prepared a positively charged chitin whisker-based hydrogel as a scaffold for bone tissue engineering, which promotes the adhesion and proliferation of osteoblast cells. The increased surface roughness enabled by chitin whiskers further contributed to cell adhesion. A cancer cell-adhesive hydrogel was designed to interact with glioblastoma cells through thiol-mediated adhesion [185], impeding the invasive motility of tumor cells and resulting in a significant decrease in tumor mass.

Studies have also found that the aligned nanofibrous topology is conducive to cell adhesion and migration [186, 187]. Subsequently, a hydrogel-based bone scaffold with anisotropic nanofibrous morphology was developed [188], where chitin nanofibers were aligned in the same direction by mechanical deformation-induced orientation. This alignment facilitated the stretching and attachment of BMSCs' filopodia to the nanofibrous surface, promoting cell spreading and alignment along the orientation direction. Expanding upon this feature, exogenously electrospun PCL fibers were coated onto GelMA hydrogel [189], facilitating the attachment of human umbilical vein endothelial cells (HUVECs) to the fibers and their growth along the fiber direction. This process thereby accelerates epithelialization and chronic wound healing. Moreover, Hou *et al* [190] utilized soft lithography to prepare GelMA hydrogel with defined surface roughness gradients and systematically studied BMSCs' sensing behavior in response to a broad range of roughness (surface texture, randomness created by polishing, etching, blasting, or anodizing). The cells exhibited deformation or remodeling on the rough hydrogel surface to balance high cellular traction force and gather more adhesive ligands. A similar phenomenon was observed when cells adhered to a microfiber network, providing further insight into the previous case. Once cells adhere to the hydrogel surface, they can sense ambient mechanical signals and transduce them into biochemical signals, leading to transcriptional regulation in the nucleus.

Consequently, the mechanical properties of the hydrogel significantly influence cell growth and differentiation. For example, BMSCs on stiff substrates (>30 kPa) tend to produce more stress fibers and focal adhesions, preferentially undergoing osteogenic differentiation. In contrast, on soft substrates (<10 kPa), cell adhesion is substantially suppressed, and cells preferentially undergo adipogenic differentiation [191]. An intricate circular-patterned screening model was designed to observe the effects of hydrogel stiffness on the regulation of stem cell adhesion, migration, and differentiation. Circular-patterned hydrogels with different stiffness ranges were fabricated via dynamic covalent crosslinking between aminated gelatin and oxidized HA. BMSCs seeded in the center of the hydrogel exhibited distinct behaviors in response to the varying stiffness levels. Stem cell differentiation directions were induced by the hydrogels, with lower stiffness promoting nerve and muscle differentiation, and higher stiffness promoting osteogenesis [192]. Additionally, Chen *et al* [193] utilized Ca^{2+} ions to continuously increase the stiffness of GelMA/Alg hydrogel, inducing a dynamic response from mesenchymal stem cells (MSCs) to changes in matrix stiffness. This led to osteogenic differentiation and accelerated repair of calvarial defects.

Furthermore, encapsulating cells in 3D hydrogel networks is another key strategy for regulating cell behavior, distinct from cell adhesion, growth, and differentiation on 2D hydrogel surfaces. For further details, interested readers can refer to the additional references provided at the end of this chapter [194–196].

4.4.6 Inhibition of protein/cell adhesion

The uncontrolled and non-specific adsorption of proteins and cells, leading to severe biofouling, remains a persistent concern in biomedical applications. It is highly detrimental for implantable biomedical devices, biosensors, and surgical and protective instruments [197, 198]. Biofouling not only reduces the efficacy of the devices, but also leads to the formation of dense collagen capsules and biofilms on the device's surface, resulting in microbial contamination, tissue infection, thrombosis-like side effects, and immune response or inflammation [199]. Until now, the most common way to reduce non-specific adsorption is the immobilization of neutral or zwitterionic hydrophilic polymers-based hydrogel coating onto the implant material surface (figure 4.13).

In this field, hydrophilic coatings such as PEG and poly(2-hydroxyethyl methacrylate) (poly(HEMA))-based hydrogels provide a hydration layer near the surface, serving as a physical barrier that hinders cell and protein adhesion according to

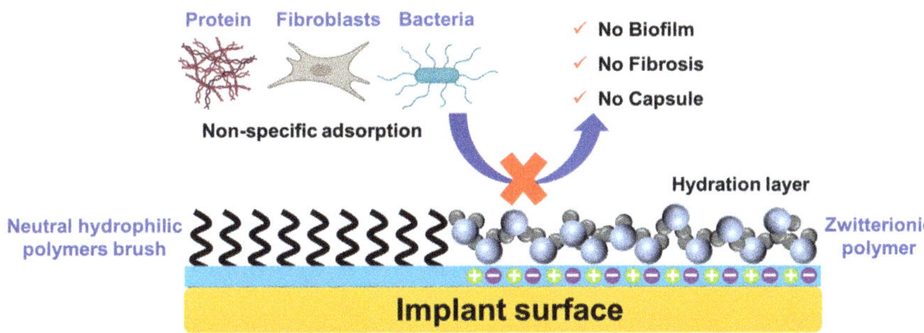

Figure 4.13. Illustration of the inhibition of unspecific protein/cell adhesion on an implant surface.

thermodynamic principles [200]. Hydrophilic polymer brushes, i.e. poly(HEMA) and poly[oligo(ethylene glycol) methyl ether methacrylate], were grafted onto the surface of CS hydrogel to inhibit protein–CS interactions [201]. The coating could significantly reduce the protein adsorption and avoid platelet activation and leukocyte adhesion. Peng *et al* [202] combined the anti-adhesive properties of PEG with the antibacterial properties of CS to prepare a blended hydrogel coating on a stainless-steel archwire for dental appliances. This dual-functional platform with integrated antifouling and antibacterial effect showed a great potential in biomedical device applications. Additionally, introducing anionic components to neutralize the cationic charge of CS [203] or reacetylate the $-NH_2$ group [204] could also confer antifouling properties, resulting in reduced bacterial/cellular adsorption and an accelerated wound healing process. However, over time, they undergo a loss of hydrophilicity attributed to oxidation *in vivo*.

Another approach to prevent biofouling involves the introduction of zwitterionic polymers and mixed-charge/pseudozwitterionic polymers into bio-based hydrogels [205]. As previously discussed, zwitterionic hydrogels with reduced net charge can interact with water molecules through electrostatically induced hydration or ionic solvation, leading to a robust hydration layer on the surface [206]. However, the majority of conventional zwitterionic polymers based on acrylate units are non-biodegradable and have limited *in vivo* applications. Therefore, zwitterionic functional groups are grafted on biopolymers to integrate the features of biodegradability and antifouling ability. Yu *et al* [207] synthesized sulfobetaine-derived dextran, together with CMCS, forming an injectable hydrogel through Schiff base crosslinking. The hydrogel showed excellent antifouling properties, i.e. strong resistance to bacteria and nonspecific proteins. Specifically, the zwitterionic hydrogels can also serve as a stem cell delivery platform to provide a sterile microenvironment for proliferation and maintenance of the stem cell function (stemness), thereby significantly accelerating wound healing and promoting non-scar skin tissue regeneration. In addition, carboxybetaine-derived dextran was synthesized to prevent specific protein adsorption, resist allogeneic repulsion and bacterial adhesion. This promotes wound healing by reducing the risk of inflammatory responses [208].

4.4.7 Hydrophobization

The exploration discussed above is mainly based on the hydrophilic properties of hydrogels, which create a moist environment at the contact interface to promote cell proliferation and tissue regeneration. However, the hydrophobization of hydrogels allows transport of hydrophobic drugs and enhances wet adhesion for tissue engineering (figure 4.14) [209].

As stated earlier, β-CD, characterized by its a hydrophobic inner cavity and hydrophilic exterior, is used as a matrix for encapsulating hydrophobic drugs in hydrogel networks through the formation of inclusion complexes [210]. Pivato *et al* explored the mechanism of drug dynamics in a β-CD hydrogel matrix and its correlation with macroscopic release kinetics at the molecular level [211]. The two steps used were Korsmeyer–Peppas kinetics driven by Fickian diffusion and a steady-state condition maintained for a long time, respectively. Together, they demonstrate the combined effect of solute-to-polymer adsorption and β-CD encapsulation within the hydrogel. Researchers have also explored β-CD for various grafting systems to form self-assembled nanoparticles. A novel β-CD was designed by grafting it with hydrophobic polyurethane at the periphery to control its hydrophilic–hydrophobic balance [212]. A hydrophobic anticancer drug (paclitaxel) would be released from the intricate superstructure through a non-Fickian diffusion process, resulting in the killing of cancer cells. β-CD was also used as a crosslinker to connect Alg via the –NH_2 group of ethylenediamine-modified β-CD and the –COOH groups of Alg through EDC/NHS chemistry [213]. The obtained nanogel was used for encapsulation of 5-fluorouracil (commonly use in the treatment of cancer) as a drug model and showed a controlled released property. Similarly, a hydrophobically modified gelatin was reported to produce hydrogels with a high content of hydrophobic segments [214]. The resulting hydrophobic interaction enhanced the adsorption of fluorescein sodium, which is a model for hydrophobic therapeutic agents.

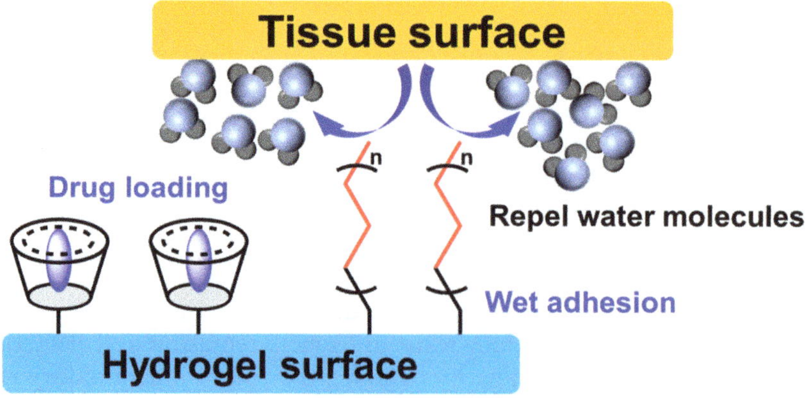

Figure 4.14. Illustration of the hydrophobic function on a bio-based hydrogel surface.

Moreover, as already mentioned, the adhesion strength of hydrogels decreases in wet or underwater conditions due to the formation of hydration layer. In addition to establishing strong covalent bonds, controlling the size of hydrophobic domains within hydrogels can enhance their adhesion capabilities. Li *et al* [215] employed gum arabic and sodium dodecyl sulfate as surfactants to produce hydrophobic association hydrogels. Owing to the hydrophobic polypeptide chains and dodecyl methacrylate repel water molecules at the interface, hydrogels showed adhesive behavior towards various substances. Furthermore, the disassociation of disulfide bonds in proteins will lead to hydrophobic amino acid eversion, thus forming a hydrophobic surface by protein assembly [216]. Liu *et al* [217] developed a hydrophobic natural sericin protein together with TA as a bioglue or sealant for wound healing. The self-hydrophobization enhanced the effect of catechol groups, such as hydrogen bonding and electrostatic interactions, which could seal skin incisions, mend internal wound defects, and prevent fluid leakage.

4.5 Conclusions and perspectives

Numerous biopolymers have been used for biomedical applications either directly or after artificial synthesis. Our objective here was to examine the current state of research and offer insights for the future development of bio-based hydrogels. As shown herein, biopolymers and their derivatives possess plenty of remarkable features including low cost, excellent biocompatibility, and controllable biodegradability.

However, many natural polymers often lose their original structural properties after complex isolation and purification, which presents challenges in adapting to the highly intricate microenvironment of the human body. Therefore, selecting a suitable carrier form is particularly crucial. Over the past few decades, various forms such as fibers, films, micro/nano-particles, sponges, and hydrogels have been developed. Among these, hydrogels have attracted much attention due to their similarity to ECM tissue. Subsequently, many preparation methods are carefully designed by regulating the physical structure or changing chemical composition, thus deriving lots of unique properties. This chapter provides an overview of biopolymers used in hydrogel preparation, and insights into general hydrogel formation mechanisms, including physical and crosslinking processes. It also discusses preparation methods, such as molecular self-assembly, 3D printing, and DN strategy, contextualizing them within the biomedical field for applications such as wound dressing, drug delivery, tissue engineering, and coating for implant devices. Finally, this chapter highlights the importance of properties at the interface of bio-based hydrogels, including antibacterial properties, hemostatic abilities, anticoagulant activities, adhesive capacities, biomolecule and cell immobilization, inhibition of protein/cell adhesion, and hydrophobization.

Advances in the bio-medical field require novel and innovative materials and devices. The development of smart and personalized biomaterials for tailored therapies in medicine signifies the future direction of medicine. Key considerations which can be addressed by scientists in the realm of bio-based materials development include formulating new strategies for efficiently introducing bioactive molecules, marker molecules, and functional groups onto biopolymers. Moreover, this research involves the innovative design of novel supramolecular structures derived from self-assembly process. These approaches aim to tailor the overall mechanical, chemical, and interfacial properties of bio-based hydrogels to create functional and personalized materials suitable for successful application at the interface of biomedical tissues (e.g. Janus structure, patterned structure, microneedle structure, and the emerging soft robot). Moreover, for every researcher, it is very important to consider current good manufacturing practices (e.g. FDA approved) during the design process, as this is essential for successful clinical translation. Here, it is worth noting that several successful examples (e.g. Hydrofiber®, Deflux™, and Keragel®) of bio-based hydrogel implementation are already in clinical use, showcasing their potential to replace synthetic polymers thanks to their superior inherent properties. Finally, through the collaborative efforts of the research community, we believe that bio-based hydrogels will find extensive applications in biomedicine, eventually transitioning to clinical use in the foreseeable future.

Bibliography

[1] Zhang Y S and Khademhosseini A 2017 Advances in engineering hydrogels *Science* **356** eaaf3627

[2] Correa S, Grosskopf A K, Lopez Hernandez H, Chan D, Yu A C, Stapleton L M and Appel E A 2021 Translational applications of hydrogels *Chem. Rev.* **121** 11385–457

[3] Muir V G and Burdick J A 2021 Chemically modified biopolymers for the formation of biomedical hydrogels *Chem. Rev.* **121** 10908–49

[4] Tunçer S 2021 Biopolysaccharides: properties and applications *Polysaccharides* (Hoboken, NJ: Wiley) pp 95–134

[5] Rajalekshmy G P, Lekshmi Devi L, Joseph J and Rekha M R 2019 An overview on the potential biomedical applications of polysaccharides *Functional Polysaccharides for Biomedical Applications* (Amsterdam: Elsevier) ch 2 pp 33–94

[6] Shen H, Sun T and Zhou J 2023 Recent progress in regenerated cellulose fibers by wet spinning *Macromol. Mater. Eng.* **308** 2300089

[7] Seddiqi H, Oliaei E, Honarkar H, Jin J, Geonzon L C, Bacabac R G and Klein-Nulend J 2021 Cellulose and its derivatives: towards biomedical applications *Cellulose* **28** 1893–931

[8] Tu H, Zhu M, Duan B and Zhang L 2021 Recent progress in high-strength and robust regenerated cellulose materials *Adv. Mater.* **33** 2000682

[9] Mahiout A, Meinhold H, Kessel M, Schulze H and Baurmeister U 1987 Dialyzer membranes: effect of surface area and chemical modification of cellulose on complement and platelet activation *Artif. Organs* **11** 149–54

[10] Yang Y, Lu Y-T, Zeng K, Heinze T, Groth T and Zhang K 2021 Recent progress on cellulose-based ionic compounds for biomaterials *Adv. Mater.* **33** 2000717

[11] Chang C and Zhang L 2011 Cellulose-based hydrogels: present status and application prospects *Carbohydr. Polym.* **84** 40–53

[12] Wang Z M, Li L, Zheng B S, Normakhamatov N and Guo S Y 2007 Preparation and anticoagulation activity of sodium cellulose sulfate *Int. J. Biol. Macromol.* **41** 376–82

[13] Ong X-R, Chen A X, Li N, Yang Y Y and Luo H-K 2023 Nanocellulose: recent advances toward biomedical applications *Small Sci.* **3** 2200076
[14] Fu J, Yang F and Guo Z 2018 The chitosan hydrogels: from structure to function *New J. Chem.* **42** 17162–80
[15] Li Z and Lin Z 2021 Recent advances in polysaccharide-based hydrogels for synthesis and applications *Aggregate* **2** e21
[16] Duan B, Huang Y, Lu A and Zhang L 2018 Recent advances in chitin based materials constructed via physical methods *Prog. Polym. Sci.* **82** 1–33
[17] Jiang X, Zeng F, Zhang L, Yu A and Lu A 2023 Engineered injectable cell-laden chitin/chitosan hydrogel with adhesion and biodegradability for calvarial defect regeneration *ACS Appl. Mater. Interfaces* **15** 20761–73
[18] Lee K Y and Mooney D J 2012 Alginate: properties and biomedical applications *Prog. Polym. Sci.* **37** 106–26
[19] Cao Y, Cong H, Yu B and Shen Y 2023 A review on the synthesis and development of alginate hydrogels for wound therapy *J. Mater. Chem.* B **11** 2801–29
[20] Hernández-González A C, Téllez-Jurado L and Rodríguez-Lorenzo L M 2020 Alginate hydrogels for bone tissue engineering, from injectables to bioprinting: a review *Carbohydr. Polym.* **229** 115514
[21] Motta; F L, Andrade C C P and Santana M H A 2013 A review of xylanase production by the fermentation of xylan: classification, characterization and applications *Sustainable Degradation of Lignocellulosic Biomass* (London: InTech) pp 252–75
[22] Han T, Song T, Pranovich A and Rojas O J 2022 Engineering a semi-interpenetrating constructed xylan-based hydrogel with superior compressive strength, resilience, and creep recovery abilities *Carbohydr. Polym.* **294** 119772
[23] Zhang B, Zhong Y, Dong D, Zheng Z and Hu J 2022 Gut microbial utilization of xylan and its implication in gut homeostasis and metabolic response *Carbohydr. Polym.* **286** 119271
[24] Jackson R L, Busch S J and Cardin A D 1991 Glycosaminoglycans: molecular properties, protein interactions, and role in physiological processes *Physiol. Rev.* **71** 481–539
[25] Karjalainen J M, Tammi R H, Tammi M I, Eskelinen M J, Ågren U M, Parkkinen J J, Alhava E M and Kosma V-M 2000 Reduced level of CD44 and hyaluronan associated with unfavorable prognosis in clinical stage I cutaneous melanoma *Am. J. Pathol.* **157** 957–65
[26] Necas J, Bartosikova L, Brauner P and Kolar J 2008 Hyaluronic acid (hyaluronan): a review *Vet. Med.* **53** 397–411
[27] Highley C B, Prestwich G D and Burdick J A 2016 Recent advances in hyaluronic acid hydrogels for biomedical applications *Curr. Opin. Biotechnol.* **40** 35–40
[28] Luo Z, Wang Y, Li J, Wang J, Yu Y and Zhao Y 2023 Tailoring hyaluronic acid hydrogels for biomedical applications *Adv. Funct. Mater.* **33** 2306554
[29] Tian X, Peng X, Long X, Lin J, Zhang Y, Zhan L and Zhao G 2022 Oxidized chondroitin sulfate eye drops ameliorate the prognosis of fungal keratitis with anti-inflammatory and antifungal effects *J. Mater. Chem.* B **10** 7847–61
[30] Sharma R, Kuche K, Thakor P, Bhavana V, Srivastava S, Mehra N K and Jain S 2022 Chondroitin sulfate: emerging biomaterial for biopharmaceutical purpose and tissue engineering *Carbohydr. Polym.* **286** 119305
[31] Cheng C, Sun S and Zhao C 2014 Progress in heparin and heparin-like/mimicking polymer-functionalized biomedical membranes *J. Mater. Chem.* B **2** 7649–72

[32] Sasisekharan R and Venkataraman G 2000 Heparin and heparan sulfate: biosynthesis, structure and function *Curr. Opin. Chem. Biol.* **4** 626–31

[33] Nazarzadeh Zare E *et al* 2024 Biomedical applications of engineered heparin-based materials *Bioact. Mater.* **31** 87–118

[34] Reizabal A, Costa C M, Pérez-Álvarez L, Vilas-Vilela J L and Lanceros-Méndez S 2023 Silk fibroin as sustainable advanced material: material properties and characteristics, processing, and applications *Adv. Funct. Mater.* **33** 2210764

[35] Ling S, Li C, Jin K, Kaplan D L and Buehler M J 2016 Liquid exfoliated natural silk nanofibrils: applications in optical and electrical devices *Adv. Mater.* **28** 7783–90

[36] Melke J, Midha S, Ghosh S, Ito K and Hofmann S 2016 Silk fibroin as biomaterial for bone tissue engineering *Acta Biomater.* **31** 1–16

[37] Zheng M, Wang X, Chen Y, Yue O, Bai Z, Cui B, Jiang H and Liu X 2023 A review of recent progress on collagen-based biomaterials *Adv. Healthcare Mater.* **12** 2202042

[38] Lin K, Zhang D, Macedo M H, Cui W, Sarmento B and Shen G 2019 Advanced collagen-based biomaterials for regenerative biomedicine *Adv. Funct. Mater.* **29** 1804943

[39] Wang P, Meng X, Wang R, Yang W, Yang L, Wang J, Wang D-A and Fan C 2022 Biomaterial scaffolds made of chemically cross-linked gelatin microsphere aggregates (C-GMSs) promote vascularized bone regeneration *Adv. Healthcare Mater.* **11** 2102818

[40] Kurian A G, Singh R K, Patel K D, Lee J-H and Kim H-W 2022 Multifunctional GelMA platforms with nanomaterials for advanced tissue therapeutics *Bioact. Mater.* **8** 267–95

[41] Xu X, Hu J, Xue H, Hu Y, Liu Y-n, Lin G, Liu L and Xu R-a 2023 Applications of human and bovine serum albumins in biomedical engineering: a review *Int. J. Biol. Macromol.* **253** 126914

[42] Ghuman J, Zunszain P A, Petitpas I, Bhattacharya A A, Otagiri M and Curry S 2005 Structural basis of the drug-binding specificity of human serum albumin *J. Mol. Biol.* **353** 38–52

[43] Caraceni P, Tufoni M and Bonavita M E 2013 Clinical use of albumin *Blood Transfus.* **11** s18–25

[44] Feroz S, Muhammad N, Ratnayake J and Dias G 2020 Keratin-based materials for biomedical applications *Bioact. Mater.* **5** 496–509

[45] Shavandi A, Silva T H, Bekhit A A and Bekhit A E-D A 2017 Keratin: dissolution, extraction and biomedical application *Biomater. Sci.* **5** 1699–735

[46] Wang L, Shang Y, Zhang J, Yuan J and Shen J 2023 Recent advances in keratin for biomedical applications *Adv. Colloid Interface Sci.* **321** 103012

[47] Debelle L and Tamburro A M 1999 Elastin: molecular description and function *Int. J. Biochem. Cell Biol.* **31** 261–72

[48] Mammi M, Gotte L and Pezzin G 1968 Evidence for order in the structure of α-elastin *Nature* **220** 371–3

[49] Goel R, Gulwani D, Upadhyay P, Sarangthem V and Singh T D 2023 Unsung versatility of elastin-like polypeptide inspired spheroid fabrication: a review *Int. J. Biol. Macromol.* **234** 123664

[50] Annabi N, Mithieux S M, Weiss A S and Dehghani F 2009 The fabrication of elastin-based hydrogels using high pressure CO_2 *Biomaterials* **30** 1–7

[51] Wang S, Wei Y, Wang Y and Cheng Y 2023 Cyclodextrin regulated natural polysaccharide hydrogels for biomedical applications—a review *Carbohydr. Polym.* **313** 120760

[52] Villiers A 1891 Sur la fermentation de la fécule par l'action du ferment butyrique *C. R. Acad. Sci.* **112** 536–8

[53] Liu Z, Ye L, Xi J, Wang J and Feng Z-g 2021 Cyclodextrin polymers: structure, synthesis, and use as drug carriers *Prog. Polym. Sci.* **118** 101408

[54] Liu G, Yuan Q, Hollett G, Zhao W, Kang Y and Wu J 2018 Cyclodextrin-based host–guest supramolecular hydrogel and its application in biomedical fields *Polym. Chem.* **9** 3436–49

[55] Li J, Harada A and Kamachi M 1994 Sol–gel transition during inclusion complex formation between α-cyclodextrin and high molecular weight poly(ethylene glycol)s in aqueous solution *Polym. J.* **26** 1019–26

[56] Li J, Ni X and Leong K W 2003 Injectable drug-delivery systems based on supramolecular hydrogels formed by poly(ethylene oxide)s and α-cyclodextrin *J. Biomed. Mater. Res.* A **65** 196–202

[57] Figueiredo P, Lintinen K, Hirvonen J T, Kostiainen M A and Santos H A 2018 Properties and chemical modifications of lignin: towards lignin-based nanomaterials for biomedical applications *Prog. Mater Sci.* **93** 233–69

[58] Lu X, Gu X and Shi Y 2022 A review on lignin antioxidants: their sources, isolations, antioxidant activities and various applications *Int. J. Biol. Macromol.* **210** 716–41

[59] Das A K, Mitra K, Conte A J, Sarker A, Chowdhury A and Ragauskas A J 2024 Lignin—a green material for antibacterial application—a review *Int. J. Biol. Macromol.* **261** 129753

[60] Gan J, Chen L, Chen Z, Zhang J, Yu W, Huang C, Wu Y and Zhang K 2023 Lignocellulosic biomass-based carbon dots: synthesis processes, properties, and applications *Small* **19** 2304066

[61] Khadem E, Ghafarzadeh M, Kharaziha M, Sun F and Zhang X 2024 Lignin derivatives-based hydrogels for biomedical applications *Int. J. Biol. Macromol.* **261** 129877

[62] Gan D, Xing W, Jiang L, Fang J, Zhao C, Ren F, Fang L, Wang K and Lu X 2019 Plant-inspired adhesive and tough hydrogel based on Ag-Lignin nanoparticles-triggered dynamic redox catechol chemistry *Nat. Commun.* **10** 1487

[63] Jing H, Huang X, Du X, Mo L, Ma C and Wang H 2022 Facile synthesis of pH-responsive sodium alginate/carboxymethyl chitosan hydrogel beads promoted by hydrogen bond *Carbohydr. Polym.* **278** 118993

[64] Xue W, Liu X, Yu W and Ma X 2006 Preparation of protein-loaded microspheres with size $\leqslant 10\ \mu m$ by electrostatic droplet generation technology *Chin. Sci. Bull.* **51** 279–86

[65] Shi L, Ding P, Wang Y, Zhang Y, Ossipov D and Hilborn J 2019 Self-healing polymeric hydrogel formed by metal–ligand coordination assembly: design, fabrication, and biomedical applications *Macromol. Rapid Commun.* **40** 1800837

[66] You J, Zhou J, Li Q and Zhang L 2012 Rheological study of physical cross-linked quaternized cellulose hydrogels induced by β-glycerophosphate *Langmuir* **28** 4965–73

[67] Xie H, Wang Z, Wang R, Chen Q, Yu A and Lu A 2024 Self-healing, injectable hydrogel dressing for monitoring and therapy of diabetic wound *Adv. Funct. Mater.* **34** 2401209

[68] Ding B, Gao H, Song J, Li Y, Zhang L, Cao X, Xu M and Cai J 2016 Tough and cell-compatible chitosan physical hydrogels for mouse bone mesenchymal stem cells *in vitro* *ACS Appl. Mater. Interfaces* **8** 19739–46

[69] Deng P, Yao L, Chen J, Tang Z and Zhou J 2022 Chitosan-based hydrogels with injectable, self-healing and antibacterial properties for wound healing *Carbohydr. Polym.* **276** 118718

[70] Zhao D, Huang J, Zhong Y, Li K, Zhang L and Cai J 2016 High-strength and high-toughness double-cross-linked cellulose hydrogels: a new strategy using sequential chemical and physical cross-linking *Adv. Funct. Mater.* **26** 6279–87

[71] Yang X, Jiang X, Yang H, Bian L, Chang C and Zhang L 2020 Biocompatible cellulose-based supramolecular nanoparticles driven by host–guest interactions for drug delivery *Carbohydr. Polym.* **237** 116114

[72] Itzhakov R, Tworowski D, Sadot N, Sayas T, Fallik E, Kleiman M and Poverenov E 2023 Nucleoside-based cross-linkers for hydrogels with tunable properties *ACS Appl. Mater. Interfaces* **15** 7359–70

[73] Liu P, Lin W, Wieduwild R, Towers R, Thomas A K, Günther M, Butdayev S, Wobus M, Bornhäuser M and Zhang Y 2021 Displaying lipid chains in a peptide–polysaccharide-based self-assembled hydrogel network *Chem. Mater.* **33** 2756–68

[74] Zhang Y, Furyk S, Bergbreiter D E and Cremer P S 2005 Specific ion effects on the water solubility of macromolecules: PNIPAM and the Hofmeister series *J. Am. Chem. Soc.* **127** 14505–10

[75] He Q, Huang Y and Wang S 2018 Hofmeister effect-assisted one step fabrication of ductile and strong gelatin hydrogels *Adv. Funct. Mater.* **28** 1705069

[76] Guo B, Liang Y and Dong R 2023 Physical dynamic double-network hydrogels as dressings to facilitate tissue repair *Nat. Protoc.* **18** 3322–54

[77] Ding R, Wei X, Liu Y, Wang Y, Xing Z, Wang L, Liu H and Fan Y 2023 Epidermal growth factor-loaded microspheres/hydrogel composite for instant hemostasis and liver regeneration *Smart Mater. Med.* **4** 173–82

[78] Chang C, Zhang L, Zhou J, Zhang L and Kennedy J F 2010 Structure and properties of hydrogels prepared from cellulose in NaOH/urea aqueous solutions *Carbohydr. Polym.* **82** 122–7

[79] Ye D, Chang C and Zhang L 2019 High-strength and tough cellulose hydrogels chemically dual cross-linked by using low- and high-molecular-weight cross-linkers *Biomacromolecules* **20** 1989–95

[80] Hu H, Wang F, Yu L, Sugimura K, Zhou J and Nishio Y 2018 Synthesis of novel fluorescent cellulose derivatives and their applications in detection of nitroaromatic compounds *ACS Sustainable Chem. Eng.* **6** 1436–45

[81] Huang J and Jiang X 2018 Injectable and degradable pH-responsive hydrogels via spontaneous amino–Yne click reaction *ACS Appl. Mater. Interfaces* **10** 361–70

[82] Lee Y, Lim S, Kim J A, Chun Y H and Lee H J 2023 Development of thiol–ene reaction-based HA hydrogel with sustained release of EGF for enhanced skin wound healing *Biomacromolecules* **24** 5342–52

[83] Qiao Y, Liu X, Zhou X, Zhang H, Zhang W, Xiao W, Pan G, Cui W, Santos H A and Shi Q 2020 Gelatin templated polypeptide co-cross-linked hydrogel for bone regeneration *Adv. Healthc. Mater.* **9** 1901239

[84] Li Z, Lu F and Liu Y 2023 A review of the mechanism, properties, and applications of hydrogels prepared by enzymatic cross-linking *J. Agric. Food Chem.* **71** 10238–49

[85] Kim B S, Kim S H, Kim K, An Y H, So K H, Kim B G and Hwang N S 2020 Enzyme-mediated one-pot synthesis of hydrogel with the polyphenol cross-linker for skin regeneration *Mater. Today Bio.* **8** 100079

[86] Deng P, Chen F, Zhang H, Chen Y and Zhou J 2021 Conductive, self-healing, adhesive, and antibacterial hydrogels based on lignin/cellulose for rapid MRSA-infected wound repairing *ACS Appl. Mater. Interfaces* **13** 52333–45

[87] Cao J, Xiao L and Shi X 2019 Injectable drug-loaded polysaccharide hybrid hydrogels for hemostasis *RSC Adv.* **9** 36858–66

[88] Jiang X, Zeng F, Yang X, Jian C, Zhang L, Yu A and Lu A 2022 Injectable self-healing cellulose hydrogel based on host–guest interactions and acylhydrazone bonds for sustained cancer therapy *Acta Biomater.* **141** 102–13
[89] Xu D, Huang J, Zhao D, Ding B, Zhang L and Cai J 2016 High-flexibility, high-toughness double-cross-linked chitin hydrogels by sequential chemical and physical cross-linkings *Adv. Mater.* **28** 5844–9
[90] Deng P, Liang X, Chen F, Chen Y and Zhou J 2022 Novel multifunctional dual-dynamic-bonds crosslinked hydrogels for multi-strategy therapy of MRSA-infected wounds *Appl. Mater. Today* **26** 101362
[91] Chen K, Wu Z, Liu Y, Yuan Y and Liu C 2022 Injectable double-crosslinked adhesive hydrogels with high mechanical resilience and effective energy dissipation for joint wound treatment *Adv. Funct. Mater.* **32** 2109687
[92] Liang Y, Li Z, Huang Y, Yu R and Guo B 2021 Dual-dynamic-bond cross-linked antibacterial adhesive hydrogel sealants with on-demand removability for post-wound-closure and infected wound healing *ACS Nano* **15** 7078–93
[93] Fleming S and Ulijn R V 2014 Design of nanostructures based on aromatic peptide amphiphiles *Chem. Soc. Rev.* **43** 8150–77
[94] Zou P, Chen W-T, Sun T, Gao Y, Li L-L and Wang H 2020 Recent advances: peptides and self-assembled peptide-nanosystems for antimicrobial therapy and diagnosis *Biomater. Sci.* **8** 4975–96
[95] Zhao X and Zhang S 2006 Molecular designer self-assembling peptides *Chem. Soc. Rev.* **35** 1105–10
[96] Fu K, Wu H and Su Z 2021 Self-assembling peptide-based hydrogels: fabrication, properties, and applications *Biotechnol. Adv.* **49** 107752
[97] Mei L, Xu K, Zhai Z, He S, Zhu T and Zhong W 2019 Doxorubicin-reinforced supramolecular hydrogels of RGD-derived peptide conjugates for pH-responsive drug delivery *Org. Biomol. Chem.* **17** 3853–60
[98] Yaguchi A, Oshikawa M, Watanabe G, Hiramatsu H, Uchida N, Hara C, Kaneko N, Sawamoto K, Muraoka T and Ajioka I 2021 Efficient protein incorporation and release by a jigsaw-shaped self-assembling peptide hydrogel for injured brain regeneration *Nat. Commun.* **12** 6623
[99] Tomadoni B, Capello C, Valencia G A and Gutiérrez T J 2020 Self-assembled proteins for food applications: a review *Trends Food Sci. Tech.* **101** 1–16
[100] Meleties M, Katyal P, Lin B, Britton D and Montclare J K 2021 Self-assembly of stimuli-responsive coiled-coil fibrous hydrogels *Soft Matter* **17** 6470–6
[101] Feng F *et al* 2023 Cooperative assembly of a designer peptide and silk fibroin into hybrid nanofiber gels for neural regeneration after spinal cord injury *Sci. Adv.* **9** eadg0234
[102] Shi L *et al* 2017 Self-healing silk fibroin-based hydrogel for bone regeneration: dynamic metal–ligand self-assembly approach *Adv. Funct. Mater.* **27** 1700591
[103] Lin X, Zhang L and Duan B 2021 Polyphenol-mediated chitin self-assembly for constructing a fully naturally resourced hydrogel with high strength and toughness *Mater. Horiz.* **8** 2503–12
[104] Zhang R, Deng L, Guo J, Yang H, Zhang L, Cao X, Yu A and Duan B 2021 Solvent mediating the *in situ* self-assembly of polysaccharides for 3D printing biomimetic tissue scaffolds *ACS Nano* **15** 17790–803

[105] Li F, Tang J, Geng J, Luo D and Yang D 2019 Polymeric DNA hydrogel: design, synthesis and applications *Prog. Polym. Sci.* **98** 101163

[106] Geng J, Yao C, Kou X, Tang J, Luo D and Yang D 2018 A fluorescent biofunctional DNA hydrogel prepared by enzymatic polymerization *Adv. Healthc. Mater.* **7** 1700998

[107] Jiang X, Li M, Guo X, Chen H, Yang M and Rasooly A 2019 Self-assembled DNA-THPS hydrogel as a topical antibacterial agent for wound healing *ACS Appl. Bio Mater.* **2** 1262–9

[108] Nayak S, Kumar P, Shankar R, Mukhopadhyay A K, Mandal D and Das P 2022 Biomass derived self-assembled DNA-dot hydrogels for enhanced bacterial annihilation *Nanoscale* **14** 16097–109

[109] Gibson I, Rosen D and Stucker B 2015 Development of additive manufacturing technology *Additive Manufacturing Technologies: 3D Printing, Rapid Prototyping, and Direct Digital Manufacturing* (New York: Springer) pp 19–42

[110] Erkoc P, Uvak I, Nazeer M A, Batool S R, Odeh Y N, Akdogan O and Kizilel S 2020 3D printing of cytocompatible gelatin-cellulose-alginate blend hydrogels *Macromol. Biosci.* **20** 2000106

[111] Compaan A M, Song K and Huang Y 2019 Gellan fluid gel as a versatile support bath material for fluid extrusion bioprinting *ACS Appl. Mater. Interfaces* **11** 5714–26

[112] Yu C, Schimelman J, Wang P, Miller K L, Ma X, You S, Guan J, Sun B, Zhu W and Chen S 2020 Photopolymerizable biomaterials and light-based 3D printing strategies for biomedical applications *Chem. Rev.* **120** 10695–743

[113] Weber P, Cai L, Rojas F J A, Garciamendez-Mijares C E, Tirelli M C, Nalin F, Jaroszewicz J, Święszkowski W, Costantini M and Zhang Y S 2023 Microfluidic bubble-generator enables digital light processing 3D printing of porous structures *Aggregate* **5** e409

[114] Li C, Zheng Z, Jia J, Zhang W, Qin L, Zhang W and Lai Y 2022 Preparation and characterization of photocurable composite extracellular matrix-methacrylated hyaluronic acid bioink *J. Mater. Chem.* **B 10** 4242–53

[115] Pereira R F, Lourenço B N, Bártolo P J and Granja P L 2021 Bioprinting a multifunctional bioink to engineer clickable 3D cellular niches with tunable matrix microenvironmental cues *Adv. Healthc. Mater.* **10** 2001176

[116] Li X, Liu B, Pei B, Chen J, Zhou D, Peng J, Zhang X, Jia W and Xu T 2020 Inkjet bioprinting of biomaterials *Chem. Rev.* **120** 10793–833

[117] Teo M Y, Kee S, RaviChandran N, Stuart L, Aw K C and Stringer J 2020 Enabling free-standing 3D hydrogel microstructures with microreactive inkjet printing *ACS Appl. Mater. Interfaces* **12** 1832–9

[118] Zhou L, Ramezani H, Sun M, Xie M, Nie J, Lv S, Cai J, Fu J and He Y 2020 3D printing of high-strength chitosan hydrogel scaffolds without any organic solvents *Biomater. Sci.* **8** 5020–8

[119] Champeau M, Heinze D A, Viana T N, de Souza E R, Chinellato A C and Titotto S 2020 4D printing of hydrogels: a review *Adv. Funct. Mater.* **30** 1910606

[120] Dong Y, Wang S, Ke Y, Ding L, Zeng X, Magdassi S and Long Y 2020 4D printed hydrogels: fabrication, materials, and applications *Adv. Mater. Technol.* **5** 2000034

[121] Lai J, Ye X, Liu J, Wang C, Li J, Wang X, Ma M and Wang M 2021 4D printing of highly printable and shape morphing hydrogels composed of alginate and methylcellulose *Mater. Des.* **205** 109699

[122] Zhang H, Shi L W E and Zhou J 2023 Recent developments of polysaccharide-based double-network hydrogels *J. Polym. Sci.* **61** 7–43

[123] Gong J P, Katsuyama Y, Kurokawa T and Osada Y 2003 Double-network hydrogels with extremely high mechanical strength *Adv. Mater.* **15** 1155–8
[124] Huang J, Frauenlob M, Shibata Y, Wang L, Nakajima T, Nonoyama T, Tsuda M, Tanaka S, Kurokawa T and Gong J P 2020 Chitin-based double-network hydrogel as potential superficial soft-tissue-repairing materials *Biomacromolecules* **21** 4220–30
[125] Choi S, Choi Y and Kim J 2019 Anisotropic hybrid hydrogels with superior mechanical properties reminiscent of tendons or ligaments *Adv. Funct. Mater.* **29** 1904342
[126] Aldana A A, Morgan F L C, Houben S, Pitet L M, Moroni L and Baker M B 2021 Biomimetic double network hydrogels: combining dynamic and static crosslinks to enable biofabrication and control cell-matrix interactions *J. Polym. Sci.* **59** 2832–43
[127] Nakajima T, Sato H, Zhao Y, Kawahara S, Kurokawa T, Sugahara K and Gong J P 2012 A universal molecular stent method to Toughen any hydrogels based on double network concept *Adv. Funct. Mater.* **22** 4426–32
[128] Zhao Y *et al* 2014 Proteoglycans and glycosaminoglycans improve toughness of biocompatible double network hydrogels *Adv. Mater.* **26** 436–42
[129] Yang F, Zhao J, Koshut W J, Watt J, Riboh J C, Gall K and Wiley B J 2020 A synthetic hydrogel composite with the mechanical behavior and durability of cartilage *Adv. Funct. Mater.* **30** 2003451
[130] Chen Q, Zhu L, Zhao C, Wang Q, Zheng J and Robust A 2013 One-pot synthesis of highly mechanical and recoverable double network hydrogels using thermoreversible sol–gel polysaccharide *Adv. Mater.* **25** 4171–6
[131] Shi X and Wu P 2021 A smart patch with on-demand detachable adhesion for bioelectronics *Small* **17** 2101220
[132] Dhand A P, Davidson M D, Galarraga J H, Qazi T H, Locke R C, Mauck R L and Burdick J A 2022 Simultaneous one-pot interpenetrating network formation to expand 3D processing capabilities *Adv. Mater.* **34** 2202261
[133] Wang Z, Wei; H, Huang; Y, Wei; Y and Chen J 2023 Naturally sourced hydrogels: emerging fundamental materials for next-generation healthcare sensing *Chem. Soc. Rev.* **52** 2992–3034
[134] Larsson D G J and Flach C-F 2022 Antibiotic resistance in the environment *Nat. Rev. Microbiol.* **20** 257–69
[135] Helander I M, Nurmiaho-Lassila E L, Ahvenainen R, Rhoades J and Roller S 2001 Chitosan disrupts the barrier properties of the outer membrane of gram-negative bacteria *Int. J. Food Microbiol.* **71** 235–44
[136] He C, Ke M, Zhong Z, Ye Q, He L, Chen Y and Zhou J 2021 Effect of the degree of acetylation of chitin nonwoven fabrics for promoting wound healing *ACS Appl. Bio Mater.* **4** 1833–42
[137] Lin Y, Xu J, Dong Y, Wang Y, Yu C, Li Y, Zhang C, Chen Q, Chen S and Peng Q 2023 Drug-free and non-crosslinked chitosan/hyaluronic acid hybrid hydrogel for synergistic healing of infected diabetic wounds *Carbohydr. Polym.* **314** 120962
[138] Zhao X, Guo B, Wu H, Liang Y and Ma P X 2018 Injectable antibacterial conductive nanocomposite cryogels with rapid shape recovery for noncompressible hemorrhage and wound healing *Nat. Commun.* **9** 2784
[139] You J, Xie S, Cao J, Ge H, Xu M, Zhang L and Zhou J 2016 Quaternized chitosan/poly (acrylic acid) polyelectrolyte complex hydrogels with tough, self-recovery, and tunable mechanical properties *Macromolecules* **49** 1049–59

[140] Qu J, Zhao X, Liang Y, Zhang T, Ma P X and Guo B 2018 Antibacterial adhesive injectable hydrogels with rapid self-healing, extensibility and compressibility as wound dressing for joints skin wound healing *Biomaterials* **183** 185–99

[141] Peng N, Wang Y, Ye Q, Liang L, An Y, Li Q and Chang C 2016 Biocompatible cellulose-based superabsorbent hydrogels with antimicrobial activity *Carbohydr. Polym.* **137** 59–64

[142] Xie F, Jiang L, Xiao X, Lu Y, Liu R, Jiang W and Cai J 2022 Quaternized polysaccharide-based cationic micelles as a macromolecular approach to eradicate multidrug-resistant bacterial infections while mitigating antimicrobial resistance *Small* **18** 2104885

[143] Hou Y, Tan T, Guo Z, Ji Y, Hu J and Zhang Y 2022 Gram-selective antibacterial peptide hydrogels *Biomater. Sci.* **10** 3831–44

[144] Gao F, Ahmed A, Cong H, Yu B and Shen Y 2023 Effective strategies for developing potent, broad-spectrum antibacterial and wound healing promotion from short-chain antimicrobial peptides *ACS Appl. Mater. Interfaces* **15** 32136–47

[145] Patrulea V *et al* 2022 Synergistic effects of antimicrobial peptide dendrimer-chitosan polymer conjugates against *Pseudomonas aeruginosa Carbohydr. Polym.* **280** 119025

[146] Zhang X, Qin M, Xu M, Miao F, Merzougui C, Zhang X, Wei Y, Chen W and Huang D 2021 The fabrication of antibacterial hydrogels for wound healing *Eur. Polym. J.* **146** 110268

[147] Cao Z *et al* 2021 Antibacterial hybrid hydrogels *Macromol. Biosci.* **21** 2000252

[148] Zhong Y, Xiao H, Seidi F and Jin Y 2020 Natural polymer-based antimicrobial hydrogels without synthetic antibiotics as wound dressings *Biomacromolecules* **21** 2983–3006

[149] Montazerian H *et al* 2022 Engineered hemostatic biomaterials for sealing wounds *Chem. Rev.* **122** 12864–903

[150] Guo B, Dong R, Liang Y and Li M 2021 Haemostatic materials for wound healing applications *Nat. Rev. Chem.* **5** 773–91

[151] Xie H, Xia H, Huang L, Zhong Z, Ye Q, Zhang L and Lu A 2021 Biocompatible, antibacterial and anti-inflammatory zinc ion cross-linked quaternized cellulose–sodium alginate composite sponges for accelerated wound healing *Int. J. Biol. Macromol.* **191** 27–39

[152] Guo S, Ren Y, Chang R, He Y, Zhang D, Guan F and Yao M 2022 Injectable self-healing adhesive chitosan hydrogel with antioxidative, antibacterial, and hemostatic activities for rapid hemostasis and skin wound healing *ACS Appl. Mater. Interfaces* **14** 34455–69

[153] Zhu Z, Zhang K, Xian Y, He G, Pan Z, Wang H, Zhang C and Wu D 2023 A Choline phosphoryl-conjugated chitosan/oxidized dextran injectable self-healing hydrogel for improved hemostatic efficacy *Biomacromolecules* **24** 690–703

[154] Chen G, Yu Y, Wu X, Wang G, Ren J and Zhao Y 2018 Bioinspired multifunctional hybrid hydrogel promotes wound healing *Adv. Funct. Mater.* **28** 1801386

[155] Park E, Lee J, Huh K M, Lee S H and Lee H 2019 Toxicity-attenuated glycol chitosan adhesive inspired by mussel adhesion mechanisms *Adv. Healthc. Mater.* **8** 1900275

[156] Wang Y, Kim K, Lee M S and Lee H 2018 Hemostatic ability of chitosan-phosphate inspired by coagulation mechanisms of platelet polyphosphates *Macromol. Biosci.* **18** 1700378

[157] Luu C H, Nguyen N-T and Ta H T 2024 Unravelling surface modification strategies for preventing medical device-induced thrombosis *Adv. Healthc. Mater.* **13** 2301039

[158] Faustino C M C, Lemos S M C, Monge N and Ribeiro I A C 2020 A scope at antifouling strategies to prevent catheter-associated infections *Adv. Colloid Interface Sci.* **284** 102230

[159] Luu C H, Nguyen N-T and Ta H T 2023 Unravelling surface modification strategies for preventing medical device-induced thrombosis *Adv. Healthc. Mater.* **13** 2301039

[160] Biran R and Pond D 2017 Heparin coatings for improving blood compatibility of medical devices *Adv. Drug Deliv. Rev.* **112** 12–23

[161] Jiang Y, Guo Y, Wang H, Wang X and Li Q 2023 Hydrogel coating based on dopamine-modified hyaluronic acid and gelatin with spatiotemporal drug release capacity for quick endothelialization and long-term anticoagulation *Int. J. Biol. Macromol.* **230** 123113

[162] Yao M *et al* 2022 Microgel reinforced zwitterionic hydrogel coating for blood-contacting biomedical devices *Nat. Commun.* **13** 5339

[163] Zhang L, Cao Z, Bai T, Carr L, Ella-Menye J-R, Irvin C, Ratner B D and Jiang S 2013 Zwitterionic hydrogels implanted in mice resist the foreign-body reaction *Nat. Biotechnol.* **31** 553–6

[164] Wang J, Sun H, Li J, Dong D, Zhang Y and Yao F 2015 Ionic starch-based hydrogels for the prevention of nonspecific protein adsorption *Carbohydr. Polym.* **117** 384–91

[165] Yao M, Sun H, Guo Z, Sun X, Yu Q, Wu X, Yu C, Zhang H, Yao F and Li J 2021 A starch-based zwitterionic hydrogel coating for blood-contacting devices with durability and bio-functionality *Chem. Eng. J.* **421** 129702

[166] Sunshine H and Iruela-Arispe M L 2017 Membrane lipids and cell signaling *Curr. Opin. Lipidol* **28** 408–13

[167] Zhao Y, Song S, Ren X, Zhang J, Lin Q and Zhao Y 2022 Supramolecular adhesive hydrogels for tissue engineering applications *Chem. Rev.* **122** 5604–40

[168] Ghobril C and Grinstaff M W 2015 The chemistry and engineering of polymeric hydrogel adhesives for wound closure: a tutorial *Chem. Soc. Rev.* **44** 1820–35

[169] Li M, Pan G, Zhang H and Guo B 2022 Hydrogel adhesives for generalized wound treatment: design and applications *J. Polym. Sci.* **60** 1328–59

[170] Li S, Cong Y and Fu J 2021 Tissue adhesive hydrogel bioelectronics *J. Mater. Chem.* B **9** 4423–43

[171] Lei X-X *et al* 2023 Multifunctional two-component *in situ* hydrogel for esophageal submucosal dissection for mucosa uplift, postoperative wound closure and rapid healing *Bioact. Mater.* **27** 461–73

[172] Pan M, Nguyen K-C T, Yang W, Liu X, Chen X-Z, Major P W, Le L H and Zeng H 2022 Soft armour-like layer-protected hydrogels for wet tissue adhesion and biological imaging *Chem. Eng. J.* **434** 134418

[173] Pan G, Li F, He S, Li W, Wu Q, He J, Ruan R, Xiao Z, Zhang J and Yang H 2022 Mussel- and barnacle cement proteins-inspired dual-bionic bioadhesive with repeatable wet-tissue adhesion, multimodal self-healing, and antibacterial capability for nonpressing hemostasis and promoted wound healing *Adv. Funct. Mater.* **32** 2200908

[174] Wu X *et al* 2022 An injectable asymmetric-adhesive hydrogel as a $GATA6^{+}$ cavity macrophage trap to prevent the formation of postoperative adhesions after minimally invasive surgery *Adv. Funct. Mater.* **32** 2110066

[175] Cao J, Wu P, Cheng Q, He C, Chen Y and Zhou J 2021 Ultrafast fabrication of self-healing and injectable carboxymethyl chitosan hydrogel dressing for wound healing *ACS Appl. Mater. Interfaces* **13** 24095–105

[176] Fisher S A, Baker A E G and Shoichet M S 2017 Designing peptide and protein modified hydrogels: selecting the optimal conjugation strategy *J. Am. Chem. Soc.* **139** 7416–27

[177] Arslan M 2020 Fabrication and reversible disulfide functionalization of PEGylated chitosan-based hydrogels: platforms for selective immobilization and release of thiol-containing molecules *Eur. Polym. J.* **126** 109543
[178] Deng Y, Li R, Wang H, Yang B, Shi P, Zhang Y, Yang Q, Li G and Bian L 2022 Biomaterial-mediated presentation of jagged-1 mimetic ligand enhances cellular activation of notch signaling and bone regeneration *ACS Nano* **16** 1051–62
[179] Lei L, Wang X, Zhu Y, Su W, Lv Q and Li D 2022 Antimicrobial hydrogel microspheres for protein capture and wound healing *Mater. Des.* **215** 110478
[180] Xiong Y, Zhang X, Ma X, Wang W, Yan F, Zhao X, Chu X, Xu W and Sun C 2021 A review of the properties and applications of bioadhesive hydrogels *Polym. Chem.* **12** 3721–39
[181] Bellis S L 2011 Advantages of RGD peptides for directing cell association with biomaterials *Biomaterials* **32** 4205–10
[182] Xu Q, Zhang Z, Xiao C, He C and Chen X 2017 Injectable polypeptide hydrogel as biomimetic scaffolds with tunable bioactivity and controllable cell adhesion *Biomacromolecules* **18** 1411–8
[183] Diaz C and Missirlis D 2023 Amyloid-based albumin hydrogels *Adv. Healthc. Mater.* **12** 2201748
[184] Huang Y, Yao M, Zheng X, Liang X, Su X, Zhang Y, Lu A and Zhang L 2015 Effects of chitin whiskers on physical properties and osteoblast culture of alginate based nanocomposite hydrogels *Biomacromolecules* **16** 3499–507
[185] Cha J and Kim P 2021 Cancer cell-sticky hydrogels to target the cell membrane of invading glioblastomas *ACS Appl. Mater. Interfaces* **13** 31371–8
[186] Li Y, Huang G, Zhang X, Wang L, Du Y, Lu T J and Xu F 2014 Engineering cell alignment *in vitro Biotechnol. Adv.* **32** 347–65
[187] Wang S, Hashemi S, Stratton S and Arinzeh T L 2021 The effect of physical cues of biomaterial scaffolds on stem cell behavior *Adv. Healthc. Mater.* **10** 2001244
[188] Lin X *et al* 2022 Anisotropic hybrid hydrogels constructed via the noncovalent assembly for biomimetic tissue scaffold *Adv. Funct. Mater.* **32** 2112685
[189] Zhong H, Huang J, Luo M, Fang Y, Zeng X, Wu J and Du J 2023 Near-field electrospun PCL fibers/GelMA hydrogel composite dressing with controlled deferoxamine-release ability and retiform surface for diabetic wound healing *Nano Res.* **16** 599–612
[190] Hou Y, Yu L, Xie W, Camacho L C, Zhang M, Chu Z, Wei Q and Haag R 2020 Surface roughness and substrate stiffness synergize to drive cellular mechanoresponse *Nano Lett.* **20** 748–57
[191] Engler A J, Sen S, Sweeney H L and Discher D E 2006 Matrix elasticity directs stem cell lineage specification *Cell* **126** 677–89
[192] Wang Q, Wang Z, Zhang D, Gu J, Ma Y, Zhang Y and Chen J 2022 Circular patterns of dynamic covalent hydrogels with gradient stiffness for screening of the stem cell microenvironment *ACS Appl. Mater. Interfaces* **14** 47461–71
[193] Chen Z and Lv Y 2022 Gelatin/sodium alginate composite hydrogel with dynamic matrix stiffening ability for bone regeneration *Composites* B **243** 110162
[194] Lou J and Mooney D J 2022 Chemical strategies to engineer hydrogels for cell culture *Nat. Rev. Chem.* **6** 726–44
[195] Dou X-Q and Feng C-L 2017 Amino acids and peptide-based supramolecular hydrogels for three-dimensional cell culture *Adv. Mater.* **29** 1604062

[196] Rizwan M, Baker A E G and Shoichet M S 2021 Designing hydrogels for 3D cell culture using dynamic covalent crosslinking *Adv. Healthcare Mater.* **10** 2100234

[197] Vaisocherová H, Brynda E and Homola J 2015 Functionalizable low-fouling coatings for label-free biosensing in complex biological media: advances and applications *Anal. Bioanal. Chem.* **407** 3927–53

[198] Liu S and Guo W 2018 Anti-biofouling and healable materials: preparation, mechanisms, and biomedical applications *Adv. Funct. Mater.* **28** 1800596

[199] Pavithra D and Doble M 2008 Biofilm formation, bacterial adhesion and host response on polymeric implants—issues and prevention *Biomed. Mater.* **3** 034003

[200] Lynn A D, Kyriakides T R and Bryant S J 2010 Characterization of the *in vitro* macrophage response and *in vivo* host response to poly(ethylene glycol)-based hydrogels *J. Biomed. Mater. Res.* A **93** 941–53

[201] Buzzacchera I, Vorobii M, Kostina N Y, de los Santos Pereira A, Riedel T, Bruns M, Ogieglo W, Möller M, Wilson C J and Rodriguez-Emmenegger C 2017 Polymer brush-functionalized chitosan hydrogels as antifouling implant coatings *Biomacromolecules* **18** 1983–92

[202] Peng L, Chang L, Si M, Lin J, Wei Y, Wang S, Liu H, Han B and Jiang L 2020 Hydrogel-coated dental device with adhesion-inhibiting and colony-suppressing properties *ACS Appl. Mater. Interfaces* **12** 9718–25

[203] Tang Y, Cai X, Xiang Y, Zhao Y, Zhang X and Wu Z 2017 Cross-linked antifouling polysaccharide hydrogel coating as extracellular matrix mimics for wound healing *J. Mater. Chem.* B **5** 2989–99

[204] Liu R-R, Shi Q-Q, Meng Y-F, Zhou Y, Mao L-B and Yu S-H 2024 Biomimetic chitin hydrogel via chemical transformation *Nano Res.* **17** 771–7

[205] Erathodiyil N, Chan H-M, Wu H and Ying J Y 2020 Zwitterionic polymers and hydrogels for antibiofouling applications in implantable devices *Mater. Today* **38** 84–98

[206] Li Q *et al* 2022 Zwitterionic biomaterials *Chem. Rev.* **122** 17073–154

[207] Yu Q *et al* 2023 Zwitterionic polysaccharide-based hydrogel dressing as a stem cell carrier to accelerate burn wound healing *Adv. Healthc. Mater.* **12** 2202309

[208] Qiu X, Zhang J, Cao L, Jiao Q, Zhou J, Yang L, Zhang H and Wei Y 2021 Antifouling antioxidant zwitterionic dextran hydrogels as wound dressing materials with excellent healing activities *ACS Appl. Mater. Interfaces* **13** 7060–9

[209] Gosecka M, Gosecki M and Jaworska-Krych D 2023 Hydrophobized hydrogels: construction strategies, properties, and biomedical applications *Adv. Funct. Mater.* **33** 2212302

[210] Wankar J, Kotla N G, Gera S, Rasala S, Pandit A and Rochev Y A 2020 Recent advances in host–guest self-assembled cyclodextrin carriers: implications for responsive drug delivery and biomedical engineering *Adv. Funct. Mater.* **30** 1909049

[211] Pivato R V, Rossi F, Ferro M, Castiglione F, Trotta F and Mele A 2021 β-cyclodextrin nanosponge hydrogels as drug delivery nanoarchitectonics for multistep drug release kinetics *ACS Appl. Polym. Mater.* **3** 6562–71

[212] Shukla A, Singh A P, Dubey T, Hemalatha S and Maiti P 2019 Third generation cyclodextrin graft with polyurethane embedded in hydrogel for a sustained drug release: complete shrinkage of melanoma *ACS Appl. Bio Mater.* **2** 1762–71

[213] Hosseinifar T, Sheybani S, Abdouss M, Hassani Najafabadi S A and Shafiee Ardestani M 2018 Pressure responsive nanogel base on alginate-cyclodextrin with enhanced apoptosis mechanism for colon cancer delivery *J. Biomed. Mater. Res.* A **106** 349–59

[214] Takei T, Yoshihara R, Danjo S, Fukuhara Y, Evans C, Tomimatsu R, Ohzuno Y and Yoshida M 2020 Hydrophobically-modified gelatin hydrogel as a carrier for charged hydrophilic drugs and hydrophobic drugs *Int. J. Biol. Macromol.* **149** 140–7

[215] Li S, Gao Y, Jiang H, Duan L and Gao G 2018 Tough, sticky and remoldable hydrophobic association hydrogel regulated by polysaccharide and sodium dodecyl sulfate as emulsifiers *Carbohydr. Polym.* **201** 591–8

[216] Wang D, Ha Y, Gu J, Li Q, Zhang L and Yang P 2016 2D protein supramolecular nanofilm with exceptionally large area and emergent functions *Adv. Mater.* **28** 7414–23

[217] Liu H *et al* 2022 Bio-inspired self-hydrophobized sericin adhesive with tough underwater adhesion enables wound healing and fluid leakage sealing *Adv. Funct. Mater.* **32** 2201108

Chapter 5

Bio-based aerogels for EMI shielding applications

Suji Mary Zachariah, Gopika G Nair, Yves Grohens and Sabu Thomas

5.1 Introduction

In an era where electronic devices are becoming more and more common and where wireless communication networks are growing, electromagnetic interference (EMI) shielding is becoming more and more important. The functionality, dependability, and safety of electronic equipment can all be negatively impacted by EMI, or undesired disturbances that interfere with electrical circuits as a result of radiation and conduction. Because of this, there is a great need for EMI shielding materials that are durable, effective, and lightweight. Aerogels are extremely lightweight, porous materials with significant porosity, low density, and limited heat conductivity [1]. Aerogels are unique solids that obtain their extraordinary qualities from their nanoporous structure. They are gels consisting of a distributed gas phase within a microporous solid. Because the gas component of an aerogel replaces the liquid component without significantly collapsing the solid network structure, aerogels have a high specific surface area and a relatively low density. Because of their translucent quality, they are also referred to as 'frozen smoke' or 'blue smoke'. These are three-dimensional networks of interconnected nanoparticles that provide special qualities for a range of research and commercial uses. The last ten years have witnessed a paradigm shift in aerogel research, with an emphasis on bio-based aerogels made from renewable resources. Traditionally, silica-based aerogels were employed for a wide range of applications. Their potential for sustainability, biodegradability, adaptability to a range of desired qualities, and reduced environmental impact in contrast to conventional, non-renewable materials make them especially interesting [2, 3]. Recent advancements in the production of conductive materials have enabled their integration, allowing them to serve as effective EMI shields without sacrificing their inherent beneficial properties. The combination of the inherent properties of aerogels with the most recent advancements in bio-based materials has opened the way for environmentally friendly EMI shielding solutions.

doi:10.1088/978-0-7503-6184-2ch5

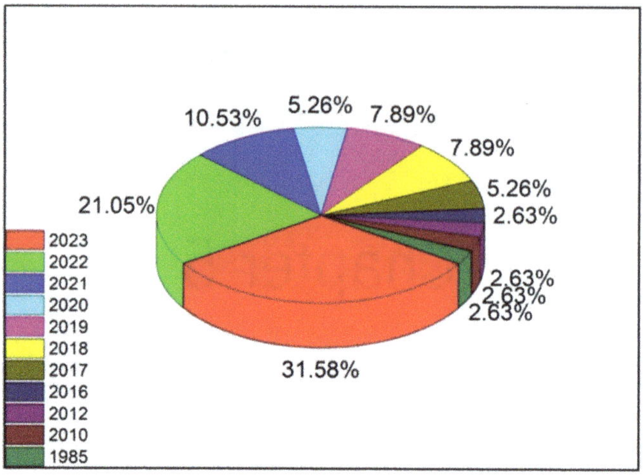

Figure 5.1. Pie chart of the number of research publications per year in the area of biomaterials for EMI shielding applications (data from Science Direct).

However, the biological breakdown of these aerogels may not always be advantageous, particularly in applications requiring long-term stability.

Figure 5.1 illustrates the percentage of publications focusing on EMI shielding materials of bio-based origin. The rate of progress over the years has been quite slow, likely due to the complexity of research in this interdisciplinary field, along with performance challenges and other factors. In this chapter, we will explore the realm of bio-based aerogels and their potential applications in EMI shielding in detail. Their intrinsic properties, combined with their environmentally benign nature, offer a unique combination of advantages. The ability to embed or disperse conductive fillers such as carbon nanotubes, graphene, and metallic nanoparticles within these aerogels has opened avenues to enhance their EMI shielding effectiveness without adding significant weight. This is particularly pivotal for industries such as aerospace, where the balance between shielding efficacy and weight can dictate the feasibility of a material's application. However, like any new technology, there are challenges to overcome. Ongoing research is focused on extending the lifespan of these materials, particularly in light of their possible biodegradability, and optimizing the trade-off between the incorporation of conductive fillers and their inherent aerogel qualities. Nevertheless, bio-based aerogels stand out as a great example of sustainable innovation as the need for EMI shielding materials grows in our rapidly advancing technological world. These materials align with the core objective of modern research: integrating sustainability with functionality. There is cause for optimism regarding the increasing role these materials will play in defining EMI shielding solutions in the future as research progresses.

5.2 Overview of aerogels

Aerogels have a long history dating back to the early 1900s, with Samuel Kistler creating the first silica-based aerogel in 1931. By substituting a gas for the gel's liquid component, Kistler's creative research produced a solid that preserved the gel's structure but was primarily made of air. The first silica aerogel was produced by this technique, and it had the characteristic low density and high porosity of aerogels. Since Kistler's discovery, researchers have added more types of aerogels to the family, such as carbon and polymer aerogels. Thanks to considerable advancements in aerogel fabrication techniques, aerogel characteristics may now be tailored to suit particular applications. Supercritical drying, sol–gel processes, and changes in precursor materials have all facilitated the evolution of aerogels with increased mechanical strength, greater flexibility, and broader applications.

Three distinct methods are used to generate aerogels: the hydrothermal, sol–gel, and template methods. Metal oxides, polymers such as cellulose and carbon, and silica may all be transferred into materials at low temperatures and low cost using the sol–gel process. However, silica gels are among the most widely utilized and easiest to make. The initial stage in synthesis is the production of a colloidal suspension, or sol, into which the precursor material is disseminated in a solution of reactants, solvents, and catalysts. This promotes events related to polymerization, hydrolysis, and polycondensation inside the gel-like structure, which leads to branching and cross-linking. The last step involves drying using either ambient pressure drying (APD), freeze-drying (FD), or supercritical drying (SCD). Depending on the method used, the products are referred to as aerogels, cryogels, or xerogels, in that order. The drying stage has a significant impact on aerogel characteristics because it can damage the porous framework. During the SCD process, the wet gel is heated in a closed container above the critical parameters, which vary based on the impregnated fluid. Since CO_2 is non-flammable, inert, and non-toxic, and has mild critical values of 31 °C and 73.8 bar of pressure, it is commonly utilized. FD involves freezing at a low temperature and sublimation in a vacuum [4]. Carbon, silica, cellulose, clay, graphene, polyvinyl alcohol, and pectin aerogels have all been successfully produced by freeze-drying. Although FD is less expensive than SCD and a more environmentally friendly method for creating porous structures with minimal shrinkage, its usage is restricted due to the risk of cracking caused by solvent crystallization within the pores [5, 6]. APD is another popular drying method that is far less expensive than SCD. Instead of evaporating under pressure and heat, the liquid trapped in the gel's pores is exchanged with a solvent having a low surface tension. This also results in a chemical change to the gel surface, where polar surface groups are replaced by non-polar ones. Following that, the solvent leaves the gel's pores at room temperature. The fourth technique is thermal drying evaporation, which creates Xerogels by directly evaporating the liquid. However, the high surface tension causes an increase in capillary pressure, which in turn causes low porosity and substantial shrinkage.

With hydrothermal synthesis, chemical reactions are carried out in a sealed environment (such as Teflon-lined autoclaves) at temperatures and pressures higher than those found outside. The process of creating functional molecules and materials one after the other *in situ* by employing a substrate is known as template synthesis. With aerogels, a network can be built on top of an existing structure or an aerogel can be used for additional functionalization once it has been formed. There are also a few known changes that provide aerogels with improved qualities. Better absorption of electromagnetic waves could be improved by both unidirectional and bidirectional freeze-drying. Third, polymer impregnation for mechanical stability and endurance; and, second, high-temperature heat treatments for improved conductivity.

5.3 General concepts, mechanisms and terminologies of EMI shielding

The process of shielding any material from the potentially harmful effects of stray radiations is known as EMI shielding. Radiation cannot get through the shield due to a shielding system that uses absorption, reflection, or numerous internal reflections. The electrical conductivity, relative permittivity, magnetic permeability, and interior surface area of a material all affect its capacity to shield. Among the different types of shielding systems, reflection is the primary mechanism for EMI shielding. This is especially apparent in materials with substantial conductivity. There should be electromagnetic field-interacting mobile charge carriers in the material. However, good conduction path connectivity is more important than high conductivity, thus it is not a necessary and sufficient condition. In materials with low electrical conductivity and electric or magnetic dipoles, another mechanism that emerges is absorption. For this, high magnetic permeability and dielectric constant materials are used. Shielding effectiveness increases with thickness because absorption is proportionate to the material's dimensions. Multiple internal reflections are an additional EMI shielding technique. It refers to the reflections generated by the shield's various surfaces or points of contact. For numerous internal reflections, a large surface area is required; porous or foam materials provide this. The mechanism under consideration is only relevant to thin films and porous media. Thick materials lose all meaning at thicknesses greater than the skin depth, which represents the distance required to reduce the wave to $1/e$, or 37% [7]. The shielding efficiency is described as the sum total of the above three processes and is given by

$$SE_{total} = SE_R + SE_A + SE_M, \tag{5.1}$$

where SE_{total} denotes the total shielding efficiency, SE_R, SE_A, and SE_M correspond to the shielding efficiency via reflection, absorption, and multiple internal reflections, respectively.

The predominant reflection or absorption loss is determined by the incident radiation and the material's properties. Electrical conductivity is the most important factor in EMI shielding. While EM wave absorbers can provide EMI absorption, only electrically conductive materials can provide an adequate level of shielding efficiency. A higher frequency results in a decrease in reflection loss but an increase in absorption loss. While the increase in absorption loss is partly caused by the decrease in skin depth, the drop in reflection loss is caused by the increase in shield impedance with frequency. The higher the dielectric polarizability of the material, the greater the EMI shielding effectiveness (SE). The conductivity mismatch between the matrix and filler is the cause of this. The interaction between the radiation's electric field and the material's electric dipoles increases absorption.

Furthermore, when they interact with the magnetic dipoles, materials having a significant magnetic permeability absorb more electromagnetic radiation. When the shield's thickness increases, absorption loss increases while reflection loss stays constant. Thicker shields with increased magnetic permeability and electrical conductivity yield better absorption of electromagnetic waves. Electrical connectivity enhances shielding by the continuous electric field lines formed in the material. This also enhances magnetic power loss by allowing the flow of eddy current induced by the magnetic field. Hence, within a composite which is formed from a conductive filler and non-conducting matrix, electrical connectivity is favorable for shielding [8, 9].

EMI mitigation can be sustainably designed using bio-based composites in various morphological forms, such as aerogels, films, and papers. Incorporating conductive biopolymers from sustainable sources into these materials enables the creation of lightweight and efficient EMI shielding solutions. Bio-based aerogels, known for their low density and excellent insulation properties, can be integrated into electronic devices, while composite films and papers provide versatile coatings and substrates for EMI shielding. Furthermore, the development of bio-based blends, combining conductive and non-conductive biopolymers, offers a customizable approach to address specific EMI challenges. This innovative use of bio-based materials not only contributes to enhanced EMI protection but also aligns with sustainable, environmentally friendly practices, presenting a significant advancement in both electronic device performance and ecological responsibility.

5.4 Research progress of bio-based aerogels for EMI shielding

Research on bio-based aerogels for EMI shielding has been steadily progressing, driven by the demand for sustainable and effective materials in electronics, telecommunications, and other industries. Bio-based aerogels, in particular, have garnered significant attention due to their unique properties, making them a promising option for both environmental sustainability and technological innovation in EMI shielding applications. Research on these bio-based aerogels includes natural biopolymers such as cellulose, lignin, chitosan, as well as synthetic biopolymers such as polyvinyl alcohol (PVA), and polycaprolactone (PCL). In the following sections,this chapter discusses the synthesis, processing techniques, and functionalization strategies of various bio-based aerogels for EMI shielding.

5.5 Natural biopolymers and their composites for EMI shielding

5.5.1 Wood-based aerogels

Wood-based composites, which are inexpensive, lightweight, sustainable, and naturally porous with a hierarchical structure, have been the subject of extensive research in recent years. The lack of interfacial compatibility and electrical conductivity along with the intrinsic defects including rotting and flammability deteriorate the practical application of wood as a material for shielding against EM radiations. An effective strategy to overcome these limitations of wood is to develop multifunctional composites of wood with the incorporation of suitable nanomaterials. These composites can combine the properties of wood as well as the characteristics of the nanomaterials. Wood-based composites are emerging as promising materials for numerous applications which include wood–plastic composites which possess resistance to weather, high strength wood ceramics and electrically conducting wood–metal composites [10–14]. Wood, composed of cellulose, hemicellulose, and lignin, is particularly accessible for chemical modification, as well as being biocompatible and biodegradable, making it an ideal candidate for the fabrication of sustainable EMI shielding and microwave absorption (MA) materials. The porous structure and large surface area of natural wood is considered to be beneficial for microwave absorption. The presence of pores within wood facilitates impedance matching and also enables the radiation to undergo multiple reflections. Moreover, the porous structure of wood allows the facile penetration of the nanomaterials into its interior and hence imparts it with properties such as electrical conductivity and shielding ability [10].

Various methods have been employed for the fabrication of wood-based EMI shielding and microwave absorbing materials (figure 5.2). Among the different techniques, direct internal synthesis is one of the most accepted methods. In this method, the wood is impregnated with appropriate particles directly. Since particles with larger size cannot be impregnated to the interior of the wood, *in situ* co-precipitation methods are now preferred. An insufficient amount of filler in the wood

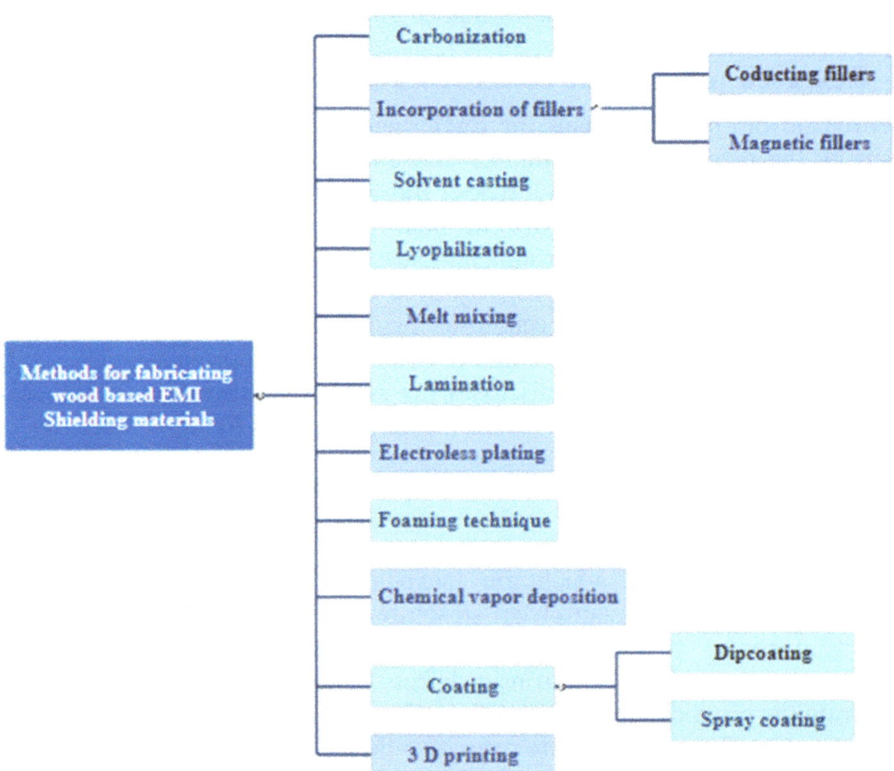

Figure 5.2. Different methods for fabricating wood-based EMI shielding materials.

leads to poor absorption properties, which presents a key limitation of this fabrication technique. Different surface deposition methods such as electroless plating, the wet chemical method, the sol–gel method, the hydrothermal method and chemical vapor deposition have been adapted for introducing conducting and magnetic nanomaterials to the surface of wood. Surface deposition methods are efficient for controlling the surface coating and thus the electromagnetic properties of the composites. The limitations of direct internal synthesis and surface coating methods can be addressed through a more effective approach known as hot pressing. This method can be used for fabricating EMI shielding and microwave absorbing materials with smaller units of wood such as wood fiber, wood flour and wood cellulose. In this method, the wood is initially combined with the desired filler components or adhesives and then molded into boards or films via hot pressing [10]. Through all these methods a large number of wood-based composites for EMI shielding have been developed with exceptional shielding efficiency and microwave absorption properties.

Metal-based composites: Metals and metal oxides are important candidates among EMI shielding materials due to their high saturation magnetization, excellent dielectric properties, and significant magnetic loss capabilities. Incorporating metal or metal oxide nanoparticles into the wood matrix is an effective strategy to develop shielding materials. The hierarchical morphology of the wood and the existence of many active groups and rough sites permit the simple inclusion of metal/metal oxides into it and also help these particles to develop continuous and interconnected conductive networks within the wood matrix [15–19]. Electroless plating serves as a good method for incorporating metal/metal oxide nanoparticles into wood. A composite of birch veneer and copper was fabricated by Sun *et al* using electroless plating, a method that uses a chemical reduction reaction rather than an external electrical power source to deposit a metal layer onto a substrate. The composite exhibited an outstanding EMI SE of 60 dB in the frequency range of 10 MHz to 1.5 GHz. A similar method was carried out by Liu *et al* to develop a Nickel/wood composite leading to a composite with promising shielding efficiency [20]. Even though metal-based composites possess high conductivity as well as SE, their corrosive nature reduces their suitability in practical applications [21]. To overcome this, Shu *et al* coated wood veneers with an Ni–Mo–P alloy, known for its superior corrosion resistance, using electroless plating. In a 3.5% NaCl solution, they observed that increasing the percentage of Mo in the alloy led to a higher corrosion potential and a lower corrosion current density, indicating improved corrosion resistance of the composite. Along with the better corrosion inhibition, the composite achieved good electrical conductivity as well as a maximum shielding efficiency of ~60 dB in the frequency range of 9 kHz to 1.5 GHz [22].

In addition to metals and their alloys, metal oxides are also promising candidates for developing EMI shielding materials. Metal oxides possess outstanding electromagnetic characteristics, chemical stability, simple synthesis routes and low cost, which enhances their suitability to be incorporated in matrices such as wood to fabricate EMI shielding and microwave absorbing materials. Gao *et al* could produce a magnetic wood composite with considerably high saturation magnetization (M_s = 25.3 emu g^{-1}) by incorporating Fe_3O_4 into wood by using an impregnation method. The strong magnetic properties of this composite contribute effectively to magnetic loss, enhancing its performance in relevant applications [23]. A magnetic carbon composite graphite/Fe_3O_4/Fe_3C was developed via an *in situ* pyrolysis of Fe_3O_4/wood shavings. The EM wave absorption ability of the composite was observed to be elevated by the phase shift from amorphous carbon to graphitic carbon through carbonization at high temperature. Moreover the continuous distribution of Fe_3O_4/Fe_3C which provided the impedance matching and superior dielectric and magnetic loss contributed to the microwave absorption performance of the lightweight composite and hence achieved a reflection loss (RL) of −26.72 dB at 10.52 GHz with a thickness of 3.15 mm [24].

Wood-based carbon composites: The EMI shielding and MA efficiency of carbon-based materials are being explored extensively due to their excellent electrical conductivity, lightweight, and outstanding resistance towards oxidation. Because of its distinct porous and hierarchical nanostructure, carbon derived from biomass, particularly that generated from wood, is highly desirable when it is used for EMI shielding and MA materials [25–27]. A porous biomass-derived pyrolyzed carbon-based microwave absorbing material was developed from natural wood by Xi *et al* with an orderly parallel channel structure. A small part of the incident EM wave was reflected off the side walls of the channels, while the majority of the radiation entered the channels and was attenuated through absorption loss. The material exhibited a maximum RL of −68.3 dB and an absorption bandwidth of 7.63 GHz. The intrinsic structure of the wood played a pivotal role in the shielding performance of the material. Additionally, the absorption loss was highly sensitive to temperature, as the samples annealed at different temperatures showed variations in carbon content, which in turn affected their electrical conductivity [28]. Ziming *et al* used natural wood to fabricate an interconnected carbon scaffold. They employed a sequential process of delignification followed by carbonization to create the scaffolds. The scaffolds were backfilled with an epoxy matrix in order to make an epoxy/carbon (EP/carbon) composite for EMI shielding. The microchannel structure and the increased graphitic degree of the carbon scaffold, both resulting from the high-temperature carbonization of wood, contributed to the composite's high SE. The EP/carbon composite possessed a conductivity of 12.5 S cm^{-1} and a thermal conductivity of 0.58 W $(m\ K)^{-1}$. The average SE of the composite was found to be 27.8 dB in the X-band with a thickness of 2 mm [28].

A flexible shielding material with better microwave absorption was introduced by Dong *et al.* They decorated porous biomass-derived carbon (PBDC) with *in situ* grown MnO nanorods. The synergistic effect between the PBDC and MnO and the porous structure of the material allowed the material to achieve a minimum reflection loss of −51.6 dB at 10.4 GHz with a thickness of 2.47 mm [29]. Another effective strategy to improve the microwave absorption of carbon composites is the incorporation of magnetic materials. Zheng *et al* incorporated Ni nanoparticles into a three-dimensional porous carbon skeleton derived from wood stalks to obtain a multifunctional bio-based EMI shielding material. The carbon composite fabricated through pyrolysis consisted of a highly ordered anisotropic porous architecture and functions as a dense 3D electrically conductive network.

The embedding of nickel nanoparticles made the composite magnetic and could impart a better EMI shielding ability. The magnetic carbon composite attained an SE of 50.8 dB in the X-band with a very small thickness of 2 mm. Moreover, the composite exhibited outstanding mechanical properties with a compressive strength of 11.7 MPa and a super-hydrophobic nature with a water contact angle of 152.1° [30]. A wood derived three-dimensional porous biochar/iron (Fe) composite fabricated by Lou *et al* could also acquire excellent microwave absorbing property and possessed a minimum reflection loss of −57.6 dB at 6.92 GHz. Both the dielectric and magnetic losses played a key role in imparting this superior shielding ability to the composite [31].

Conducting polymer-based composites: Integrating the unique properties of conducting polymers such as electrical conductivity to the mechanical and structural advantages of wood is a promising technique to develop sustainable electromagnetic shielding and microwave absorbing materials. *In situ* polymerization is a facile method to incorporate conducting polymers into wood. He *et al* used this method to graft polyaniline (PANI) on wood and hence developed an EMI shielding composite. The composite could achieve a maximum electrical conductivity of 9.23 S cm^{-1} and a maximum SE of −60 dB [32]. In similar method Zhang *et al* also synthesized flexible poplar composite membranes coated with PANI for EMI shielding. Poplar wood was treated chemically and by ultrasonication to produce homogeneous cellulose nanofibers. These nanofibers were subsequently used to create a flexible cellulose nanofiber/polyaniline (CNF/PANI) membrane with a core–shell heterostructure. A 0.28 mm composite membrane with 50% PANI demonstrated an SE of 25.2 dB. The shielding ability was ascribed to the porous and core–shell heterostructure of the composite [33].

Gan *et al* investigated the EMI shielding efficiency of a wood/polypyrrole (PPy) composite. They impregnated an interconnected layer of PPy into wood channels by *in situ* chemical vapor deposition, which imparted a high electrical conductivity of 39 S cm^{-1} to the composite. The composite achieved an SE of ~58 dB with a thickness of 3.5 cm in the X-band. The process of fabricating this composite did not involve the carbonization of the wood and thus retains up to ~28.7 times more compressive and tensile strength compared to conventional carbonized wood-based conducting and EMI shielding composites [34]. A non-carbonized nanostructured wood composite was prepared by Chen *et al*. A porous wood aerogel with a 3D network structure was fabricated by removing lignin and hemicellulose from wood, followed by *in situ* polymerization of aniline, which resulted in the formation of a conducting aerogel with 22.07 S cm^{-1} electrical conductivity. The SE of the composite aerogel was found to be 27.63 dB in the X-band with a thickness of 2–3 mm. More interestingly, the aerogel showed outstanding flame retardancy and the ability of photothemal conversion under near-infrared (NIR) irradiation, thus making it a multifunctional material [35].

5.5.2 Cellulose-based aerogels

Cellulose is derived from plant sources, making it renewable and abundant. It is a linear polymer made up of repeating glucose units linked together by

β-1,4-glycosidic bonds. Its unique structure, characterized by long and straight chains, enables the formation of strong fibers. These fibers are highly versatile, offering tailorability, cost-effectiveness, and compatibility with other materials. Additionally, their potential for multifunctionality such as providing mechanical strength, thermal insulation, or flame retardancy makes them suitable for a wide range of applications, including EMI shielding. Aerospace and automobile industries usually require lightweight materials. Owing to these advantages, cellulose is used as a substrate, binder, and derived carbon material for EMI shielding materials. Regardless of the matrix used, five primary types of aerogel-based composites, carbon-based (carbon nanotubes (CNTs), graphene, reduced graphene oxide (rGO)), metal-based (MXene), biomass-based, polymer-based, and hybrid composites (combining two or more different fillers), are being researched extensively for this application.

Carbon-based composites: Using a urea/sodium hydroxide/CNT/cellulose aqueous suspension and freeze-drying, Huang *et al* [36] constructed an environmentally friendly conductive aerogel. The structure was created by varying the cellulose concentration in the NaOH/urea solution, with the continuous water phase serving as a pore-foaming agent. The reported shielding effectiveness ranged from approximately 12.8–14.6 dB for a 10 wt% CNT loading, and 18.7–22.5 dB for a 12 wt% CNT loading (figure 5.3). Further the specific shielding effectiveness (SSE) values were reported to be 183 and 219 dB cm^3 g^{-1}, which is significantly higher than for typical metals such as copper, which has an SSE of 10 dB cm^3 g^{-1}. Given that cellulose composite aerogels have extremely high porosities of roughly 94%, EM microwaves can easily penetrate the aerogel's interior with minimal reflection. As incoming microwaves reflect and scatter between the cell walls and CNTs within the cellulose composite aerogel, they become increasingly attenuated and confined due

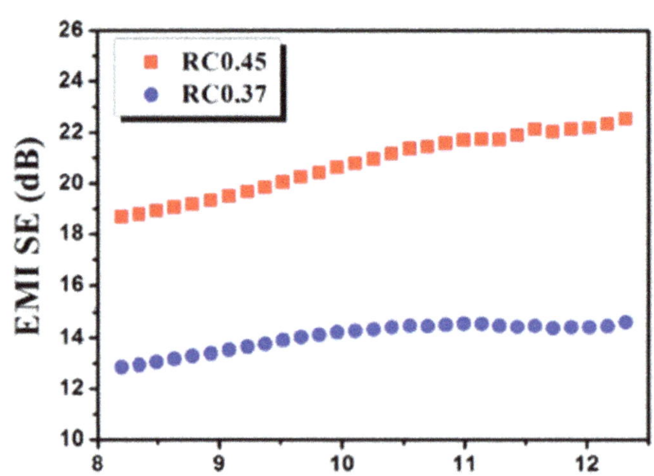

Figure 5.3. EMI shielding effectiveness of regenerated cellulose (RC) 0.45 (12 wt% CNT loading) and RC 0.37 (10 wt% CNT loading).

to the aerogel's limited exits. This 'trap' makes it difficult for the microwaves to escape, ultimately leading to their absorption and conversion into heat.

Li *et al* [37] reported a lightweight cellulose nanofibril/reduced graphene oxide carbon aerogel fabricated by unidirectional freeze-drying and pyrolysis. The results show that the aerogels with unidirectionally aligned pores possess better compression resilience and EMI shielding performance in the radial direction. The unidirectional aerogel with low density (≈ 0.0058 g cm^{-3}) exhibits a high EMI SE of ≈ 33 dB at the X-band and a specific EMI SE of 5759 dB cm^3 g^{-1}. Therefore, the fabricated aerogel demonstrated a promising potential in the field of next-generation EMI shielding materials.

Through simple processing methods, Pai *et al* [38] created a range of biodegradable THz absorbers that resulted in highly conductive CNF/SBC aerogels that were exclusively made from biomass. First, they treated sugarcane bagasse in a hydrothermal autoclave at 200 °C for 24 h to fabricate conductive biochar. This was followed by freeze-drying and pyrolysis at 900 °C in a N_2 environment. Following this, CNFs and sustainable biochar (SBC) were combined in weight ratios of 1:1, 1:2.5, and 1:5. The aerogels showed a maximum EMI SE of 70 dB at 0.8 THz at a 1:5 ratio. It is worth noting that these functional THz shields exhibit minimal reflection of incoming signals, primarily absorbing the THz waves. Due to the high porosity of aerogels, the incoming THz radiation becomes scattered within the multiple interfaces of the conductive porous network and is absorbed without generating any secondary reflection or further EM pollution. Further, graphite-like structures in the biochar significantly enhance conductivity and contribute to improved absorption.

Conducting polymer-based composites: Pai *et al* [39] developed conductive aerogels via *in situ* polymerizing PANI over CNF followed by freeze-drying. Figure 5.4 shows the reaction scheme for the *in situ* polymerization. The aerogel showed a shielding ability of 32 dB (for 1:1 ratio of PANI:CNF) in the X-band with 0.019 25 g c.c.$^{-1}$ density, 98.6% porosity, specific EMI SE of $\sim$1667 dB cm^3 g^{-1}, and ultra-fast heat dissipation capability upon high power microwave exposure. The dielectric parameters also confirmed the better absorption capability of PANI/CNF aerogel with large dielectric losses. The formation of a well defined porous network (figure 5.4(B)) with the strong inter-molecular hydrogen bonding between OH groups of CNF and nitrogen groups of PANI resulted in excellent shielding values.

A lightweight composite made of polypyrrole/cellulose aerogel (PPy/CA) was created using freeze-drying and *in situ* polymerization methods. The findings demonstrate that varying the compression ratios from 0% to 65% can successfully regulate the complex permittivity of the PPy/CA composite. At a compression ratio of 65%, the composite shows optimal microwave absorption performance. At 8.53 GHz, with a thickness of 5 mm, the minimum RL value hits -12.24 dB. Furthermore, by varying the thickness within a range of 4–5 mm, the PPy/CA composite's effective bandwidth (RL less than -10 dB) can span the whole X-band (8.2–12.4 GHz), as shown in figure 5.5. It could prove to be a promising microwave absorber in the future due to its well-controlled microwave absorption capability, green nature, lightweight design, and strong absorption [40].

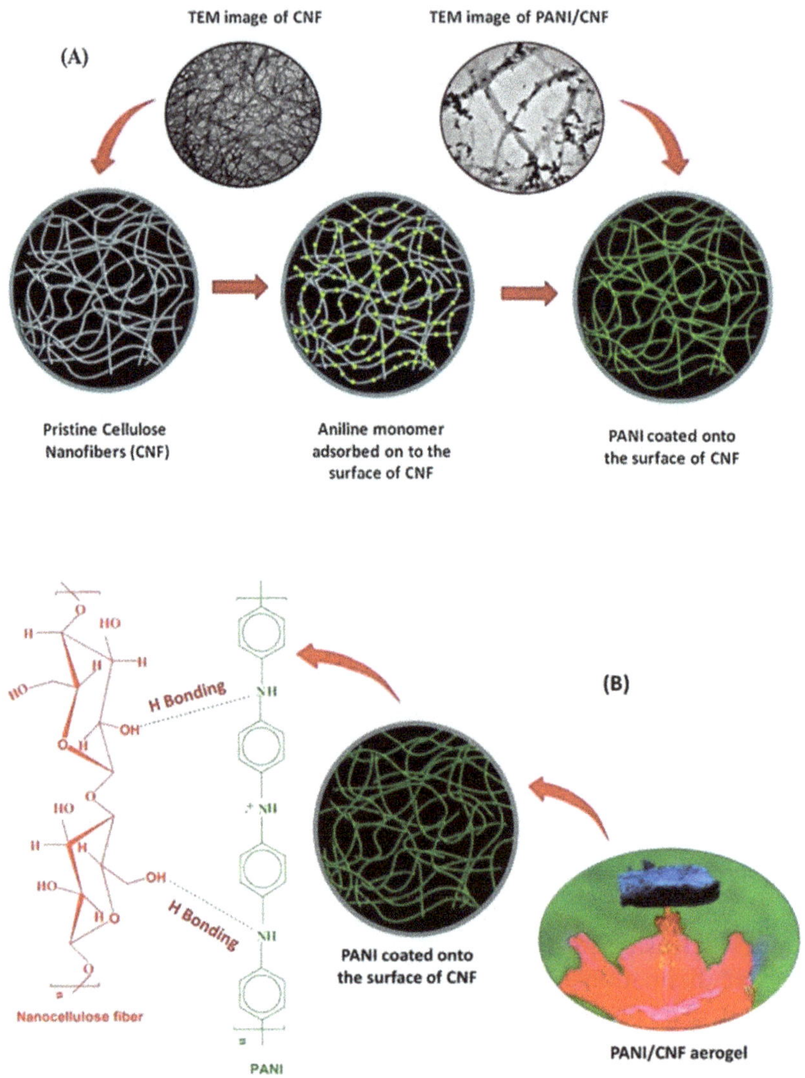

Figure 5.4. (A) Scheme for *in situ* polymerization of aniline monomer onto the surface of cellulose nanofibers. (B) Specific secondary interactions between PANI and CNF and a photograph depicting the lightweight nature of PANI/CNF aerogel. (Reproduced with permission from [39]. Copyright 2020 Elsevier.)

Metal-based composites: Fei *et al* [41] fabricated a novel lightweight Co/C@CNF aerogel by freeze-drying CNF anchored Zeolitic Imidazolate Framework-67 (ZIF-67) followed by annealing between 700 °C and 900 °C. The CNF provided a 3D structure which acted as a continuous and interconnected conductive network, with Co/C nanoparticles uniformly dispersed throughout the aerogel, thereby imparting an effective shielding property. Enhancing the annealing temperature also had an apparent effect on shielding ability, due to the graphitization degree and improved magnetic performance. The shielding effectiveness was reported to be 35.1 dB under

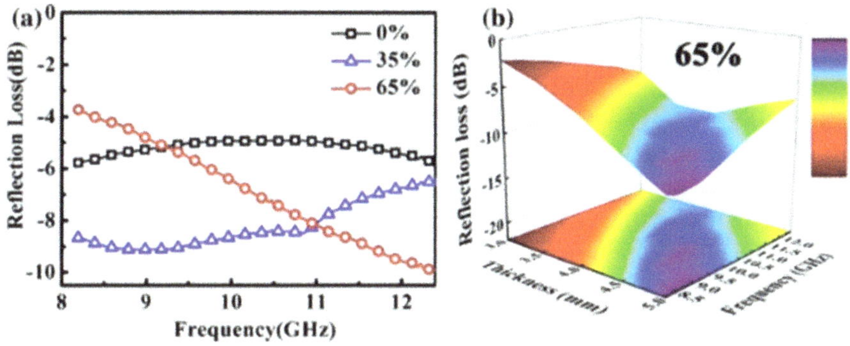

Figure 5.5. (a) Reflection loss properties of PPy/CA under different compression ratios. (b) Three-dimensional reflection loss curves of PPy/CA with the compression ratio of 65%. (Reproduced with permission from [40]. Copyright 2020 Elsevier.)

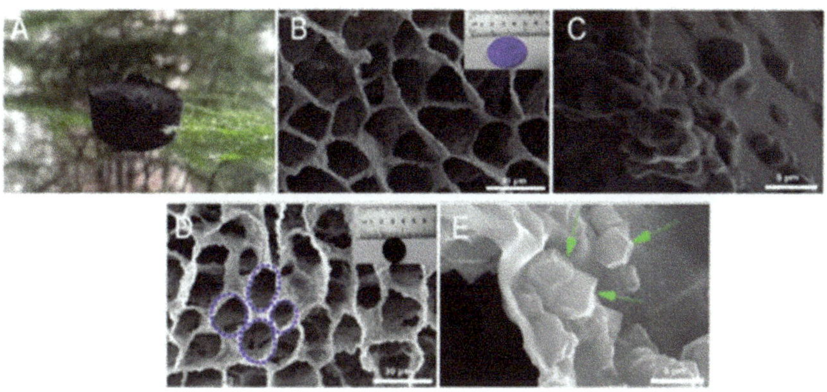

Figure 5.6. (A) Photos of Co/C@CNF aerogel, (B), (C) SEM images of ZIF-67/CNF aerogel, and (D), (E) microstructure of Co/C@CNF-900.

the conditions of 900 °C, with ultra-high absorption performance in the X-band. Its corresponding density was as low as 1.74 mg cm^{-3}, and the specific SE was 20 172.4 dB cm^3 g^{-1}. The SEM images in figures 5.6(B)–(E) show a continuous and interconnected network after pyrolysis, which made the aerogel conducive to attenuate much electromagnetic wave energy. Figure 5.6(D) shows the open porous structure with size ranging from 20 to 30 μm.

Hybrid filler-based composites: Chen *et al* [42] used a co-precipitation approach to create aerogels based on cellulose, rGO, and Fe_3O_4. To create the cellulose/GO hydrogels, a solution of NaOH, urea, and water was first pre-cooled to 12 °C. Then, cellulose and GO were added, and the material was cast and subsequently coagulated with H_2SO_4. To create the cellulose/rGO composite aerogels, the hydrogel was lyophilized after being submerged in a vitamin C solution for GO reduction. To create cellulose/rGO/Fe_3O_4 aerogels, the material was subsequently submerged in a solution of NaOH, $FeCl_3{\cdot}6H_2O$, and $FeCl_2{\cdot}4H_2O$ and then washed. The influence of rGO loading was studied which revealed that addition of rGo enhanced the total shielding ability. Aerogel with (5 wt% rGO and 8 wt% Fe_3O_4) and without (0 wt% rGO and 8 wt% Fe_3O_4) rGO showed a shielding ability of 32.4–40.1 and 25.3–29.7 dB, respectively. This improvement is attributed to the conductive rGo network, which effectively attenuates incoming electromagnetic waves. Second, numerous micro-capacitors could be developed that altered dispersion of Fe_3O_4 and enhanced the real part of permittivity and the absorbing capability.

5.5.3 Lignin-based aerogels

Zeng *et al* [43] fabricated extremely elastic, lightweight, reduced graphene oxide (rGO)/lignin-derived carbon (LDC) composite aerogels with aligned micron-sized holes and cell walls. The inclusion of a tiny proportion of LDC in the cell walls increased the interfacial polarization effect while retaining the number of charge carriers and conductivity of the cell walls, significantly increasing the cell walls' wave absorption capacity. rGO/LDC aerogels also have bigger cell walls with higher integrity than rGO aerogels, which improves the multiple reflection ability of the aligned cell walls. The synergistic effects of the multiphase cell walls and favored microstructures of the rGO/LDC aerogels resulted in a high shielding efficiency of 21.3–49.2 dB at an ultralow density of 2.0–8.0 mg cm^{-3}. This corresponds to an SSE of up to 53 250 dB cm^2 g^{-1}, which exceeds the values reported for other carbon or metal-based shields. Furthermore, the crucial roles that microstructures play in influencing the EMI shielding performance becomes particularly evident when comparing the shielding effectiveness in directions parallel and perpendicular to the cell wall, as well as during an *in situ* compression process.

5.5.4 Alginate-based aerogels

Zhou *et al* [44] reported an ultrathin $Ti_3C_2T_x$/calcium alginate (CA) aerogel films with a sponge-like structure of micron-scale thickness, fabricated via vacuum filtration and freeze-drying. Alginate served as an interlayer spacer between the MXene sheets to prevent stacking and to form the sponge-like structure. This structure facilitated dissipation of incident electromagnetic waves via scattering and multiple reflection, achieving a shielding effectiveness of 54.3 dB and an SSE of 17 586 dB cm^2 g^{-1}.

5.5.5 Chitosan-based aerogels

Chitosan (CS) is the second most abundant aminopolysaccharide biopolymer and, as such, it exhibits excellent chemical compatibility with inorganic materials. Chitosan is a derivative of chitin, which is a natural polymer obtained from the exoskeletons of crustaceans such as shrimp and crab. Chitosan aerogels are typically produced through a sol–gel process which involves creating a colloidal suspension or sol of chitosan, followed by gelation to form a three-dimensional network. The gel is then subjected to a solvent exchange process, where the liquid within the gel is replaced with a solvent such as ethanol. Finally, the solvent is removed using a supercritical drying technique to obtain the chitosan aerogel with a highly porous structure [46].

Metal-based composites: A hierarchically porous MXene/CS-based hybrid carbon aerogel with durably high EMI shielding performance and thermal insulation performance was fabricated by Wu *et al* via a freeze-drying method followed by a controlled heat treatment. Owing to the special MXene–TiO_2–C heterostructure, the synergistic effects of the hybrid composite showed (0.1322 vol% MXene) a high absolute EMI SE of ~61.4 dB and an SSE of 5155.46 dB cm^3 g^{-1} in the X-band at a thickness of ~3 mm. Notably, the significant hydrophobicity and strong mechanical performance endowed with durable high EMI shielding and thermal insulation performance. The hybrid carbon aerogel exhibited a high EMI SE of ~40.5 dB and an ultralow thermal conductivity of ~25.5 mW m^{-1} K^{-1} after storage in a hygrothermal environment for 30 days [45].

Heteroatom doping of aerogels is a promising strategy to dissipate electromagnetic waves. Wang *et al* [46] fabricated N-doped biomass-based carbon aerogel by freeze-drying and a specific two-stage pyrolysis using chitin and chitosan as carbon and nitrogen source. The content and type of doping elements and electrical conductivity was manipulated by controlling the content of precursors and carbonization temperature. The N-doping induced a dipole polarization, while the unique three-dimensional porous architecture of the biomass-based carbon aerogel facilitated multiple reflections and scattering of electromagnetic waves (EMWs). The as-prepared biomass-based carbon aerogel, prepared at 1300 °C, exhibited an optimal EMI performance of 82 dB. The absorption coefficients could be tuned in the range of 0.31–0.88. This outstanding EMI performance originated from a variety of mechanisms including conduction loss, dipole polarization loss, and interface polarization loss.

Conducting polymer-based composites: Zhang *et al* [47] reported a work on structurally enhanced cellulose framework with PANI composite aerogel, cellulose–chitosan/PANI aerogel (CCPA). Notably, this aerogel showed brilliant microwave absorption properties which reached −54.76 dB in reflection loss, with the effective absorption bandwidth of 6.04 GHz. More importantly, the infrared thermal imaging showed that those samples, coated with PANI, were fully insulated for efficient temperature transfer (figure 5.7). At the same time, with an increase of chitosan, the structural stability significantly improved. The 3D mesh structure and excellent circuit network are beneficial for strong conductivity loss. Meanwhile, the existence of porous structure also promoted the space charge polarization, and these loss factors integrate with each other to promote EMW absorption. With a filler ratio of 35%, the RL_{min} reaches −54.76 dB and the widest f_e is 5.12 GHz. When the filler ratio is 40%, the maximum bandwidth can also achieve 6.04 GHz. The characteristic superiority of both structure and function makes cellulose–chitosan/PANI aerogel (CCPA) a promising multifunctional material for practical applications.

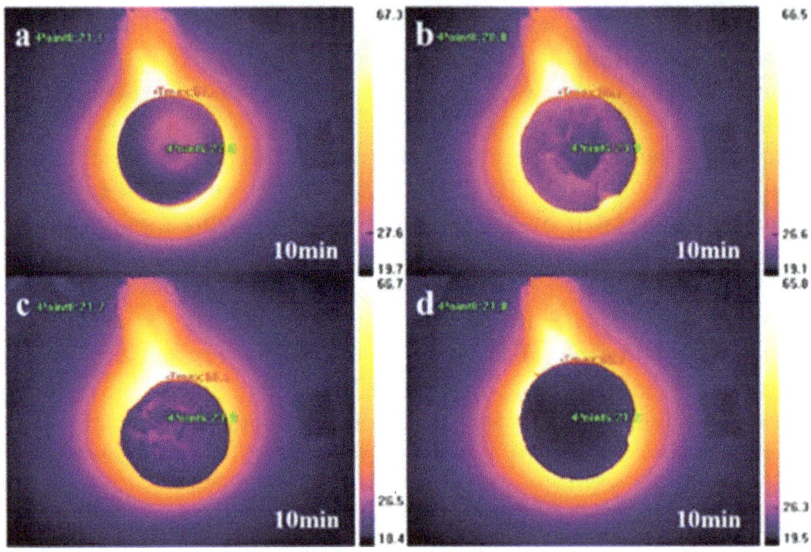

Figure 5.7. Thermal images of cellulose–chitosan/PANI aerogel composite. (Reproduced with permission from [47]. Copyright 2020 Elsevier.)

5.6 Synthetic biopolymers and composites for EMI shielding

5.6.1 Polyvinyl alcohol-based aerogels

PVA aerogels are a class of lightweight materials derived from polyvinyl alcohol, which is known for its biocompatibility, low toxicity, and remarkable mechanical strength. It is a water-soluble polymer and is derived from the polymerization of vinyl acetate, followed by partial or complete hydrolysis of the acetate groups. The resulting PVA polymer consists of repeating vinyl alcohol units, and its properties can be modified by adjusting the degree of hydrolysis. The use of PVA, a biodegradable and non-toxic material, aligns with the growing demand for sustainable and eco-friendly technologies.

Carbon-based composites: Yuan *et al* [48] fabricated PVA/carbon-based aerogels via a simple, low cost method which offers a promising solution to the pollution problems associated with the traditional incineration and landfill disposal of leather wastes (LWs). The obtained composite, consisting of Cr containing leather waste (LW)/PVA carbon aerogels achieved an SE value of 62.66 dB and a conductivity of 1.9436 S mm^{-1}. The interaction between LWs and PVA promoted the formation of a carbon skeleton, effectively embedding chromium. Additionally, the heteroatom doping and interconnected three-dimensional porous structure contributed to an enhanced conductivity and improved electromagnetic shielding performance.

Similarly, Vazhayal *et al* [49] developed carbon aerogels (CAs) from waste tissue paper (WTP) and PVA. Initially, the WTP was washed and dried to remove contaminants followed by adding this pulp (2, 4, 6, 8 wt%) to 100 ml PVA solution, crosslinking agents (formaldehyde) and catalyst (HCl). The resulting pulp is left overnight for polymerization followed by freeze-drying at −80 °C and carbonization at 900 °C to form 3D carbon aerogels. The average shielding values were reported to be 20.4, 25.5, 38.7, 40.0, and 31.4 dB for 0, 2, 4, 6, and 8 wt%, respectively. Because of the extremely high porosities and surface area, a considerable portion of the incident EM waves was absorbed within the porous structure, leaving only a minimal amount of waves to pass through the material. The high conductivity of WTP-PVA CA contributes to conduction loss, while the functional groups, defects, and interfaces between graphitic and non-graphitic regions cause polarization, relaxation loss, and thus improved absorption capacity via dipole formation and charge accumulation at porous interfaces under the influence of an electric field.

Conducting polymer-based composites: Zhang *et al* [50] fabricated an aerogel by combining leather solid waste with PVA and PANI to form LSW/PVA/PANI aerogels with high mechanical robustness, flame retardancy, and EMI shielding capability. Amino carboxyl groups in LSW, generated by solid-state shear milling (S3M) technology, formed strong hydrogen-bond interactions with the PVA molecular chains. This interaction enhanced the aerogel's compressive strength to 15.6 MPa and increased the temperature at which it began to deform to 112.7 °C, at a thickness of 2.5 cm. Moreover, it contained a high concentration of nitrogen (N) atoms, which contribute to its flame-retardant properties. This ensures an effective flame-retardant mechanism and increases the limiting oxygen index (LOI) value of the LSW/PVA aerogel to 32.0% at a thickness of 2.5 mm, indicating improved resistance to combustion. Notably, by the cyclic coating method, a conductive PANI layer could be polymerized on the surface of LSW/PVA aerogel, which led to the construction of a sandwich structure with impressive EMI shielding capability. The EMI SE reached more than 40 dB, and the specific shielding effectiveness (SSE) reached 73.0 dB cm^3 g^{-1}. The inherent dipoles in collagen fibers and the conductive PANI synergistically produced an internal multiple reflection and absorption mechanism.

Metal-based composites: Xu *et al* [51] fabricated porous Ti_2CT_x MXene/poly (vinyl alcohol) composite foams constructed by few-layered Ti_2CT_x (f-Ti_2CT_x) MXene and poly(vinyl alcohol) (PVA) via a facile freeze-drying method. As superior EMI shielding materials, their calculated specific shielding effectiveness reached up to 5136 dB cm^2 g^{-1} with an ultralow filler content of only 0.15 vol% and a reflection effectiveness (SER) of less than 2 dB, representing the excellent absorption-dominated shielding performance. A contrast experiment revealed that the good impedance matching derived from the multiple porous structures, internal reflection, and polarization effect (dipole and interfacial polarization) played a synergistic role in the improved absorption efficiency and superior EMI shielding performance.

Hybrid filler-based composites: Sambyal *et al* [52] fabricated a three-dimensional (3D) porous $Ti_3C_2T_x$/CNT hybrid aerogel via a bidirectional freezing method for lightweight EMI shielding applications. The synergism of the lamellar and porous structure of the MXene/CNT hybrid aerogels contributed extensively to their excellent electrical conductivity (9.43 S cm^{-1}) and superior EMI SE value of 103.9 dB at 3 mm thickness at the X-band frequency, the latter of which is the best value reported for synthetic porous nanomaterials.

Li *et al* [53] fabricated Ni foam/graphene oxide/polyvinyl alcohol (Ni/GO/PVA) composite aerogels using a freeze-drying method. With the increase in the GO content, the overall EMI shielding performance is improved. Ni/GO/PVA-20 has excellent EMI shielding performance and reached 87 dB at a thickness of 2.0 mm. The EMI shielding performance decreased with the decrease in thickness. The EMI shielding performance of Ni/GO/PVA-20 with thicknesses of 1.0 and 1.5 mm was 60 and 70 dB, respectively. The porous structure of the disordered GO and aerogels increased the propagation distance of electromagnetic waves. The dielectric loss, magnetic loss, and multiple reflection and scattering are the main mechanisms of electromagnetic energy attenuation.

5.6.2 Polycaprolactone-based aerogels

Polycaprolactone (PCL) is a biodegradable polyester which is widely explored for developing sustainable EMI shielding materials. It is an aliphatic semi-crystalline polymer with a melting temperature in the range of 59 °C–64 °C and a glass transition temperature of −60 °C. The rubbery structure attained by PCL at physiological temperatures imparts high roughness and good mechanical properties including high strength and elasticity to it, depending on its molecular weight [54]. Its non-toxicity and superior mechanical properties along with great biocompatibility make it suitable for developing EMI shielding materials.

Metal-based composites: Ahmad *et al* studied the microwave absorbing properties of a nickel oxide reinforced PCL composite. They utilized a melt blend technique to fabricate the composite. The energy dissipation resulting from the interaction between electromagnetic waves and the composite particles explained the mechanism of electromagnetic wave absorption in the NiO–PCL composite. Therefore, the behavior is significantly influenced by factors such as frequency range, dielectric characteristics, thickness, and filler loading. The study found that the shielding efficiency of the composite increased with increasing filler concentration and exhibited an absorption-dominated shielding mechanism [55].

Sun *et al* investigated metal alloy and flexible polymer composites in order to substitute inorganic particle-reinforced composites for electromagnetic protective shell products. They analysed the influence of metal alloy fillers in the mechanical and absorbing performance of 3D printed PCL composites. From the comprehensive evaluation of the composites, it was observed that the 3D printed sample with a 2 mm thickness and 30% filer content could effectively reduce the electric field radiation from 281 to 41 $V\,m^{-1}$ [56]. Hou *et al* proposed an inventive approach combining alternate casting and electroless plating techniques to create asymmetric hierarchical PCL composites made of conductive and impedance matching layers. The electromagnetic waves are attenuated by the polarization loss brought on by the two layers with different electrical conductivities. The PCL composite, which consisted of Fe_3O_4@rGO/Ni/Ag as the filler material, exhibits an outstanding EMI SE of 47.6 dB along with a remarkably reduced power coefficient of reflectivity (0.27). The gradient distribution of the fillers provided both a positive conductive and a negative magnetic gradient. The greater the gradient, the more dielectric and magnetic losses are induced, leading to an improved absorption mechanism that may overcome the limitations of non-adjustable reflective properties [57].

Carbon-based composites: Morphological modification, such as the creation of double percolation structures, can only occur in polymer blends through the substitution of components. Nevertheless, the weight percentage of either polymer in the blends often falls between 40% and 60%, which is not sufficient to develop a densely conductive route. A PLA/PCL mix containing only 30 weight percent PCL was used by Liu *et al* to successfully produce a double percolation structure. In the blend, silicon dioxide (SiO_2) was added to the PLA phase which effectively modified the viscosity and elasticity of the PLA phase and thus resulted in the modification of the morphology of the PCL phase from a segregated to a continuous structure. In the polymer blend, multi-wall carbon nanotubes (MWCNTs) are dispersed in the PCL phase to develop a conductive network and could impart enormous electrical conductivity to the composite. The PLA/PCL composite could achieve an increase in conductivity from 1.19×10^{-14} S cm^{-1} to 4.57×10^{-4} S cm^{-1} with the addition of 1 wt% MWCNT and 6 wt% SiO_2. With only 3.0 wt% MWCNTs, the nanocomposites exhibited a shielding effectiveness of approximately 26 dB, blocking 99.997% of microwaves. This research presents a crucial approach to develop environmentally friendly nanocomposites with exceptional EMI shielding and an ultralow percolate threshold [58].

Thomassin *et al* developed PCL foams incorporated with MWCNTs by using supercritical CO_2. They employed melt mixing and co-precipitation methods for the fabrication of the composite. The resulting composite, with a uniform open cell structure, exhibited an EMI SE of 60–80 dB with a very low volume percentage of MWCNT (0.25%). An absorption-dominated shielding mechanism with very negligible reflectivity and low filler content make this composite an efficient material for EMI shielding [59].

Li *et al* used layer-assembly coextrusion to create a unique multilayer structure made of PLA and a co-continuous PLA/PCL/MWCNT (ALM) composite with a double-percolated conducting network. While maintaining an outstanding strength of over 46.0 MPa, the nacre-like structure with alternately layered rigid PLA and flexible ALM enhanced the fracture strain to 354.4%, approximately quadrupling that of the PLA/PCL/MWCNT traditional blending composite with the same composition. Furthermore, a unique frequency-selective EMI shielding performance was demonstrated by the multilayer composites, with layer number determining the adjustable shielding peak positions. Two factors contributed to their maximum shielding effectiveness, which was nearly 49.8 dB with prominent absorption loss: first, the high electrical conductivity provided by the double-percolated distribution of MWCNTs; second, the multiple wave attenuation effect that happened at the interfaces between ALM and PLA layers as well as the blend interfaces in ALM layers. This work opens up new avenues for the development of composites that may be applied to various polymeric composite systems and have exceptional mechanical and EMI shielding qualities [60].

Conducting polymer-based composites: Because of the synergistic effect of the fillers, hybrid nanocomposites have the unique ability to enhance material characteristics. An environmentally friendly hybrid nanocomposite for EMI shielding, made of PCL and reduced graphene oxide (rGO) and polyaniline (PAni) was developed by Ponnamma *et al.* Initially a PAni-PCL blend was prepared which attained better processability due to the incorporation of PAni. rGO, reduced at different temperatures, was added into the blend to make the final hybrid composites. PAni/rGO filler systems impart high electrical conductivity to PCL and the fabricate flexible composites exhibited good mechanical properties, improved thermal stabilities and promising EMI SE of up to 42 dB at 13 GHz. This composite can be used commercially as an efficient lightweight EMI shielding material [61].

5.7 Challenges of bio-based aerogels for EMI shielding

The development and application of bio-based aerogels for EMI shielding pose several challenges that researchers and engineers are actively addressing. These challenges encompass material properties, manufacturing techniques, performance optimization, and broader issues related to sustainability and industrial integration. To successfully develop and apply bio-based aerogels for EMI shielding, researchers and engineers have to encounter specific challenges, of which highly important examples are mentioned in the following.

A large number of bio-based materials are not intrinsically conducting enough to provide adequate EMI shielding. It is difficult to increase these materials' electrical conductivity without sacrificing their biocompatibility or other desired qualities. A simple and effective method to overcome this challenge is to incorporate conducting fillers into these materials. Achieving a balance between lightweight properties and mechanical strength is challenging. Bio-based aerogels must maintain structural integrity over time, resisting mechanical stresses, vibrations, and environmental factors to ensure long-term durability. Optimal design of the porous structure is crucial for achieving high EMI shielding effectiveness. Fine-tuning the porosity and pore size distribution while maintaining structural integrity is a challenging task. Better mechanical properties and lightweight characteristics can be achieved while maintaining their biocompatibility by chemical crosslinking using non-contaminant chemicals, such as citric acid, or through radiation crosslinking using gamma rays [62–66].

A major difficulty involves transitioning from laboratory-scale production to large-scale manufacturing while keeping costs low. In order to meet industrial demands, bio-based aerogel production procedures must be scalable without sacrificing the final product's economic viability. This suggests the evolution of new cost effective techniques such as 3D printing, which can effectively tune the size and shape of the pores of the aerogels. Developments may be impeded by the absence of uniform testing procedures to assess the EMI shielding performance of bio-based aerogels. Consistency in measurement procedures is vital for accurate comparisons between different research and laboratories. It is crucial to provide a sustainable supply chain for commodities derived from biomass. Careful consideration must be given to issues pertaining to ethical implications in material extraction, potential competition with food sources, and consistent sourcing. Even though bio-based materials are frequently thought to be more environmentally friendly, a thorough life cycle study is required. The assessment ought to take into account the industrial process's environmental impact, energy usage, and waste generation. Aerogels made from bio-based materials must be compatible with other materials and production methods in order to be successfully integrated into current electronic and communication systems. Providing smooth adoption across multiple businesses is a difficult task.

Obtaining regulatory approval and public trust requires addressing safety, biodegradability, and potential health impact concerns. The public's acceptability and the ability to overcome scepticism are critical to the widespread use of bio-based aerogels. Specific requirements may apply to different uses of EMI shielding. One of the most difficult aspects of developing bio-based aerogels is tailoring them to these various purposes while upholding performance requirements.

Researchers are actively working on overcoming these challenges through interdisciplinary collaborations and innovative approaches. As the field progresses, addressing these hurdles will contribute to the successful integration of bio-based aerogels into EMI shielding applications.

5.8 Conclusion and future outlook

In summary, this chapter on bio-based aerogels for EMI shielding gave insights into the compelling advancements of sustainable materials for addressing electromagnetic interference challenges. The research results discussed in this chapter affirm the viability and effectiveness of bio-based aerogels as efficient scaffold for EMI shielding materials. By emphasizing the integration of eco-friendly alternatives, this research contributes not only to the enhancement of electromagnetic compatibility but also to the broader goal of environmentally conscious technologies. The exploration of bio-based aerogels represents a promising research field towards the development of sustainable EMI shielding solutions, marking a significant step forward in the intersection of materials science and environmental responsibility.

Bibliography

[1] Husain F M *et al* 2021 Bio-based aerogels and their environment applications: an overview *Advances in Aerogel Composites for Environmental Remediation* (Amsterdam: Elsevier) pp 347–56

[2] Khan A A P *et al* (ed) 2021 *Advances in Aerogel Composites for Environmental Remediation* (Amsterdam: Elsevier)

[3] Mohammad A *et al* 2021 Aerogel and its composites for sensing, adsorption, and photocatalysis *Advances in Aerogel Composites for Environmental Remediation* (Amsterdam: Elsevier) pp 125–44

[4] Zuo L *et al* 2015 Polymer/carbon-based hybrid aerogels: preparation, properties and applications *Materials* **8** 6806–48

[5] Syeda H I and Yap P-S 2022 A review on three-dimensional cellulose-based aerogels for the removal of heavy metals from water *Sci. Total Environ.* **807** 150606

[6] Mary S K, Rose Koshy R, Arunima R, Thomas S and Pothen L A 2022 A review of recent advances in starch-based materials: bionanocomposites, pH sensitive films, aerogels and carbon dots *Carbohydr. Polym. Technol. Appl.* **3** 100190

[7] Sýkora R *et al* 2016 Rubber composite materials with the effects of electromagnetic shielding *Polym. Compos.* **37** 2933–9

[8] Chung D D L 2020 Materials for electromagnetic interference shielding *Mater. Chem. Phys.* **255** 123587

[9] Geetha S *et al* 2009 EMI shielding: methods and materials—a review *J. Appl. Polym. Sci.* **112** 2073–86

[10] Zhou M, Gu W, Wang G, Zheng J, Pei C, Fan F and Ji G 2020 Sustainable wood-based composites for microwave absorption and electromagnetic interference shielding *J. Mater. Chem.* A **8** 24267–83

[11] Ashori A 2008 Wood–plastic composites as promising green-composites for automotive industries! *Bioresour. Technol.* **99** 4661–7

[12] Rasouli D *et al* 2016 Effect of nano zinc oxide as UV stabilizer on the weathering performance of wood-polyethylene composite *Polym. Degrad. Stab.* **133** 85–91

[13] Hofenauer A *et al* 2003 Dense reaction infiltrated silicon/silicon carbide ceramics derived from wood based composites *Adv. Eng. Mater.* **5** 794–9

[14] Li J, Wang L and Liu H 2010 A new process for preparing conducting wood veneers by electroless nickel plating *Surf. Coat. Technol.* **204** 1200–5

[15] Trey S *et al* 2014 Controlled deposition of magnetic particles within the 3-D template of wood: making use of the natural hierarchical structure of wood *RSC Adv.* **4** 35678–85

[16] Merk V *et al* 2014 Hybrid wood materials with magnetic anisotropy dictated by the hierarchical cell structure *ACS Appl. Mater. Interfaces* **6** 9760–7

[17] Persson P V *et al* 2004 Silica nanocasts of wood fibers: a study of cell-wall accessibility and structure *Biomacromolecules* **5** 1097–101

[18] Hui B *et al* 2021 Boosting electrocatalytic hydrogen generation by a renewable porous wood membrane decorated with Fe-doped NiP alloys *J. Energy Chem.* **56** 23–33

[19] Hui B *et al* 2020 Natural multi-channeled wood frameworks for electrocatalytic hydrogen evolution *Electrochim. Acta* **330** 135274

[20] Liu H, Li J and Wang L 2010 Electroless nickel plating on APTHS modified wood veneer for EMI shielding *Appl. Surf. Sci.* **257** 1325–30

[21] Sun L, Li J and Wang L 2012 Electromagnetic interference shielding material from electroless copper plating on birch veneer *Wood Sci. Technol.* **46** 1061–71

[22] Shi C, Wang L and Wang L 2015 Fabrication of a hydrophobic, electromagnetic interference shielding and corrosion-resistant wood composite via deposition with Ni–Mo–P alloy coating *RSC Adv.* **5** 104750–5

[23] Gao H L *et al* 2012 *In situ* preparation and magnetic properties of Fe_3O_4/wood composite *Mater. Technol.* **27** 101–3

[24] Lou Z *et al* 2019 Synthesis of porous carbon matrix with inlaid Fe_3C/Fe_3O_4 micro-particles as an effective electromagnetic wave absorber from natural wood shavings *J. Alloys Compd.* **775** 800–9

[25] Zhang Y *et al* 2015 Broadband and tunable high-performance microwave absorption of an ultralight and highly compressible graphene foam *Adv. Mater.* **27** 2049–53

[26] Zhu G *et al* 2019 Engineering the distribution of carbon in silicon oxide nanospheres at the atomic level for highly stable anodes *Angew. Chem. Int. Ed.* **58** 6669–73

[27] Zhao H *et al* 2019 Biomass-derived porous carbon-based nanostructures for microwave absorption *Nano-Micro Letters* **11** 1–17

[28] Xi J *et al* 2017 Wood-based straightway channel structure for high performance microwave absorption *Carbon* **124** 492–8

[29] Dong S *et al* 2019 Achieving excellent electromagnetic wave absorption capabilities by construction of MnO nanorods on porous carbon composites derived from natural wood via a simple route *ACS Sustain. Chem. Eng.* **7** 11795–805

[30] Zheng Y *et al* 2020 Lightweight and hydrophobic three-dimensional wood-derived anisotropic magnetic porous carbon for highly efficient electromagnetic interference shielding *ACS Appl. Mater. Interfaces* **12** 40802–14

[31] Lou Z *et al* 2018 Synthesis of porous 3D Fe/C composites from waste wood with tunable and excellent electromagnetic wave absorption performance *ACS Sustain. Chem. Eng.* **6** 15598–607

[32] He W *et al* 2018 Characteristics and properties of wood/polyaniline electromagnetic shielding composites synthesized via *in situ* polymerization *Polym. Compos.* **39** 537–43

[33] Zhang K *et al* 2019 Flexible polyaniline-coated poplar fiber composite membranes with effective electromagnetic shielding performance *Vacuum* **170** 108990

[34] Gan W *et al* 2020 Conductive wood for high-performance structural electromagnetic interference shielding *Chem. Mater.* **32** 5280–9

[35] Chen J *et al* 2021 Wood-derived nanostructured hybrid for efficient flame retarding and electromagnetic shielding *Mater. Des.* **204** 109695

[36] Huang H D, Liu C Y, Zhou D, Jiang X, Zhong G J, Yan D X and Li Z M 2015 Cellulose composite aerogel for highly efficient electromagnetic interference shielding *J. Mater. Chem. A* **3** 4983–91

[37] Li M, Han F, Jiang S, Zhang M, Xu Q, Zhu J, Ge A and Liu L 2021 Lightweight cellulose nanofibril/reduced graphene oxide aerogels with unidirectional pores for efficient electromagnetic interference shielding *Adv. Mater. Interfaces* **8** 2101437

[38] Pai A R, Lu Y, Joseph S, Santhosh N M, Degl'Innocenti R, Lin H, Letizia R, Paoloni C and Thomas S 2023 Ultra-broadband shielding of cellulose nanofiber commingled biocarbon functional constructs: a paradigm shift towards sustainable terahertz absorbers *Chem. Eng. J.* **467** 143213

[39] Pai A R *et al* 2020 Ultra-fast heat dissipating aerogels derived from polyaniline anchored cellulose nanofibers as sustainable microwave absorbers *Carbohydr. Polym.* **246** 116663

[40] Feng S, Deng J, Yu L, Dong Y, Zhu Y and Fu Y 2020 Development of lightweight polypyrrole/cellulose aerogel composite with adjustable dielectric properties for controllable microwave absorption performance *Cellulose* **27** 10213–24

[41] Fei Y, Liang M, Yan L, Chen Y and Zou H 2020 Co/C@cellulose nanofiber aerogel derived from metal–organic frameworks for highly efficient electromagnetic interference shielding *Chem. Eng. J.* **392** 124815

[42] Chen Y, Pötschke P, Pionteck J, Voit B and Qi H 2020 Multifunctional cellulose/rGO/Fe_3O_4 composite aerogels for electromagnetic interference shielding *ACS Appl. Mater. Interfaces* **12** 22088–98

[43] Zeng Z, Wang C, Zhang Y, Wang P, Seyed Shahabadi S I, Pei Y and Lu X 2018 Ultralight and highly elastic graphene/lignin-derived carbon nanocomposite aerogels with ultrahigh electromagnetic interference shielding performance *ACS Appl. Mater. Interfaces* **10** 8205–13

[44] Zhou Z, Liu J, Zhang X, Tian D, Zhan Z and Lu C 2019 Ultrathin MXene/calcium alginate aerogel film for high-performance electromagnetic interference shielding *Adv. Mater. Interfaces* **6** 1802040

[45] Wu S *et al* 2022 Ultralight and hydrophobic MXene/chitosan-derived hybrid carbon aerogel with hierarchical pore structure for durable electromagnetic interference shielding and thermal insulation *Chem. Eng. J.* **446** 137093

[46] Wang M-L *et al* 2022 Tunable high-performance electromagnetic interference shielding of intrinsic N-doped chitin-based carbon aerogel *Carbon* **198** 142–50

[47] Zhang Z *et al* 2020 Cellulose-chitosan framework/polyailine hybrid aerogel toward thermal insulation and microwave absorbing application *Chem. Eng. J.* **395** 125190

[48] Yuan B *et al* 2021 Trash into treasure: stiff, thermally insulating and highly conductive carbon aerogels from leather wastes for high-performance electromagnetic interference shielding *J. Mater. Chem.* C **9** 2298–310

[49] Vazhayal L, Wilson P and Prabhakaran K 2020 Waste to wealth: lightweight, mechanically strong and conductive carbon aerogels from waste tissue paper for electromagnetic shielding and CO_2 adsorption *Chem. Eng. J.* **381** 122628

[50] Zhang T *et al* 2021 Leather solid waste/poly (vinyl alcohol)/polyaniline aerogel with mechanical robustness, flame retardancy, and enhanced electromagnetic interference shielding *ACS Appl. Mater. Interfaces* **13** 11332–43

[51] Xu H *et al* 2019 Lightweight Ti_2CT_x MXene/poly (vinyl alcohol) composite foams for electromagnetic wave shielding with absorption-dominated feature *ACS Appl. Mater. Interfaces* **11** 10198–207

[52] Sambyal P *et al* 2019 Ultralight and mechanically robust $Ti_3C_2T_x$ hybrid aerogel reinforced by carbon nanotubes for electromagnetic interference shielding *ACS Appl. Mater. Interfaces* **11** 38046–54

[53] Li D *et al* 2022 Lightweight and hydrophobic Ni/GO/PVA composite aerogels for ultrahigh performance electromagnetic interference shielding *Nanotechnol. Rev.* **11** 1722–32

[54] Dwivedi R *et al* 2020 Polycaprolactone as biomaterial for bone scaffolds: review of literature *J. Oral Biol. Craniofac. Res.* **10** 381–8

[55] Ahmad A F *et al* 2018 Synthesis and characterisation of nickel oxide reinforced with polycaprolactone composite for dielectric applications by controlling nickel oxide as a filler *Results Phys.* **11** 427–35

[56] Sun Z, Chen B, Wang Y, Tuo X, Gong Y and Guo J 2022 Study on metal alloy-reinforced polycaprolactone 3D printed composites for electromagnetic protection *Compos. Sci. Technol.* **225** 109516

[57] Hou M *et al* 2023 Multi-hierarchically structural polycaprolactone composites with tunable electromagnetic gradients for absorption-dominated electromagnetic interference shielding *Langmuir* **39** 6038–50

[58] Liu Y *et al* 2021 Morphology evolution to form double percolation polylactide/polycaprolactone/MWCNTs nanocomposites with ultralow percolation threshold and excellent EMI shielding *Compos. Sci. Technol.* **214** 108956

[59] Thomassin J-M *et al* 2008 Foams of polycaprolactone/MWNT nanocomposites for efficient EMI reduction *J. Mater. Chem.* **18** 792–6

[60] Li X *et al* 2024 Excellent mechanical and electromagnetic interference shielding properties of polylactic acid/polycaprolactone/multiwalled carbon nanotube composites enabled by a multilayer structure design *RSC Adv.* **14** 20390–7

[61] Ponnamma D *et al* 2016 Eco-friendly electromagnetic interference shielding materials from flexible reduced graphene oxide filled polycaprolactone/polyaniline nanocomposites *Polym.-Plast. Technol. Eng.* **55** 920–8

[62] Kaya M 2017 Super absorbent, light, and highly flame retardant cellulose-based aerogel crosslinked with citric acid *J. Appl. Polym. Sci.* **134** 45315

[63] Wang C *et al* 2021 Preparation and characterization of carboxymethylcellulose based citric acid cross-linked magnetic aerogel as an efficient dye adsorbent *Int. J. Biol. Macromol.* **181** 1030–8

[64] Wang C *et al* 2022 Characterization of antibacterial aerogel based on ε-poly-l-lysine/nanocellulose by using citric acid as crosslinker *Carbohydr. Polym.* **291** 119568

[65] Rosli N A *et al* 2022 Hydrophobic-oleophilic gamma-irradiated modified cellulose nano-crystal/gelatin aerogel for oil absorption *Int. J. Biol. Macromol.* **219** 213–23

[66] Ishak W H W *et al* 2021 Drug delivery and *in vitro* biocompatibility studies of gelatin–nanocellulose smart hydrogels cross-linked with gamma radiation *J. Mater. Res. Technol.* **15** 7145–57

Chapter 6

Hemicellulose based materials

Jun Rao, Ziwen Lv, Siyu Jia and Feng Peng

6.1 Introduction

Hemicellulose, a natural polysaccharide, is the second most abundant renewable component of lignocellulosic biomass, next to cellulose; its global annual production is approximately 60 billion tons [1, 2]. Hemicellulose belongs to a group of heterogeneous polysaccharides, whose biosynthesis routes differ from that of cellulose, and it typically constitutes 15%–35% of the dry mass of annual and perennial plants [3]. The molecular chain of hemicellulose is composed of pyranose and furanose sugar units, including D-xylose, D-mannose, D-glucose, D-galactosyl, L-arabinose, galacturonic acid, and glucuronic acid [4]. Unlike cellulose, most hemicellulose is short-chain, with a degree of polymerization (DP) value of 80–200 [5]. The type of hemicellulose depends on the plant species, cell type and location, and developmental stage.

In hardwood, hemicellulose is divided into three classes, namely, *O*-acetyl-4-*O*-methylglucuronoxylan (MeGlcp-Xylan), glucomannan (GM), and xyloglucan (XG), based on its structure. *O*-acetyl-4-*O*-methylglucuronoxylan is the predominant hemicellulose component of hemicellulose (about 15%–30%). Its backbone consists of β-D-xylopyranose residues linked by (1 $\rightarrow$ 4)-glucosidic bonds [6–9]. In softwood, hemicellulose consists of *O*-acetylgalactoglucomannan (AcGGM), arabinoglucuronoxylan (AGX), and arabinogalactan (AG) (figure 6.1(a)). *O*-acetylgalactoglucomannan is the principal component of hemicellulose in softwood species such as spruce (*Picea abies*) and pine (*Pinus sylvestris*), accounting for up to 10%–25% of their dry mass [3, 10]. The backbone is a linear or slightly branched chain comprising (1 $\rightarrow$ 4)-linked β-D-mannopyranose and β-D-glucopyranose units, whereas the (1 $\rightarrow$ 6)-linked α-galactopyranosyl units are attached only to the mannose units. Arabinoxylan is the major hemicellulose in *Gramineae* plants such as wheat, rye, and corn, and it consists of a linear main chain of (1 $\rightarrow$ 4)-β-D-xylopyranose units, in which single α-L-arabinose residues are 1,4-linked

doi:10.1088/978-0-7503-6184-2ch6

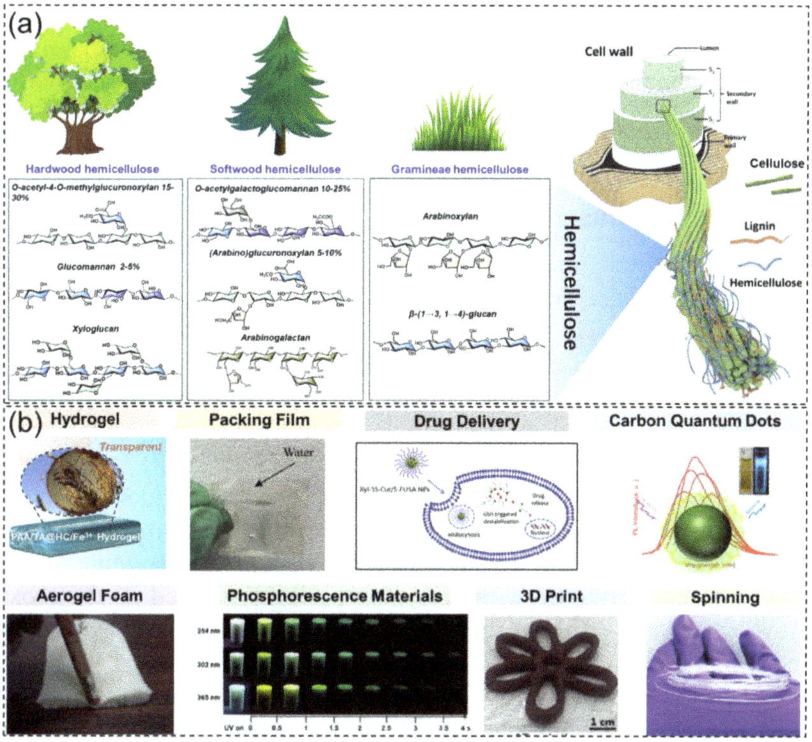

Figure 6.1. (a) Schematic representation of the plant cell wall and structural formulae of hemicellulose in hardwood (*O*-acetyl-4-*O*-methylglucuronoxylan, glucomannan, and xyloglucan); softwood (*O*-acetylgalacto-glucomannan, glucuronoxylan, and arabinogalactan); and Gramineae (arabinoxylan and β-(1 $\rightarrow$ 3, 1 $\rightarrow$ 4) glucan). (Reproduced from [13]. CC BY 4.0.) (b) The functional materials (hydrogel, films, carbon quantum dots, etc) derived from hemicellulose. (Reproduced with permission from [14]. Copyright 2023 Elsevier.)

D-xylopyranose residues attached as side residues via α-(1 $\rightarrow$ 2) and/or α-(1 $\rightarrow$ 3) linkages [11, 12].

Owing to the near unlimited supply of this sustainable polysaccharide and its excellent physical and chemical properties, hemicellulose is used in the production of various materials (emulsifiers, films, and hydrogels, etc) and fine chemicals (xylitol, ethanol, and furfural, etc) for food, medical, and energy storage applications (figure 6.1(b)) [14–16]. However, the large number of hydroxyl groups in the hemicellulose molecular chain leads to strong hydrogen bonding interactions, mainly within the molecule, making it difficult to dissolve in common solvents and limiting its processability. Meanwhile, this also makes it highly hygroscopic, leading to a decrease in the durability of the material. In contrast, the active hydroxyl groups on the main chain and side chain of hemicellulose offer opportunities for chemical modification. Chemical modification of hemicellulose offers numerous possibilities to control and tailor the properties of hemicellulose, enhance their processability and introduce new functionalities such as thermoplasticity, hydrophobicity, conductivity, and stimuli-responsiveness. Therefore, chemical modification extends the application scope of hemicellulose and facilitates its industrial production. This chapter summarizes the recent research progress on the application of hemicellulose and its derivatives in functional materials, including hydrogels, aerogels, forms, films, emulsions, and carbon quantum dots, etc.

6.2 Hemicellulose-based hydrogels

Hydrogels are a special type of soft material that possesses a continuous three-dimensional (3D) polymer network and exhibit the rheological properties of solids on a macroscopic scale [17]. The hydrogel network structures, formed through physical aggregation or chemical cross-linking, possess the capability to retain significant amounts of water. Hemicellulose has the unique physiological properties, such as inhibiting cell mutation, promoting cell adhesion and proliferation, innate immunological defense, and an anticancer effect which make it suitable for the preparation of hydrogels used in drug release and biomedical engineering. Moreover, the presence of hydroxyl and carboxylic groups on the hemicellulosic chains provides more opportunity for chemical modifications while keeping their native structure, biodegradability, and biocompatibility. Based on the cross-linking mechanisms, the design strategies for preparing hemicellulose-based hydrogels can be divided into two categories: chemical and physical hydrogels. In recent years, hemicellulose-based hydrogels have been widely used in adsorption [18, 19], drug delivery [20, 21], tissue engineering [22, 23], and wearable-electronic sensing [24, 25] applications. Herein, the preparation strategy–structure–property relationships of hemicellulose-based hydrogels are reviewed.

6.2.1 Physical crosslinking of hemicellulose-based hydrogels

The physical crosslinking in hydrogels is derived from the non-covalent interactions between polymer chains, such as hydrogen bonds, ionic bonds, hydrophobic interactions, and physical entanglement of polymer chains. The advantages of physically crosslinked hydrogels are biocompatibility and biodegradability, ascribed to the absence of chemical crosslinking agents, thereby avoiding potential cytotoxicity due to any unreacted chemical cross-linkers. Moreover, physical crosslinked hemicellulose-based hydrogels are stimuli-responsive with self-healing and injectable properties at room temperature. As for hemicellulose, its molecular chain contains a large number of hydroxyl groups that provide active sites for physical crosslinking. Green, facile, and sustainable methods for fabricating physically crosslinked hemicellulose-based hydrogels have been developed rapidly in recent years, including polyelectrolyte complexes, the freeze–thaw technique, ultrasonication, and ionic cross-linking [26].

In the plant cell wall, hemicellulose and cellulose are tightly bound together by hydrogen bonds and hydrophobic interactions. Therefore, hemicellulose can be utilized as a macromolecular physical crosslinking agent or reinforcing agent, adsorbed on the surface of polysaccharides, and forming a physical crosslinked composite hydrogel. Taking arabinoxylan (AX) as an example, when AX is added to

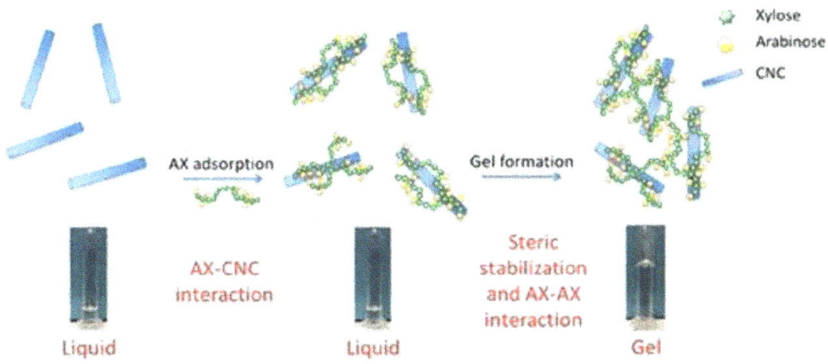

Figure 6.2. AX/CNC composite hydrogel material formation mechanism diagram. (Reproduced with permission from [27]. Copyright 2019 American Chemical Society.)

a suspension of cellulose nanocrystals (CNCs), its molecular chains undergo rapid and irreversible adsorbtion onto the surface of CNCs. This adsorption process is controlled by both thermodynamic principles and kinetics [27]. When the concentration of AX and CNC is low, it is difficult to form a gel. Further increasing the concentration of AX and CNC, AX can be approximated as a macromolecular physical crosslinking agent, which promotes the formation of a three-dimensional crosslinked network structure by adsorbing onto CNCs (figure 6.2). Significantly, the formation process of this physically adsorbed gel is affected by temperature and molecular chain structure. When the temperature is increased from 20 °C to 60 °C, the gelation time can be shortened from one week to 3 h. Additionally, the structure of hemicellulose affects both the gelation ability and gel mechanical properties. By reducing the branching degree of AX, the gelation time can be significantly shortened from three days to just 1 h, while concurrently enhancing the mechanical strength of the resulting gel. Although the decrease of the degree of branching of hemicellulose can improve its gelation ability with other polysaccharides. However, the decrease of branching degree will increase the crystallization ability of hemicellulose, which makes it difficult to effectively form gel materials.

To effectively utilize xylan (Xyl) with high crystallinity, researchers introduced positively charged chitosan (CS) and used the strong adsorption between Xyl and CS molecular chains to weaken the crystallization ability of Xyl [28]. For example, Xyl extracted from hardwood was mixed with CS in an aqueous solution and subsequently condensed into a gel film. In the process of film formation, only part of the crystalline region was formed. This is because the addition of CS greatly inhibits the crystallinity of Xyl, and the retained part of the crystalline region acts as the cross-linking point [29]. After swelling in water for a period of time, a Xyl/CS composite hydrogel with elasticity and toughness can be obtained.

One possible method to improve the mechanical properties of physical cross-linked hemicellulose-based hydrogels can be found in the freeze–thaw cycle method (figure 6.3). The freeze–thaw technique, which effectively promotes the formation of hydrogen bonds in gels, is a facile and efficient strategy to fabricate hydrogels with high mechanical strength and excellent elasticity. Guan *et al* [30, 31] first reported

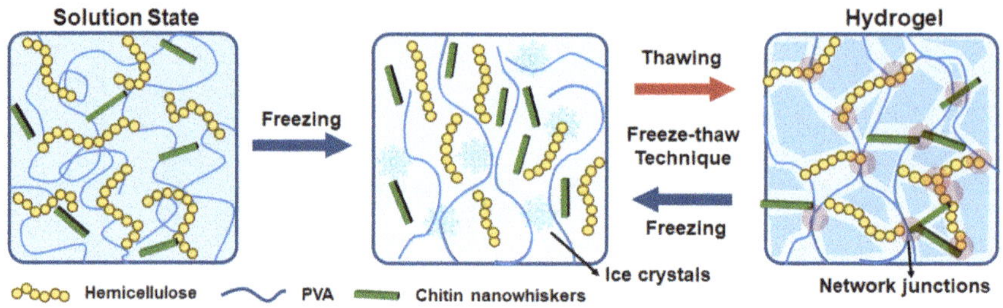

Figure 6.3. Schematic diagram of hemicellulose hydrogel formation by the freezing–thawing method.

the fabrication of physically cross-linking hemicellulose-based hydrogels using the freeze–thaw technology. The repeated freeze–thaw cycles induce physically cross-linked chain packing among the polymers, and the hydrogen bonds cause phase separation. The mechanical properties and thermal stability of the hydrogels are enhanced by increasing the number of freeze–thaw cycles. The highest compressive strength is achieved after nine freeze–thaw cycles, with the hydrogels exhibiting a compressive stress of 10.5 MPa. Therefore, they emerge as promising candidate material for applications in the tissue engineering and medical-biology fields.

6.2.2 Chemically crosslinked hemicellulose-based hydrogels

Compared with the physical crosslinking method, the preparation of hemicellulose hydrogel materials by the chemical crosslinking method has been studied more extensively and deeply. Additionally, the resulting hydrogels typically demonstrate superior mechanical strength. To date, several cross-linking methods have been reported, including small molecule cross-linking [32], free-radical polymerization induced cross-linking [33], enzymatic induced cross-linking [34], the Diels–Alder click reaction [35], and Schiff's base reaction [36].

The hemicellulose molecular chain contains potentially reactive hydroxyl groups, which can be directly crosslinked with small molecule cross-linkers, such as epichlorohydrin [32, 37, 38], ethylene glycol diglycidyl ether [39, 40], and diethylenetriaminepentaacetic dianhydride [41]. For instance, epichlorhydrin was used to generate a hydrogel with conductive properties via a one-pot reaction in an alkaline medium (figure 6.4) [32]. The hydrogel network was formed by the etherification of hemicellulose and epichlorohydrin. With increasing aniline pentamer content, the thermal stability of the electrically conductive hemicellulose-based hydrogels is enhanced, and the conductivity increases by three orders of magnitude; however, the equilibrium swelling ratios decrease.

Owing to the weak reactivity of hydroxyl groups on the hemicellulose molecular chain, the chemical modification of hemicellulose is necessary, including oxidation, hydrolysis, reduction, esterification and etherification, etc, which can introduce new active groups such as the aldehyde group, carboxyl group, sulfhydryl group, and epoxy group. For example, hydrogel materials can be efficiently and rapidly prepared by the click reaction using the high reactivity of thiolated hemicellulose. Using potassium persulfate/sodium bisulfite (KPS/SHS) as an initiator, thiolated

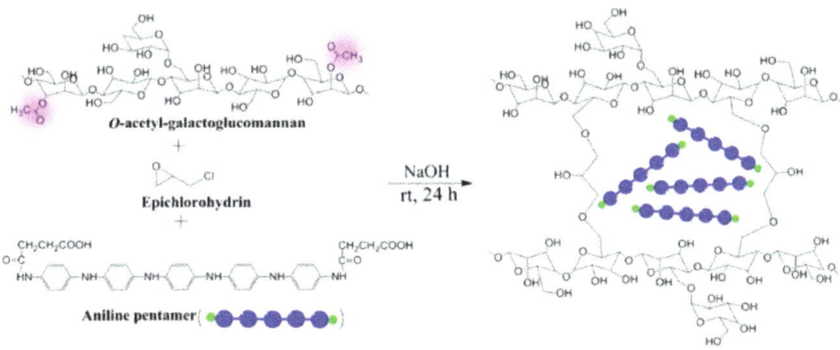

Figure 6.4. Schematic synthesis of *O*-acetylgalactoglucomannan-based hydrogels using epichlorohydrin as a cross-linker in basic media. (Reproduced with permission from [32]. Copyright 2014 American Chemical Society.)

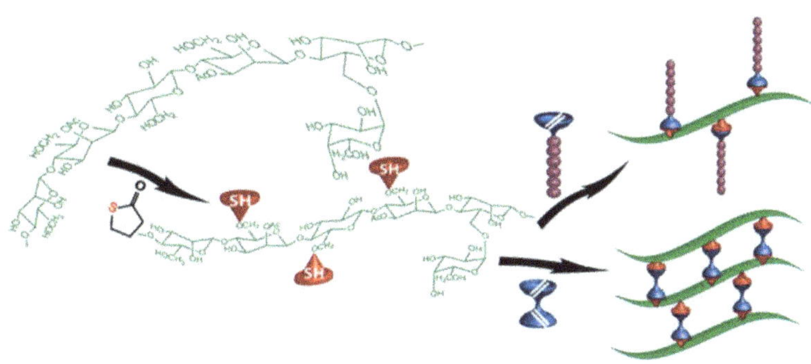

Figure 6.5. Schematic synthesis of *O*-acetylgalactoglucomannan-based hydrogels via thiol–ene reaction. (Reproduced with permission from [42]. Copyright 2015 American Chemical Society.)

hemicellulose can be reacted with bifunctional olefin monomers such as *N,N*-methylenebis-acrylamide (MBA), and the hemicellulose-based hydrogel can be obtained by the thiol–ene click reaction after heating for 15 min (figure 6.5) [42]. Due to the high activity of sulfhydryl groups, it is also possible to directly use the sulfur radicals generated by KPS/SHS without adding olefin monomers to induce sulfhydryl-substituted hemicelluloses to form a disulfide-crosslinked gel network via free-radical coupling. Oxidized hemicellulose is also one of the important raw materials for the preparation of hemicellulose-based hydrogels [43]. Partially oxidized hemicellulose can be rapidly crosslinked with macromolecules containing amino or hydroxyl groups, such as chitosan, in water by Schiff's base reaction. For instance, a hydrogel network can be achieved by cross-linking dialdehyde hemicelluloses with chitosan by the Schiff base, and silver ions were reduced by the excess aldehyde group of dialdehyde hemicellulose [44]. Importantly, the dialdehyde hemicellulose/chitosan/Ag hydrogel possessed antimicrobial activity against the model microbes *Escherichia coli* (gram-negative) and *Staphylococcus aureus* (gram-positive), which is a promising antimicrobial material for applications in the biomedical field.

Click chemistry, also known as link chemistry or dynamic combinatorial chemistry, is a novel modification strategy for synthesizing polymers with complex architectures, such as block and graft copolymers and branched, comb shaped polymers; it has emerged as a powerful method for the chemical synthesis and fabrication of tailorable materials. The types of click chemistry reactions include Cu (I)-catalyzed azide–alkyne cycloaddition (CuAAC), strain-promoted azide–alkyne cycloaddition (SPAAC), tetrazine–trans–cyclooctene ligation, and thiol–ene click reactions [45–47].

The thiol–ene reaction has gained much attention in chemical synthesis. The reaction can usually be performed under mild reaction conditions giving high conversion and selectivity, using water as a solvent. In addition, no toxic metal catalysts are needed, making the use of thiol–ene chemistry tempting for the modification of polysaccharides. Thiols and amines are valuable functional groups having high reactivity. Pahimanolis and co-workers designed a simple method for functionalizing xylan with thiols, amines and amino acids by combining traditional etherification and thiol–ene reactions [48]. First, the hydroxyl group on xylan reacts with allyl chloride under alkaline conditions at 40 °C to introduce the allyl group into the backbone of xylan, and then thiols are introduced into the backbone of xylan by click chemistry. This method provides a broad possibility for the development of new polysaccharide-based materials, and through thiol-thiol oxidative coupling, free thiol groups can be used to form hydrogels, and the shape of the resulting hydrogels can be well controlled, thus the fields of application were expanded.

Radical polymerization is the most common method for preparing hemicellulose-based hydrogels, and the commonly used monomers are acrylic acid (AA), acrylamide (AM), 2-hydroxyethyl methacrylate (HEMA), and *N*-isopropylacrylamide (NIPAM). This method offers the advantages of rapid formation of hydrogel networks at ambient temperature under mild conditions and tuneability of hydrogel by controlling the cross-linking reactions, i.e. monomer content and initiator. There are two preparation strategies. (i) The first is monomer cross-linking copolymerization, including free-radical homopolymerization and copolymerization monomer polymerization. After monomer polymerization, the network structure is formed by connecting with crosslinking agents, and then the second network structure is formed by hydrogen bonding with hemicellulose molecular chains (figure 6.6(a)). (ii) The second is graft copolymerization, modifying hemicellulose with unsaturated groups, often employing a vinyl group, to initiate free-radical polymerization, resulting in the formation of hydrogels (figure 6.6(b)) [26].

In recent years, the rapid gelation of hemicellulose-based hydrogels has attracted extensive attention. For instance, tannic acid-modified hemicellulose nanoparticles (TA@HC) and Fe^{3+} were used in the preparation of PAA/TA@HC/Fe^{3+} hydrogels. The dynamic auto-redox system consisting of TA@HC and Fe^{3+} can activate persulfate, effectively producing hydroxyl radicals and generating abundant hydroxyl groups in a short period of time (figure 6.7), consequently resulting in a fast gelation process that was completed in as short as 30 s at room temperature. [49].

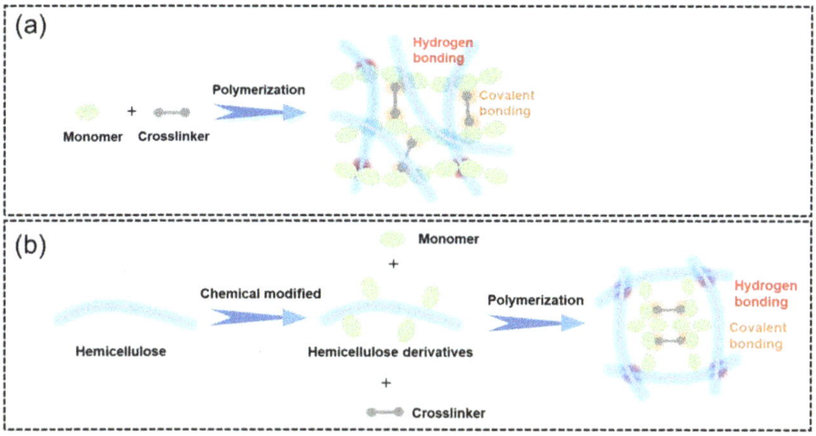

Figure 6.6. Formation of chemically cross-linking hydrogels based on radical polymerization.

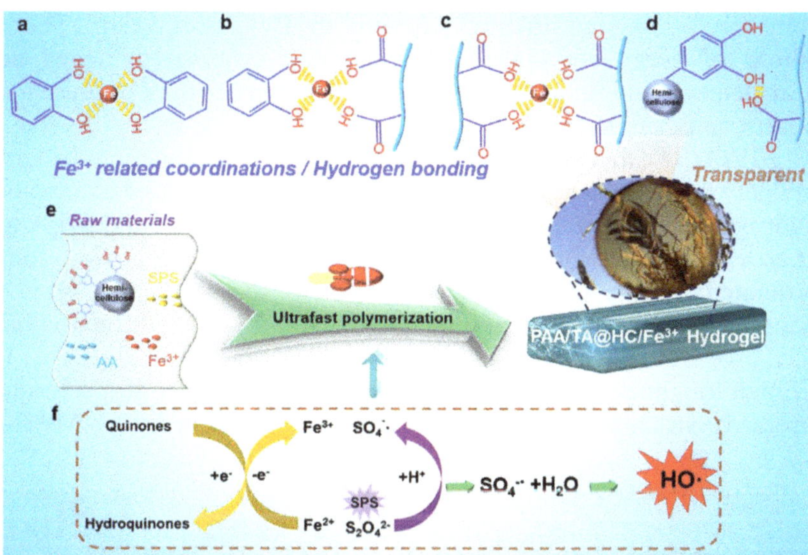

Figure 6.7. Schematic representation of hydrogel formation from PAA/TA@HC/Fe^{3+}. (Reproduced with permission from [49]. Copyright 2022 American Chemical Society.)

In addition, bromoisobutyryl bromide can also be used to modify the hydroxyl group on hemicellulose molecular chain. The modified hemicellulose is used as an atom transfer radical polymerization (ATRP) macroinitiator to initiate the cross-linking polymerization of comonomers under the catalysis of cuprous bromide/pentamethyldiethylenetriamine (CuBr/PMDETA) to form a hydrogel material [50]. However, due to the harsh reaction conditions, narrow monomer coverage and relatively complex process of ATRP, there are few reports on the preparation of hemicellulose hydrogel by this method.

6.2.3 Composite hydrogel

Hemicellulose-based hydrogels show promise for applications in biomedicine, devices, and adsorption. However, hemicellulose-based hydrogels always suffer from poor mechanical properties, a poor swelling/deswelling response, a low loading capacity, and low electrical conductivity, limiting their extensive application. To address these issues, the strategy of creating composites is adopted for fabricating high-performance hemicellulose-based hydrogels. For example, the combination of hemicellulose and filler (such as polymer, bentonite, carbon nanotubes, graphene, and MXene) can improve the mechanical properties of hydrogels, and can also give gels a variety of properties, enabling them to have a wide range of applications in many fields.

The schematic procedure for preparing a 3D crosslinked composite PVA/borax/xylan (PBX) hydrogel using a scalable freeze–thaw method is illustrated in figure 6.8 [51]. In the designed hydrogel architecture, the physically and chemically crosslinked network as the first network is formed by hydrogen bonding interactions between PVA and xylan, PVA and PVA, xylan and xylan, as well as the complexation of ester groups in borax and hydroxyl groups in PVA and xylan. The multiple combinations of borax crosslinked complexation, xylan-strengthened networks, chain entanglement, and inter/intramolecular hydrogen bonding produce a mechanically robust composite hydrogel.

A Xyl-*g*-PAA/Fe^{3+}/hydroxyl carbon nanotube (CNT) functional hydrogel, as shown in figure 6.9, exhibited excellent mechanical properties and conductivity [52]. In this hydrogel system, carboxylated Xyl was grafted with AA and MBA to form a chemical crosslinking network. Subsequently, the introduction of Fe^{3+} ions, which are coordinated by the carboxyl groups, led to the formation of a physical crosslinking network. CNTs are used as a nano-filler to further enhance the mechanical properties of the hemicellulose-based hydrogel. The composite hydrogel exhibited high toughness and elasticity. When the compressive strain reached 85%, the compressive strength of the composite hydrogel reached 10.4 MPa, and the

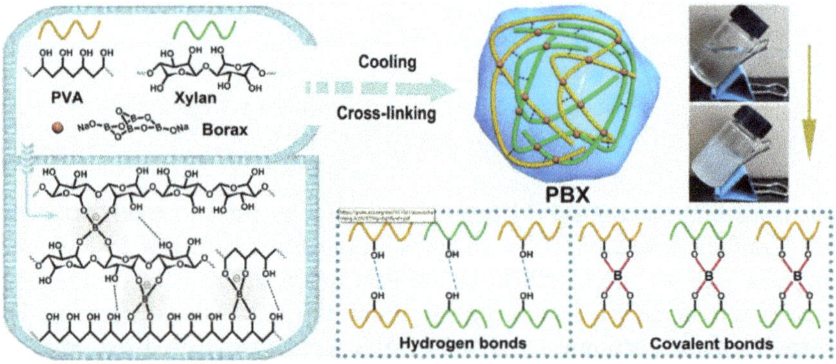

Figure 6.8. Schematic illustration of the synthesis of the PVA/borax/xylan hydrogel with multiple hydrogen bonds and covalent bonds. (Reproduced with permission from [51]. Copyright 2024 American Chemical Society.)

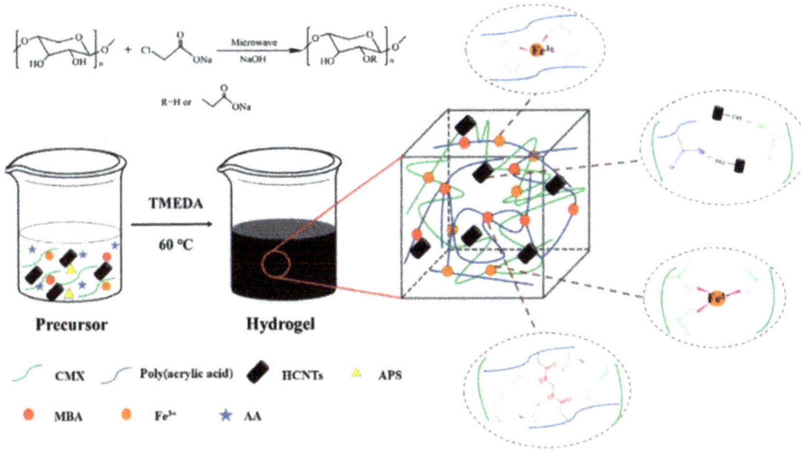

Figure 6.9. Schematic representation of hemicellulose-based double network hydrogels. (Reproduced with permission from [52]. Copyright 2019 Elsevier.)

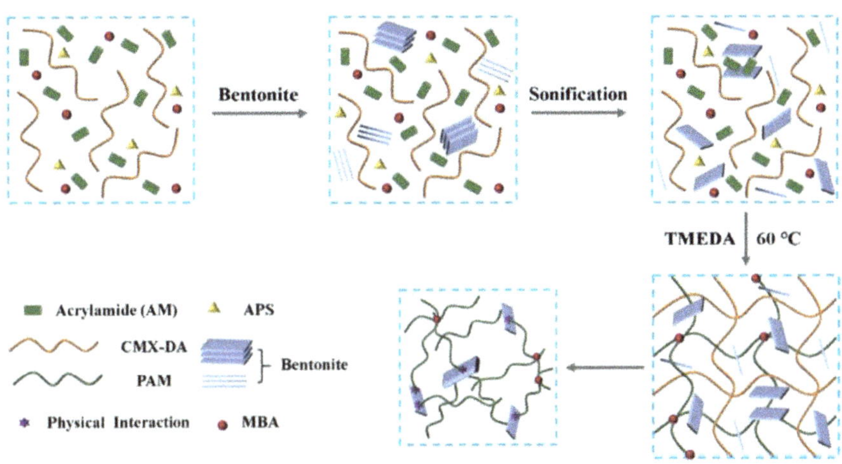

Figure 6.10. Synthesis process of xylan/PAM/bentonite composite hydrogels. (Reproduced with permission from [54]. Copyright 2022 Elsevier.)

maximum elongation at break was 1032%. After 30 compression cycles, it can quickly return to the initial state.

Bentonite is a natural non-metallic mineral with montmorillonite as the main component. It has the structure of two-dimensional nanomaterials and has the characteristics of expandability, hydrophilicity, stability, non-toxicity, and adsorption (figure 6.10) [53]. Liu and co-workers introduced bentonite into hydrogels prepared by cross-linking dopamine grafted carboxymethyl xylan with polyacrylamide (PAM) [54]. The exfoliation of bentonite improved the compatibility between inorganic phase and organic phase. The bentonite formed physical interactions with the polymer chains. The PAM chains could be integrated with the neighboring bentonite sheets by a mutual combination of polymer chains, and they could be crosslinked through non-covalent bonding such as hydrogen bonding and polymer chain entanglement, resulting in a nanocomposite hydrogel 3D network formation. With the introduction of bentonite, the Young's modulus of the hydrogel increased to 1449.3 kPa, indicating that the addition of bentonite could enhance the cross-linking density and improve the mechanical strength of the hydrogel.

6.2.4 Hemicellulose-based hydrogel adsorption material

Currently, environmental pollution has attracted global attention. The heavy metal ions (such as cadmium, plumbum, zinc, nickel, and chromium), along with organic dyes, pose a serious threat to humans, living organisms, and ecological resources worldwide, causing numerous severe diseases and significant damage to the natural environment [55, 56]. The methods for removal of metal ions from aqueous solutions include chemical precipitation, ion exchangers, chemical oxidation/reduction, reverse osmosis, electrodialysis, ultra filtration, and adsorption etc [57]. Among them, adsorption as a method with biocompatibility, flexibility, affordability, and high efficiency, has been widely studied [58]. For adsorption, the selection of adsorbent is crucial. Hydrogels are slightly crosslinked polymeric networks that enable the adsorption of many pollutants and recently, this type of adsorbent has attracted more attention due to its high-capacity and fast-response—within a few minutes [59]. Studies have shown that hemicellulose-based hydrogels as adsorbents have good adsorption effects on metal ions and dyes (figure 6.11) [60].

Polyacrylic acid (PAA) resin is a kind of super absorbent material, which contains a large number of ionizable hydrophilic ion groups. It can generate osmotic pressure inside and outside the resin by effectively ionizing cations in water, thereby exhibiting strong water absorption and water retention properties. For this reason,

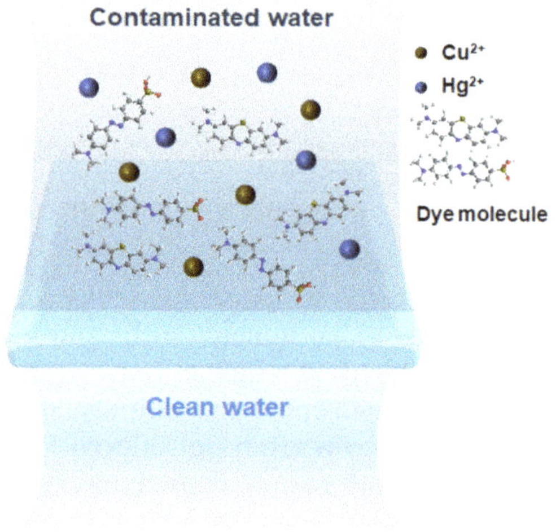

Figure 6.11. Schematic representation of hemicellulose-based hydrogel adsorption.

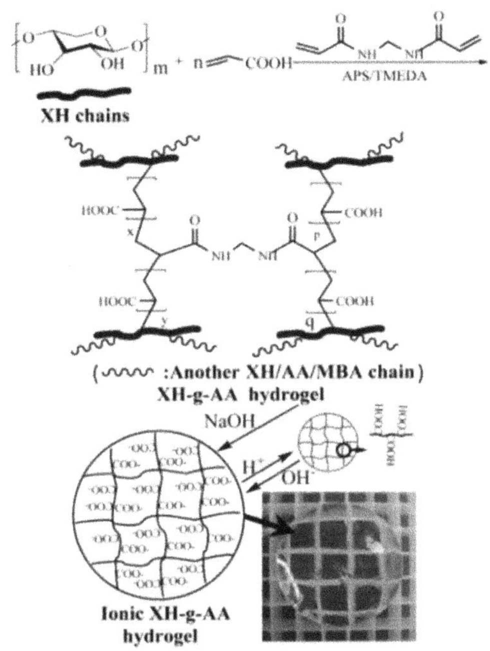

Figure 6.12. Proposed mechanistic pathway in the formation of xylan-rich hemicelluloses-graft-acrylic acid hydrogels. (Reproduced with permission from [61]. Copyright 2011 American Chemical Society.)

novel ionic hydrogels based on hemicelluloses were prepared by free-radical graft copolymerization of AA and hemicelluloses (XH) by using MBA as a cross-linker and APS/TMEDA as a redox initiator system (figure 6.12) [61]. The hydrogel has different response behaviors to different cations (such as Ga^{2+} and Na^{+}). The swelling ratio was only 7.2 at 0.005 M ion strength in $CaCl_2$ solution, which was much lower than that in NaCl solution (90.9) at 0.005 M ion strength. This indicates that the network of the ionic hydrogels is more sensitive to divalent Ca^{2+}, which is due to stronger ionic cross-linking and a more prominent charge screening effect.

Analogous to heavy metal ions, organic dyes are another category of pollutants in wastewater, and small amounts can devastate aquatic ecosystems because of their carcinogenic effects. Methylene blue (MB), a typical model of organic dye, is usually converted into organic cations in an aqueous solution. As a result it can be adsorbed by hemicellulose-based hydrogel containing anionic functional groups. For example, xylan/gelatin composite hydrogels are prepared in different molar ratios using ethylene glycol diglycidyl ether (EGDE) cross-linker, as shown in figure 6.13 [39]. Owing to the electrostatic interactions, van der Waals interactions, and hydrogen bonding, the dye molecules interact and are adsorbed on the surface of the hydrogel. The highest adsorption capacity of MB is 26.04 mg g^{-1} at pH = 5.84 °C and 25 °C. The kinetics of the adsorption process follows the pseudo-second-order model, and monolayer adsorption is adequately represented by the Langmuir isotherm model.

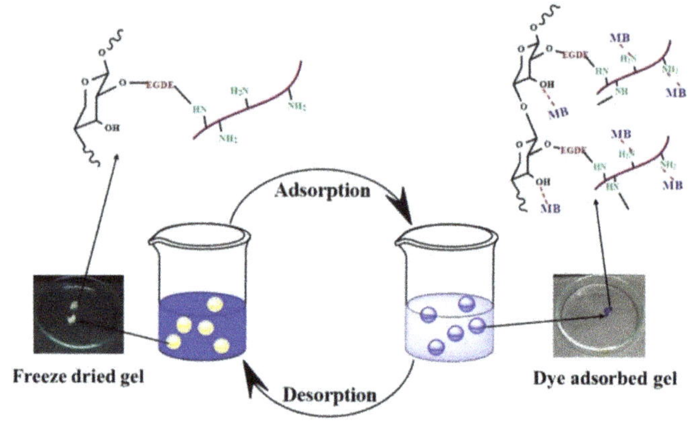

Figure 6.13. Schematic representation of xylan/gelatin composite hydrogel. (Reproduced with permission from [39]. Copyright 2021 Elsevier.)

6.2.5 Hemicellulose-based intelligent responsive hydrogel

Intelligent hydrogels can adjust their chemical structure to exhibit stimulation-responsive, injectable, self-healing, and conductive properties. Intelligent hydrogels exhibit good biocompatibility and hydrophilicity, and the introduction of stimulus-response effects can enhance their functionality and increase their range of applications [62]. The most important part of the intelligent hydrogel is the stimulus-responsive polymer, which is sensitive to external stimuli and can cause reversible or irreversible changes in the molecular structure after receiving stimulus signals and then reflecting the corresponding function [63]. According to the corresponding type, hemicellulose stimuli-responsive hydrogels can be classified as temperature-responsive [64], pH-responsive [21], magnetically responsive [38], and electrically responsive [65]. Currently, hemicellulose-based intelligent hydrogels are widely used in drug release and sensing.

Temperature-sensitive hydrogels as biocompatible materials have promising applications in drug delivery and engineering tissue. Polyisopropylacrylamide (PNIPAM) is a typical temperature-sensitive polymer. The hemicellulose-based temperature-sensitive hydrogels were prepared by the crosslinking copolymerization of xylan with N-isopropylacrylamide (NIPAm) and AA using *N,N'*-methylenebis-acrylamide (MBA) as a cross-linker and 2,2-dimethoxy-2-phenylacetophenone as a photoinitiator via ultraviolet irradiation [66]. The results indicated that the low critical solution temperature (LCST) of the hydrogels emerged at around 34 °C and increased with increasing the AA content. This is due to the introduction of a large number of hydrophilic groups to promote the formation of more hydrogen bonds in the system, resulting in the need for higher external energy to promote the phase transition of the system. The drug encapsulation efficiency of the as-prepared hydrogels reached 97.60% and the cumulative release rate of acetylsalicylic acid was 90.12% and 26.35% in the intestinal and gastric fluid, respectively. Additionally, hemicellulose-based temperature-sensitive hydrogels can be prepared by chemical modification [67]. First, reactive azide groups were incorporated into the hemicellulose backbone through the etherification of 1-azido-2,3-epoxypropane in an alkaline water/isopropanol mixture at ambient temperature, yielding degree of substitution (DS) values up to 0.28. Subsequently, the azide groups were reacted with propargyl bifunctional poly(ethylene glycol)-*b*-poly(propylene glycol)-*b*-poly (ethylene glycol) (PEG-PPG-PEG) utilizing CuAAC, leading to the formation of hydrogels. The thermo-responsive properties can be adjusted by the polymer used for crosslinking. When PEG was used for the crosslinking, the water absorption was affected by only a small amount when the temperature was increased from 7 °C to 70 °C. In contrast, the thermo-responsive PEG-PPG-PEG gels showed a considerable drop in water content when heated. These hydrogels have potential applications as drug delivery systems, or as part of separation, fractionation or self-cleaning membranes.

Additionally, pH-responsive hydrogels can be used in biomedical fields, in particular for controlled drug release. For instance, the pH-sensitive hydrogels were synthesized by radical copolymerization of the hemicellulosic polymer with AA [21]. Here, AA acts as a pH-sensitive monomer and is crosslinked with the hemicellulose chain in the presence of MBA. The results showed that the resulting hydrogels had good biodegradability. Moreover, the swelling ratios, water uptake, and degradation of hydrogels could be adjusted by the amount of acrylic acid, crosslinkers, and initiator. The cumulative release of acetylsalicylic acid as drug model was approximately 85% and 25% in simulated intestinal and gastric fluid. Comparing the drug release of acetylsalicylic acid with theophylline, as shown in figure 6.14, the difference between the cumulative release amounts of acetylsalicylic acid in the stimulated gastrointestinal solutions was more obvious than that of theophylline. The maximum drug release of acetylsalicylic acid in gastric solution was about 24%, far less than the 50% of theophylline.

Magnetic field-sensitive hydrogels generally embed ferromagnetic nanoparticles within the matrix. This allows for the remote manipulation of the hydrogel movement by an external magnetic field. Through control of the crosslinking density and swelling degree of the hydrogel, the *in situ* growth of ferromagnetic nanoparticles can be controlled, and then the size, magnetism and catalytic activity of the

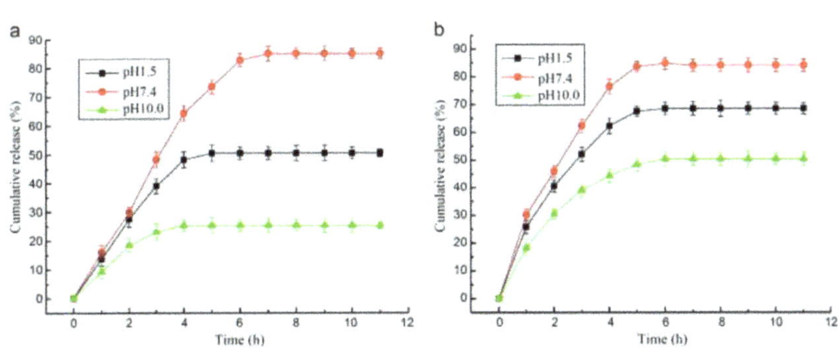

Figure 6.14. *In vitro* cumulative drug release from the drug-loaded hydrogels in various pH buffer solutions at 37 °C: (a) acetylsalicylic acid and (b) theophylline. (Reproduced with permission from [21]. Copyright 2013 Elsevier.)

nanoparticles can be adjusted. For example, when the crosslinking density of hydrogel material is low, Fe_3O_4 nanoparticles with small size and irregular shape can be obtained. With an increase of cross-linking density, the nanoparticles grow anisotropically, and finally exhibit an octahedral geometry. However, further increasing the degree of crosslinking, the anisotropic growth will slowly disappear, and the Fe_3O_4 nanoparticles will eventually exhibit a spherical structure [68]. More importantly, the introduction of magnetic particles can improve the mechanical properties of the hydrogels. In another work, the xylan/poly(acrylic acid) magnetic nanocomposite hydrogel was prepared from wheat straw xylan and Fe_3O_4 nanoparticles [69]. The composite hydrogel had a semi-interpenetrating network structure and exhibited a macro-porous structure with interconnected porous channels. After introduction of Fe_3O_4 nanoparticles, the compressive strength can reach 0.16 MPa. The thermal stability of the hydrogels was enhanced because of an increase in the contents of xylan and Fe_3O_4 nanoparticles.

Electroresponsive hydrogels can not only change their physical and chemical properties through discontinuous volume phase transitions of polyelectrolytes, but also achieve electrical conductivity by adding conductive polymers and inorganic conductive fillers. Albertsson and co-workers prepared hemicellulose-based conductive hydrogels via two strategies. One is to blend a conductive aniline pentamer (AP) with epichlorohydrin-crosslinked *O*-acetyl-galactoglucomannan (AcGGM) [32], and the other is to graft a epoxy-modified AcGGM hydrogel with an aniline tetramer, as shown in figure 6.15 [70]. In the conductive hemicellulose-based hydrogel, the equilibrium swelling ratios can also be tuned from 9.6 to 6.0 by changing the AP content level from 10% to 40% (w/w) while simultaneously altering conductivity from 9.05×10^{-9} to 1.58×10^{-6} S cm^{-1}. These conductive hemicellulose-based hydrogels with controllable conductivity, tunable swelling behavior, and acceptable mechanical properties have great potential for biomedical applications, such as biosensors, electronic devices, and tissue engineering.

Composite hydrogels with excellent mechanical properties and good frost resistance were obtained using hemicellulose as a hydrophilic carrier and polypyrrole as a conductive matrix, as shown in figure 6.16 [65]. The addition of glycerol

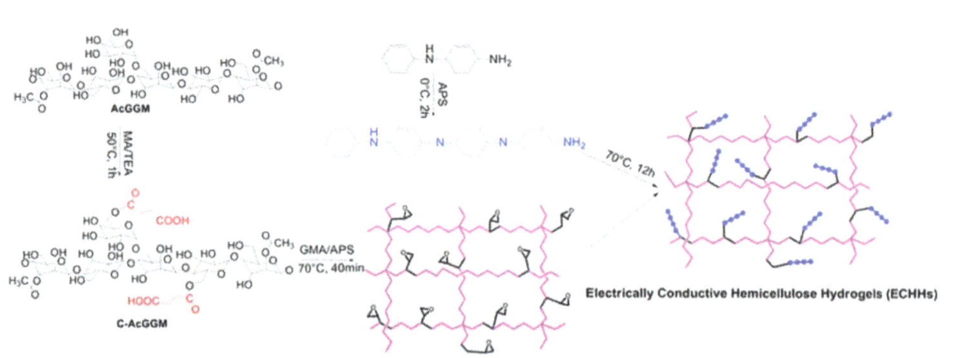

Figure 6.15. Schematic diagram of conductive gel prepared by AP and epoxy-modified AcGGM. (Reproduced with permission from [70]. Copyright 2014 Elsevier.)

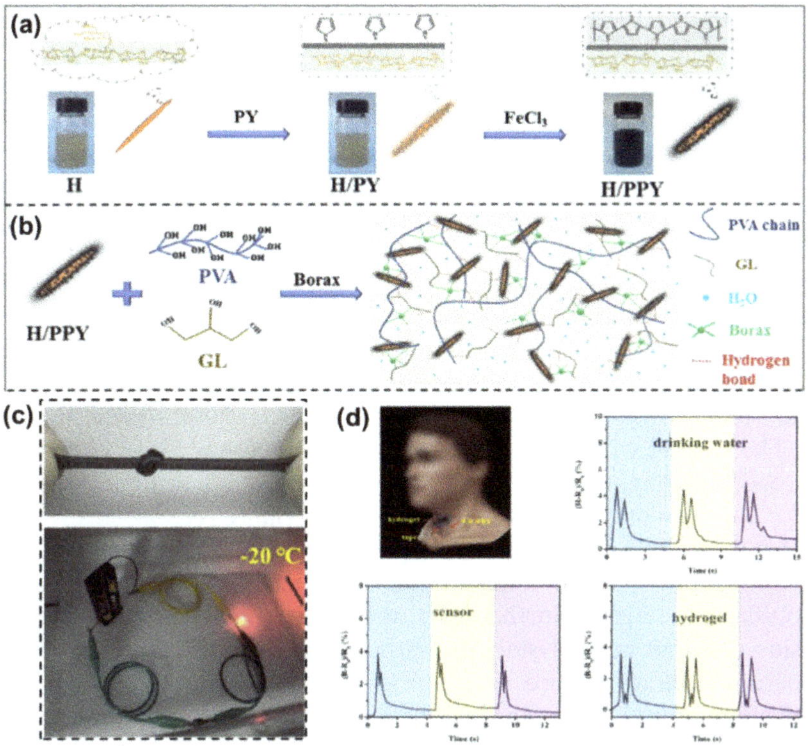

Figure 6.16. (a),(b) Construction of hemicellulose/polypyrrole hydrogel with frost resistance. (c) A physical map of the conductive hydrogel. (d) A schematic diagram of the hydrogel as a sensor and the relative resistance change of the sensor when drinking water and speaking English words. (Reproduced with permission from [65]. Copyright 2021 Elsevier.)

and sodium chloride reduced the freeze–thaw temperature of the hydrogel and increased the conductivity. At −20 °C, the hydrogel still has good electrical conductivity (figure 6.16(c)). The tensile strain, stress, compressive strength and toughness of the hydrogel reached 1094.9%, 480.6 kPa, 1790.2 kPa and 2.82 MJ m^{-3}, respectively. At the same time, the hydrogels exhibited excellent remoldability and durability, and still retained 71.8% of their after storage for 7 days. Notably, they still had good repeatability after 250 cycles. The sensors assembled by hydrogels can quickly and accurately detect motion signals including finger bending, wrist bending, and throat deformation during drinking and speaking. They are expected to be used in the fields of sports monitoring, medical monitoring, and soft intelligent robots (figure 6.16(d)).

6.3 Hemicellulose-based aerogels and foams

Aerogels are characterized by their ultra-low density (0.003–0.500 g·cm^{-3}), large specific surface area (100–1600 m^2 g^{-1}), and high-porosity (>90%) with the latter dominated by mesopores (2–50 nm) and macropores (>50 nm) [71, 72]. The performance of aerogels largely depends on the drying method because the as-obtained porous structure varies with the drying method. Lyophilization (freeze-drying) is a straightforward and versatile strategy for preparing aerogels. However, the porous structures of aerogels prepared by freeze-drying are significantly damaged, relative to those of hydrogels [73, 74]. Although supercritical CO_2 drying overcomes the problems associated with the traditional drying methods, e.g. freeze-drying and air drying, and retains the porous structure of the initial hydrogel, it is expensive [75, 76]. Thus, the choice of drying method plays a crucial role in determining the performance and characteristics of aerogels. Building upon the exploration of hemicellulose-based hydrogels in the preceding section 6.1, we find a current interest in hemicellulose-based aerogels derived from their hydrogel analogons as well. In recent years, hemicellulose-based aerogels have garnered significant attention owing to their unique properties. The unique properties exhibited by hemicellulose hydrogels also endow the derived aerogels with similarly fascinating features. The section will discuss the properties and applications of hemicellulose-based aerogels, highlighting their unique traits and potential applications.

Köhnke and co-workers prepared anisotropic xylan/CNC physically crosslinked aerogel using an ice-templating technique, where the pore morphology of the material is controlled by the solidification conditions and the molecular structure of the polysaccharide [77]. Furthermore, the reinforcement of these biodegradable foams using cellulose nanocrystals shows potential for strongly improved mechanical properties. Mikkonen and co-workers presented a novel method to prepare aerogels from polysaccharides, using specific enzymatic oxidation as a route to form hydrogels [78]. This is the first report on the use of plain galactomannan (GM) and galactoxyloglucan (XG) as aerogel matrices. The compressive moduli of the aerogels depended greatly on the oxidation, polysaccharide type, freezing method, and ambient moisture. Ice crystal templated, oriented aerogels from oxidized XG (XG-OX) showed the highest compressive modulus, 359 kPa, when determined parallel to the freezing and drying direction. In another work, freeze-casted CA crosslinked aerogel was formed using lignin-containing arabinoxylan (AX), where water and cellulose nanofibers (CNFs) were used as a solvent and as reinforcing building blocks, respectively [79]. This provides an alternative route for the preparation of a natural, barley-residue-based aerogel with superior swelling and mechanical behaviors compared to those formed with AX alone.

Currently, aerogels are promising for wearable pressure sensors due to their superior flexibility and high sensitivity. Yan and co-workers reported a series of xylan-based anisotropic aerogels with ultralight, flexible, and super-stable properties via freeze-casting technology (figure 6.17) [80]. The obtained anisotropic conductive aerogel exhibits excellent mechanical performance, with a great compressibility (undergoing a strain of 70%) and elasticity (95% height retention after 100 cycles at a strain of 20%). Moreover, the sensor demonstrated successful application as a wearable-electronic device for monitoring diverse human movements. As a pressure sensor, it shows excellent sensing property, including a high sensitivity ($S = 2.17$ kPa^{-1}) and remarkable long-term stability (2000 cycles).

Foams are used in thermal insulation and flame retardant materials due to their unique micro- and nano-porous network structure, good barrier properties and very

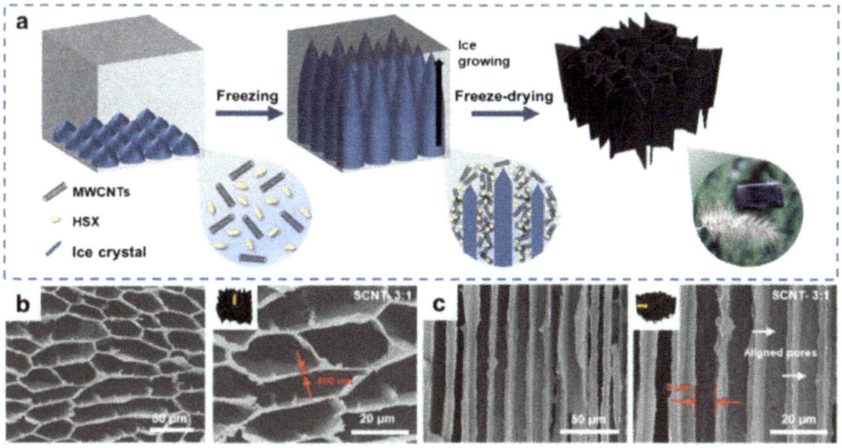

Figure 6.17. Schematic diagram of the fabrication process of xylan-based aerogels and their SEM images. (Reproduced with permission from [80]. Copyright 2024 Elsevier.)

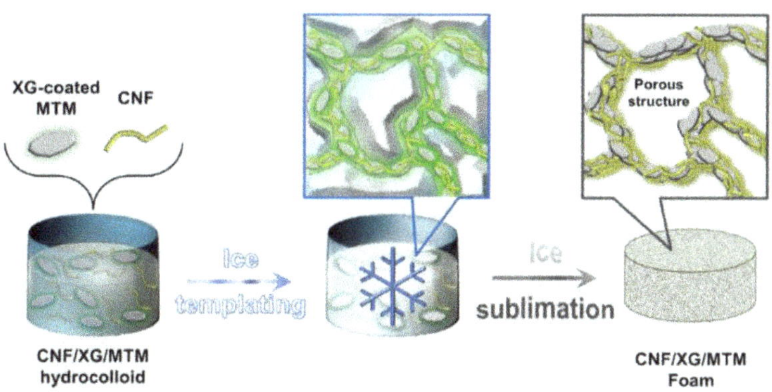

Figure 6.18. Schematization of the procedure adopted for the preparation of the CNF/XG/MTM foam. (Reproduced from [81]. John Wiley and Sons. CC BY 4.0.)

low thermal conductivity. Carosio and co-workers prepared a series of fire resistant foams with cell wall structure comprising bio-based components (cellulose nanofibers and xyloglucan-type hemicellulose), as shown in figure 6.18 [81]. Here, cellulose nanofibers were combined with 2D montmorillonite nanoplatelets and a native xyloglucan hemicellulose binder, using a water-based freeze-casting approach. The thermal conductivity of these aerogels was 0.012–0.024 W mK^{-1}, which was 2–3 orders of magnitude lower than that of traditional thermal insulation materials. The limiting oxygen index is as high as 31.5% and in the same range as the best fire-retardant synthetic foams commercially available. For self-extinguishing properties, the 33 wt% of clay nanoplatelets have a critical role in the CNF/XG/MTM cell wall nanocomposite, by acting as a barrier which delays the release of organic volatiles to feed the flame and hinder oxygen diffusion, which would otherwise facilitate combustion.

6.4 Construction and application of hemicellulose-based functional films

The increasing production and consumption of petroleum-based plastics inevitably lead to their accumulation in the natural environment, resulting in plastic pollution, which becomes poorly reversible when weathering processes fragment the plastics into micro-and nano-plastics [82]. To address these issues, the development of recyclable, biodegradable, high-performance bioplastics derived from the biomass generated by the agriculture and forestry sectors has been investigated extensively. Ever since Smart first reported hemicellulose acetate films in 1949 [83], hemicellulose-based films and coatings have attracted significant attention owing to their inherent properties such as low oxygen permeability (OP), good biocompatibility, and good biodegradability. Owing to these excellent properties, hemicellulose-based films and coatings are promising alternatives to petroleum-based synthetic polymers in food and medicine applications, such as food packaging, wound dressing, and drug delivery. Nevertheless, hemicellulose-based films and coatings have the disadvantages of water/moisture sensitivity, high hydrophilicity, poor mechanical properties, and brittleness, which limit their practical application [84–86]. Furthermore, hemicellulose with high molecular weight (extracted under highly alkaline conditions), referred to as water-insoluble hemicellulose, is crystalline, resulting in poor processability for film formation [87–89]. To this end, different strategies have been proposed for addressing these limitations, including molecular engineering and blending. In this section, the preparation strategy–structure–properties relationships of hemicellulose-based films and coatings are reviewed.

6.4.1 Physical blending strategy

The physical blending of the hemicellulose with small molecule plasticizers (such as glycerol, sorbitol, and xylitol) [90, 91], polymers (such as polyvinyl alcohol, cellulose, and chitosan) [92, 93], and 2D materials (such as montmorillonite and graphene oxide) [94, 95] can greatly enhance the film-forming property, reduce the brittleness, and provide flexibility to hemicellulose-based composite films.

Small molecule plasticizers are required for enhancing the flexibility of hemicellulose-based films; propylene glycol (PG), glycerol (Gly), sorbitol (Sor), and xylitol (Xyl) are the most commonly used plasticizers. Compared to the pure AX film, the films with an added plasticizer (Gly or Sor) have lower strength and stiffness but better flexibility [96, 97]. The plasticized AX films have lower water vapor transmission (WVPs) than the unplasticized films, attributable to antiplasticization. However, a high plasticizer content results in films with poor barrier properties, in particular a low WVP, indicating that the presence of excess plasticizers and a large amount of affinitive water weakens the chain-to-chain interactions, thereby introducing additional free volume in the AX film matrix. Furthermore, the high plasticizer content results in highly hydrophilic films, limiting their practical application [91]. Gröndahl *et al* [98] used xylitol and sorbitol as plasticizers to prepare hemicellulose-based films. The enhancement mechanism of the small molecule plasticizer in film was analyzed by dynamic thermomechanical analysis and x-ray diffraction. The stress at break and Young's modulus of the films decreased, whereas the strain at break increased with increasing plasticizer content. Sorbitol and xylitol gave rise to films exhibiting similar mechanical properties, but the xylitol plasticized films were slightly less extensible at high plasticizer levels. Additionally, the glass transition temperature of the film decreased from 40 °C–50 °C to −45 °C to −20 °C with the increase of xylitol content. When the added amount of xylitol reached 50%, the storage modulus of the films was about 1 GPa at room temperature. This is because the addition of small molecule plasticizers endows the film with crystallization ability and exhibits semi-crystalline characteristics. The films with 35 wt% sorbitol exhibited excellent oxygen barrier properties at 50% relative humidity (RH). The oxygen permeability was lower than that of plasticized starch and in the same range as that of the commercially used ethylen vinyl alcohol. Thus, glucuronoxylan has a potential in film applications, such as food packaging.

The incorporation of natural polymers, such as chitosan (CS), CNCs, CNFs, and carboxymethyl cellulose (CMC) is an effective approach for addressing the drawback of strength reduction of the hemicellulose-based films caused by the addition of small molecule plasticisers. For instance, the xylan/sulfonated nanocrystalline cellulose/Sor composite film exhibits an excellent OP of 0.1799 cm^3 μm m^{-2} d kPa [99]. In another work, the films were prepared by mixing a CNF dispersion with an AX solution without any plasticizer, resulting in a modulus of up to 7.2 GPa, tensile strength of 143 MPa, and strain-to-failure of 7.3%. Moreover, the oxygen barrier properties of the AX films are improved; they show an OP of 0.8 cm^3 μm m^{-2} d kPa at 50% RH [100]. Hemicellulose-based films and coatings must have not only low oxygen permeability, high mechanical strength, and excellent flexibility, but also excellent antibacterial properties to ensure food quality. To this end, the antibacterial properties of hemicellulose-based films can be effectively improved by introducing CS and some specific inorganic nanoparticles such as TiO_2, SiO_2, or ZnO [101–104]. In a recent work, composite films based on CS/hemicellulose were prepared by direct immersion of a CS film in a hemicellulose solution. [102]. The dried CS films were immersed in hemicellulose solutions of various concentrations for 2 h to prepare composite films with different hemicellulose loading. The cross-section of the composite film is continuous and dense (figure 6.19(a)), which is attributed to the strong hydrogen bonding between the hemicellulose and CS, resulting in a composite film with a tensile strength of 32.81 MPa and a tensile strain of 39.63%. The composite films exhibit higher mechanical strength than polyhydroxybutyrate, polypropylene, and poly (vinyl chloride) films (figure 6.19(c)). More importantly, all the films inhibit the growth of *E. coli* and *S. aureus*. The inhibition ability of the films increases with increasing TiO_2 content up to 25%, with maximum inhibition zones of 19.7 ± 1.45 and 17.1 ± 1.1 mm against *E. coli* and *S. aureus*, respectively (figure 6.19(d)). Considering these properties, the prepared CS/hemicellulose composite films can be applied to functional packaging. In another work, a drug-loading film was prepared by the cross-linking reaction of quaternized hemicellulose (QH) and CS with epichlorohydrin as the cross-linker [105]. The drug-loading film has excellent mechanical properties with a high tensile strength of up to

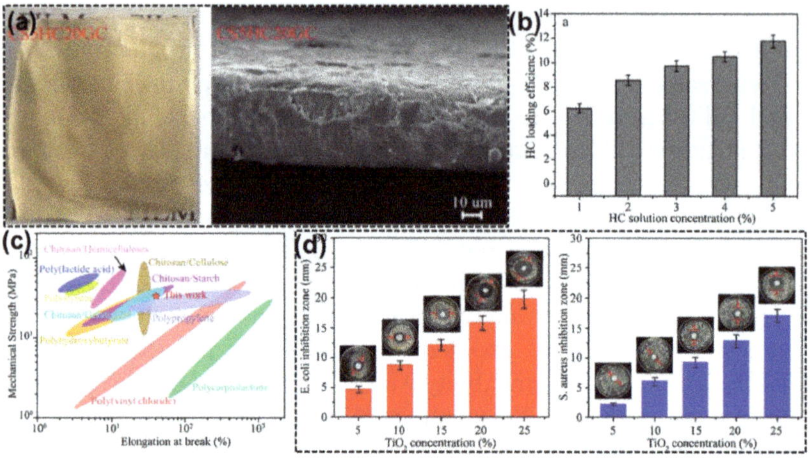

Figure 6.19. (a) Photographs and scanning electron microscopy images of CS/hemicellulose composite films. (b) Hemicellulose loading efficiency of the composite films. (c) Mechanical properties of the CS/hemicellulose composite film, compared with those of existing polymer films. (d) Antibacterial properties of the composite films for different TiO_2 concentrations. (Reproduced with permission from [102]. Copyright 2020 Elsevier.)

37 MPa. In this study, ciprofloxacin was used as a model compound to investigate the loading behavior of the film; the highest loading concentration was ~18%, and a drug release of ~20% in 48 h could be achieved. Furthermore, the results of a 293 T cell viability assay showed the excellent biocompatibility and non-toxicity of the film. These results indicate that hemicellulose-based films can be applied in biomedical applications, such as wound dressing.

In recent years, the preparation of high-performance composite films and coatings by blending hemicellulose with two-dimensional (2D) nanomaterials has attracted significant attention. Montmorillonite (MTM), bentonite (BNT), and graphene oxide (GO) are the most commonly used 2D materials [106–109]. Owing to their special anisotropic properties, excellent mechanical properties, and large specific surface area, 2D nanomaterials are special additive materials with excellent properties, which can be used widely in the preparation of composite materials. Bioinspired and highly oriented xyloglucan/montmorillonite (XG/MTM) composite films were prepared by water casting or by coating onto an oriented polyester film (figure 6.20(a)) [109]. The hydrophobic interactions between XG and the MTM surface is a potential driving mechanism for the adsorption of XG. As shown in figure 6.20(b), the cross-section of the composite film reveals an in-plane orientation of the platelets. Compared to the pure XG film, the XG/MTM composite films have significantly better mechanical properties, with the tensile strength, tensile strain, and modulus being 123 MPa, 2.1%, and 11.6 GPa, respectively, at 50% RH and 23 °C. Even in a moist state (92% RH, 23 °C), the composite films exhibit excellent mechanical properties, with the tensile strength, tensile strain, and modulus being 81 MPa, 3.0%, and 6.8 GPa, respectively; the stress is efficiently transferred from the matrix to the stiffer MTM platelets. Moreover, the above-mentioned films have much better mechanical performance than the

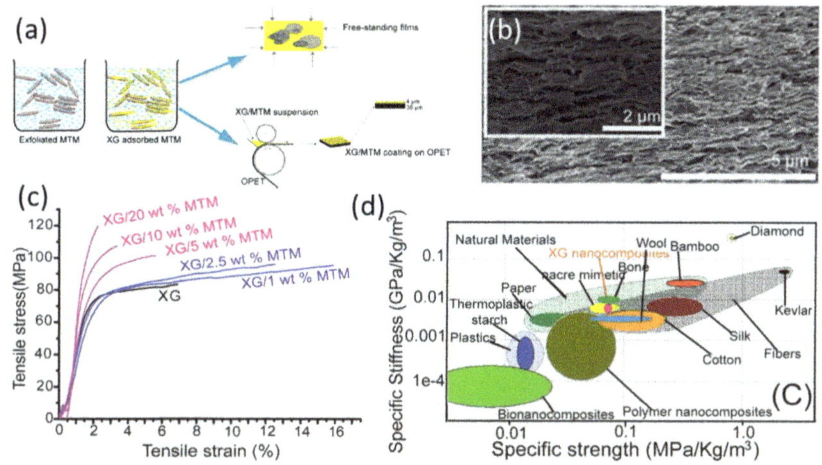

Figure 6.20. (a) Schematic of xyloglucan/montmorillonite nanocomposite preparation and film formation. (b) Mount of xyloglucan and modified xyloglucan adsorbed on the montmorillonite model surface in comparison with that on polyvinyl alcohol. (c) Typical stress–strain curves of xyloglucan/montmorillonite films conditioned at 50% RH and 23 °C. (d) Mechanical properties of the xyloglucan/montmorillonite composite film compared with those of other nanocomposites and polymer composites. (Reproduced with permission from [109]. Copyright 2013 American Chemical Society.)

conventional composites based on starch, PLA, and PCL. The excellent mechanical properties of XG–clay nanocomposites rely on the strong molecular interaction between the matrix polymer and the inorganic reinforcement. Even in a moist state, this interaction ensures that stress can be transferred efficiently from the matrix to the stiffer MTM platelets. Furthermore, the thermostability and oxygen barrier properties are significantly enhanced upon incorporating MTM, the composites with 12 vol% MTM have an OP of 1.44 $cm^3\ \mu m\ m^{-2}$ d kPa. These results provide a new strategy for preparing hemicellulose barrier films and coatings.

Peng and co-workers proposed a facile method to prepare composite films with a nacre structure based on the self-assembly of hemicellulose and MTM [108, 110]. The results show that the montmorillonite layer can be closely combined with the quaternized hemicellulose through hydrogen bonds and ionic bonds to form a continuous and dense composite film. The mechanical properties, thermal stability and barrier properties of the composite films are greatly improved. It shows great application potential in the fields of high-temperature resistant packaging materials and coatings. Additionally, the hydrophobicity and mechanical properties of the wood hydrolysate (WH) based films and coatings can be effectively improved by incorporating 2D materials such as MTM and GO [95, 111]. For example, a hydrophobic, bioinspired composite film with high strength and excellent fire retardancy was prepared by the blending of WH, MTM, and GO [95]. The nanocomposite films comprise an ordered multilayer structure, which is formed by the cross-linking of the assembled MTM sheets with WH and GO (figure 6.21). After reduction by hydroiodic acid (HI), the contact angle of the film increases from 36.6° to 88.8°. This composite film has excellent mechanical properties, with a tensile

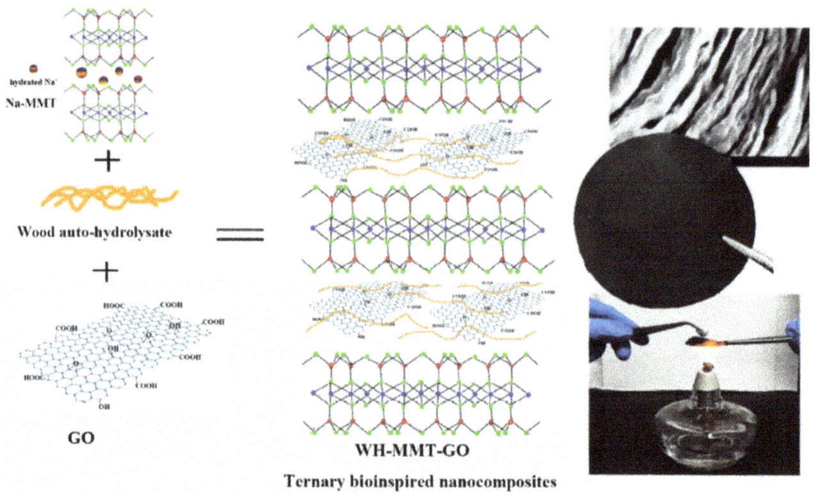

Figure 6.21. Montmorillonite and graphite oxide synergistically enhance hemicellulose-based films. (Reproduced with permission from [95]. Copyright 2018 Elsevier.)

strength of up to ~124 MPa, which is superior to that of other hemicellulose-based films. The heat release rate (HRR) curves demonstrate that the addition of MTM and reduced graphene oxide results in the highest HRR reduction of the WH films, improving the fire retardancy. These results provide an avenue for the high value-added utilization of WH in the pulping industry, and promote the application of these films in packaging, coating, and flame retardant materials.

Liquid metal (LM), a new type of multifunctional material with deformability, high conductivity, and low toxicity, has been widely applied in many fields such as electronics, mechanical engineering, and energy [112, 113]. However, embedding liquid metal directly into a polymer matrix will cause problems such as poor compatibility and liquid metal leakage. Also, liquid metals exhibit high surface tension. For example, the surface tension of EGaIn is 624 mN m^{-1}, resulting in inevitable aggregation, which limits its wide application [114]. To address these problems, the development of natural products as stabilizers for liquid metal micro/nano-droplets has attracted great attention. Hu and co-workers developed a stable and green LM aqueous colloidal ink by wrapping eutectic gallium–indium alloys (EGaIn) with carboxymethyl glucomannan (CGM) derived from radiata pine chip, which is capable of being processed into a free-standing, photothermal-actuating, and motion-monitoring Janus film, as shown in figure 6.22 [115]. The stable CGM/EGaIn inks can be patterned on different substrates to form coating layers or self-assemble into free-standing Janus films with high mechanical strength and modulus (~94 MPa and ~3.8 GPa) by density deposition. The biocompatible film demonstrated both high conductivity and large resistance variation in response to strain change (gauge factor > 500), allowing for human motion monitoring. This work provides a new prospect for the highly valued utilization of hemicelluloses and the development of biocompatible and high-performance nano-LM materials.

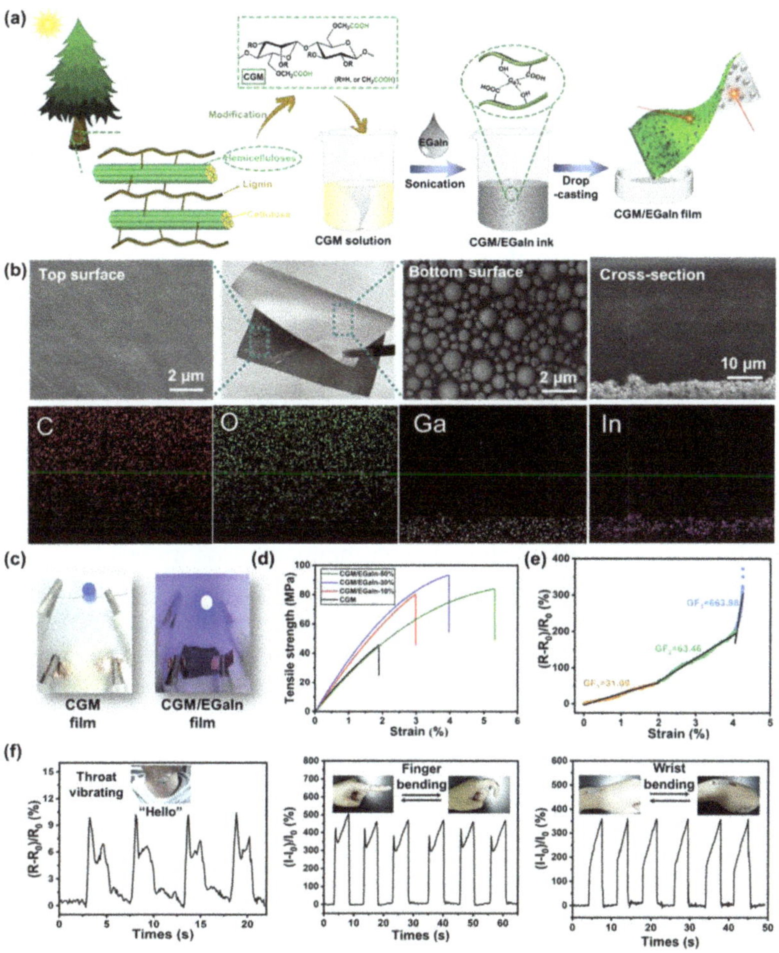

Figure 6.22. (a) Preparation flow chart of carboxymethyl glucomannan/liquid metal composite film. (b) Optical image, surface SEM image, cross-section SEM image and corresponding energy spectrum of the composite film. (c) Conductivity of the composite film. (d) Mechanical properties of the composite film. (e) Sensitivity of the composite film (GF value). (f) Composite film for monitoring human behavior (throat vibration, finger bending and wrist bending). (Reproduced with permission from [115]. Copyright 2022 American Chemical Society.)

6.4.2 Molecular engineering

Despite the film formation and flexibility of hemicellulose-based films being significantly enhanced by blending strategy, the blending strategy does not address the shortcomings of high hydrophilicity and water/moisture sensitivity, which are critical in fabricating high-performance hemicellulose-based films for practical applications. Regarding chemical modification, esterification [116, 117], etherification [118, 119], oxidation [120, 121], and graft copolymerization [122, 123] have been developed for fabricating special architectures that improve the film-forming property, hydrophobicity, mechanical properties, and thermal stability of the hemicellulose-based films and coatings.

For small molecule modification, especially esterification and etherification, alkyl chains can be introduced to the hemicellulose chain to convert the hemicellulose-based film from hydrophilic to hydrophobic, thereby overcoming the limitations that impede its practical applications. In homogeneous systems, hemicellulosic esters containing alkyl side chains of different lengths can be prepared by esterification modification [124]. Chloroform serves as an effective solvent for hemicellulosic alkyl-esters, so continuous and dense hemicellulose-based films can be prepared using a casting method. The results of contact angle experiments showed that the hydrophobic property of the hemicellulose-based film was significantly improved with the increase in length of the alkyl side chain (up to 99° water contact angle). The mechanical test results show that the maximum fracture stress and strain of the film are 29 MPa and 44%, respectively. The fracture stress and modulus decreased with the increase of the alkyl side chain length. Conversely, the fracture strain of the film increased. This is mainly because the increase in the length of the alkyl side chain reduces the bonding strength of the adjacent hemicellulose molecular chain, and the side chain acts as a toughening agent to make the film exhibit more flexibility. The results of synchronous thermal analysis showed that the thermal stability of the film increased with the increase of the length of the alkyl side chain. Furthermore, it was shown that the hemicellulosic esterified derivative can be used to prepare nanofibers by electrospinning, showing great potential in the fields of surgical sutures and filtration. The long alkyl chain can be introduced into the hemicellulose molecular chain by the etherification reaction, which greatly increases the hydrophobicity of the hemicellulose film, as shown in figure 6.23 [125]. The maximum water contact angle observed for hemicellulose film in this study was about 120°. The barrier performance of the hemicellulose film is better than that of the commercial polyethylene terephthalate. The oxygen permeability and water vapor permeability of the film are 1.21 $cm^3\ \mu m\ m^{-2}\ d^{-1}\ kPa^{-1}$ and $1.59 \times 10^{-10}\ g\ m^{-1}\ s^{-1}\ Pa^{-1}$, respectively. The excellent barrier performance makes them promising for applications in food packaging and coatings.

Graft copolymerization is also an important way to introduce long alkyl chains into hemicellulose. For instance, galactoglucomannan (GGM) film with high transparency, excellent hydrophobicity and flexibility can be prepared by the combination of etherification and emulsion polymerization, as shown in figure 6.24 [126]. First, allyl glycidyl ether (AGE) was introduced into the chain of GGM by etherification. Then, GGM latex was obtained by emulsion polymerization of etherified GGM and n-butyl acrylate in an SDS/APS initiator system. Finally, the GGM film was obtained by casting. The particle size of the prepared GGM latex increased with the increase of the side chain length and showed a single distribution, as shown in figure 6.24(d). The optimal GGM-based film not only exhibited high transmittance of 91.3% and haze of 0.43%, but also demonstrated superior hydrophobicity, achieving a 117.2° water contact angle while maintaining excellent stretchability, with elongation at break of 31.2% and an ultimate tensile stress of 30.9 MPa. This newly designed bio-based hemicellulose-modified latex with

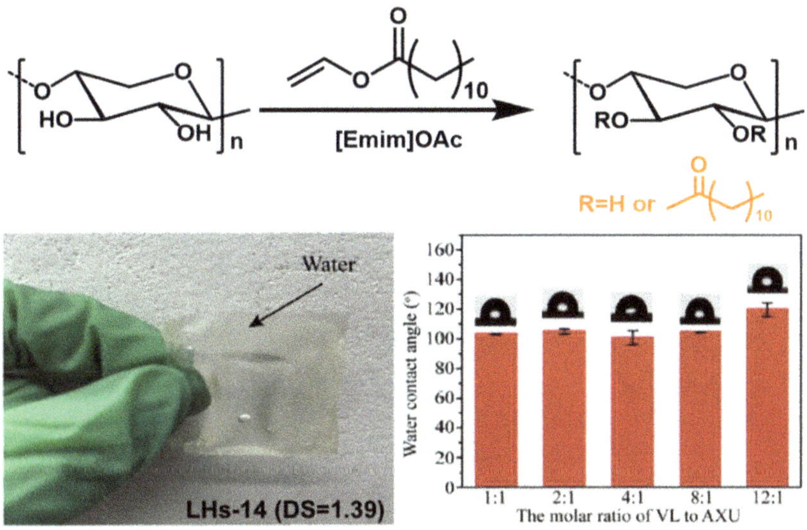

Figure 6.23. Transesterification of hemicellulose with vinyl laurate in [Emim]OAc. Flexible pouch of LHs-14 film with a DS value of 1.39 and the water contact angle of the LHs films. (Reproduced with permission from [125]. Copyright 2020 Elsevier.)

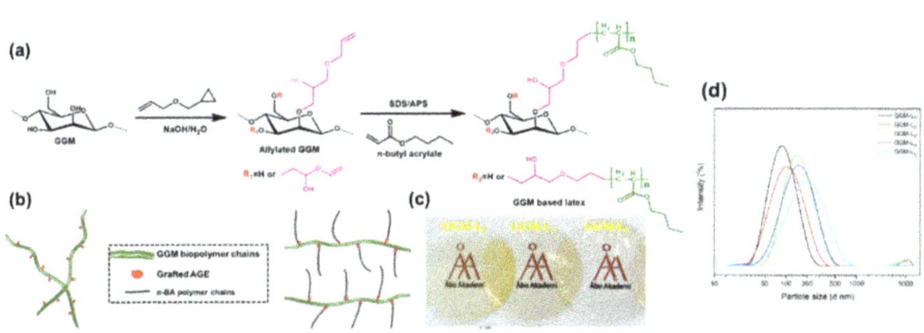

Figure 6.24. Preparation and properties of GGM membrane materials made using an emulsion polymerization strategy. (Reproduced from [126]. CC BY 4.0.)

bio-based content of up to 99 wt% can be industrially applied to replace the fossil-based polymer packaging materials, which simultaneously boosts the exploitation and utilization of hemicellulose wastes.

In a recent work, a xylan plastic (XP) was fabricated by a strategy of double cross-linking through etherification combined with hot pressing, as shown in figure 6.25 [127]. The mechanical properties, particularly the toughness of XP, were significantly enhanced by the incorporation of chemical and physical cross-linking domains. The tensile strength, toughness, and modulus of XP could reach up to 55 MPa, 2.2 MJ m^{-3}, and 1.7 GPa, respectively, which are superior to most traditional plastics. Dynamic mechanical analysis (DMA) characterizations confirmed that XP is a thermoplastic and can be hot formed. Additionally, the reversible hydrogen bond interaction between xylan chains could be simply regulated by water

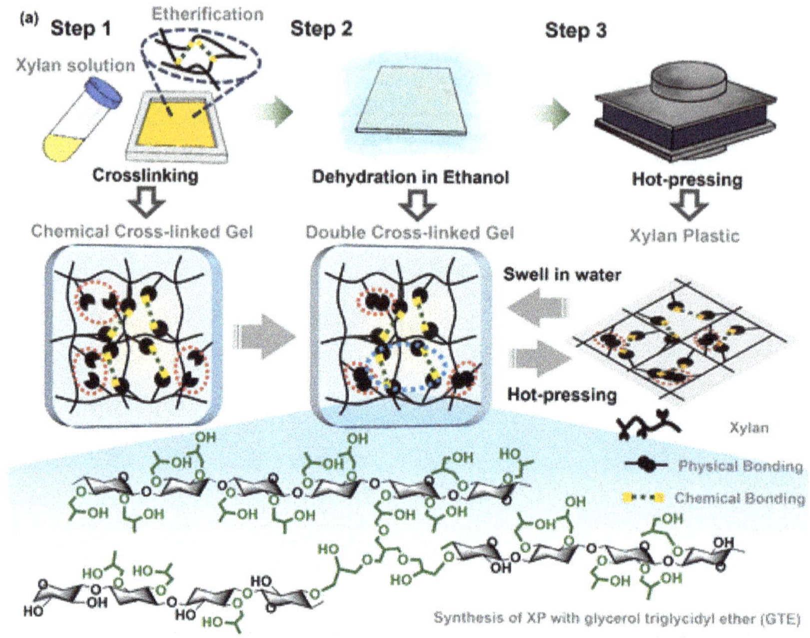

Figure 6.25. Schematic demonstrating the preparation of xylan plastic. (Reproduced with permission from [127]. Copyright 2023 American Chemical Society.)

molecules, allowing XP to be readily transformed and repeatedly reprogrammed into versatile 2D/3D shapes. Moreover, XP showed a low thermal expansion coefficient and excellent optical properties. Cytotoxicity and degradability tests demonstrated that XP had excellent non-toxicity and can be biodegraded with in 60 days. This work thus suggests an avenue for the scalable production of high-performance xylan-based plastics, in which the raw material comes from industrial wastes and exhibits great potential in response to plastic pollution.

6.5 Hemicellulose-based carbon materials

As the average percentages of carbon, oxygen, and hydrogen in hemicellulose are 44.4, 49.4, and 6.2 wt%, respectively, hemicellulose is a suitable precursor for preparing carbon materials. The conversion of hemicellulose into carbon materials (such as carbon dots and porous carbons) not only renders hemicellulose highly versatile for industrial applications but also offers a sustainable development strategy as a green energy source. To date, various cost-effective, size-controlling, and large-scale synthesis methods (such as hydrothermal carbonization, microwave-assisted hydrothermal carbonization, and pyrolysis carbonization) have been developed for fabricating hemicellulose-based carbon materials [14].

6.5.1 Hemicellulose-based carbon quantum dots

Carbon quantum dots (CQDs), a new type of small carbon nanoparticle with a particle size of less than 10 nm, were first discovered in 2004 during the purification of single walled carbon nanotubes [128, 129]. CQDs possess very strong and tunable fluorescence properties which enable their applications in biomedicine, optronics, sensors, and catalytic applications. Compared with traditional semiconductor quantum dots, lignocellulose-based CQDs have better photoluminescence characteristics, a more adjustable emission wavelength, a higher water solubility, a lower toxicity, better biocompatibility, easier surface-modification ability, and more stable optical and chemical properties [130]. Recently, hemicellulose has been used as a raw material for producing CQDs, and their application has been explored.

In 2014 Wang and co-workers prepared nitrogen-doped xylan carbon quantum dots by a one-step hydrothermal method, as shown in figure 6.26 [131]. The best quantum yield (QY) of the nitrogen-doped biomass CDs was up to 16%, which was higher than the non-doped CDs (only 2%) and those passivated by organic molecules (7%). The CDs not only had strong fluorescence, a wide band gap and low cytotoxicity, but also exhibited excitation-, pH-, concentration- and solvent-dependent optical properties, providing an excellent utilization potential in bio-imaging, photovoltaic conversion and drug delivery. In a subsequent work, Han *et al* [132] prepared nanocomposites consisting of xylan-based CQDs, silver nanoparticles, and rGO for dopamine detection. In another study, the CQDs derived from xylan via hydrothermal carbonization were utilized as a green, *in situ* reducing agent to prepare gold-nanoparticle-capped CQDs [133]. Then, an MXene was used as the immobilized matrix to fabricate Au/CQDs/MXene nanocomposites for nitrite detection. The nitrite detection results showed that the linear detection range of the

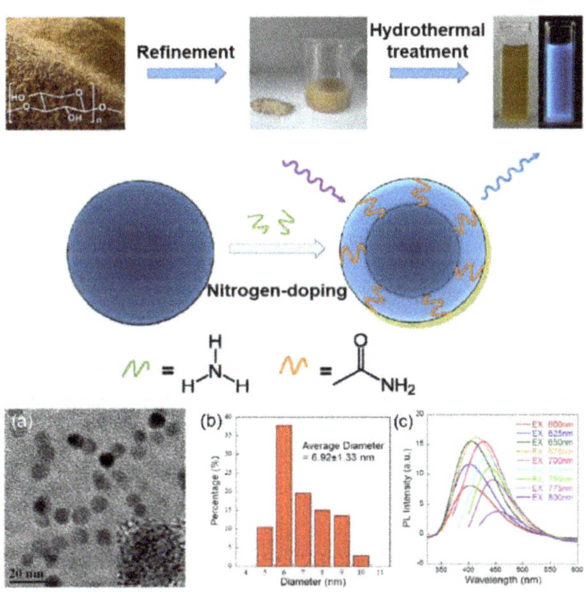

Figure 6.26. The one-step hydrothermal synthesis of *N*-doped xylan carbon quantum dots, and the size distribution and wavelength dependence of carbon quantum dots. (Reproduced with permission from [131]. Copyright 2012 Elsevier.)

sensor is from 1 to 3200 μM with a detection limit of 0.078 μM, and the sensor is superior to most reported sensors, as determined by differential pulse voltammetry method.

Currently, heavy metal contaminants such as Cu^{2+} pose a great threat to ecological security, and the discharge of sewage containing Cu^{2+} into water and soil will cause damage to environment and even human health. At present, fluorescent solutions are generally used for Cu^{2+} detection. However, they have several drawbacks such as easy leakage, difficult storage, poor photostability and inaccurate Cu^{2+} detection. To address these problems, a biomass solid-state fluorescence carbon fiber membrane detection platform (NFE-CDs) was first fabricated by encapsulating xylan carbon dots (CDs) into a metal–organic framework (E-CDs@ZIF-8) using *in situ* growth methods and electrospinning technology, as shown in figure 6.27 [134]. First, N-doped xylan carbon quantum dots (E-CDs) were prepared by the hydrothermal method, and then E-CDs@ZIF-8 materials were synthesized *in situ*. Next, in order to prepare a solid-state matrix as a platform for fluorescence detection, carbon fiber membranes (PLA/PCL) were prepared using electrospinning technology. Finally, the E-CDs@ZIF-8 solution and the carbon fiber membrane were placed together in a high-pressure reactor for reaction using a one-step hydrothermal synthesis process. Owning to its excellent fluorescence

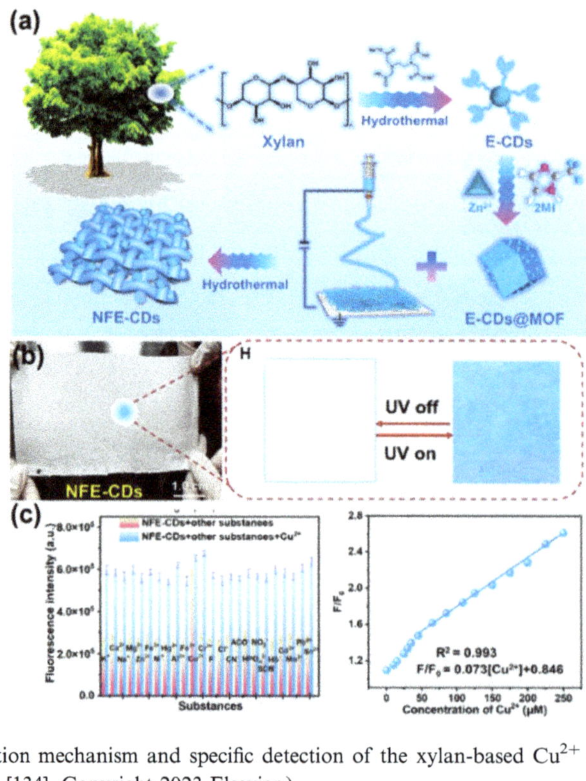

Figure 6.27. Preparation mechanism and specific detection of the xylan-based Cu^{2+} detector. (Reproduced with permission from [134]. Copyright 2023 Elsevier.)

properties, NFE-CDs showed excellent selectively for Cu^{2+} based on a 'turn-on' fluorescence response and ultra-short fluorescence response time (<1 s). The Cu^{2+} detection results showed that the linear detection range of the sensor is from 50 to 250 μM with a detection limit of 3.48 nM.

6.5.2 Hemicellulose-based porous carbon materials

Porous carbon spheres are attractive functional materials applied in blood purification, chemical protection, catalyst supports, gas storage and supercapacitor electrodes due to their uniformity, high mechanical strength, high thermostability, good fluidity, low pressure drop, large packing density, high micropore volume and large specific surface area [135, 136]. Owing to its simplicity, low cost, and environmental friendliness, hydrothermal carbonization is one of the most promising methods for preparing hemicellulose-based porous carbons. The preparation of hemicellulose-based porous carbons by hydrothermal carbonization can be divided into three steps: (i) hemicellulose is decomposed into monosaccharides (xylose) at a low pH value, high temperatures (160 °C–240 °C), and high pressure; (ii) monosaccharides are dehydrated into furfural; and (iii) furan compounds react with each other via polymerization to form the carbon microspheres, as shown in figure 6.28 [14].

Alternatively, hemicellulose porous carbon materials can be prepared by a one-step high-temperature carbonization, which is a process of high-temperature treatment of hemicellulose under an inert gas (N_2, Ar). One-step high-temperature carbonization has the advantages of simple operation and low cost. $ZnCl_2$ was introduced in the carbonization process to effectively promote the dehydration and condensation of hemicellulose, as shown in figure 6.29 [137]. $ZnCl_2$ acts as a skeleton during carbonization, providing structure to the nascent carbon when the raw material is carbonized. The porous activated carbon materials possess a large specific surface area of $1361\,m^2\,g^{-1}$ and exhibit excellent electrochemical performance.

Hemicellulose porous carbon materials exhibit a strong adsorption capacity for organic dyes and heavy metal ions in wastewater due to their large specific surface area, abundant pore structures, and numerous polar groups on the surface. As shown by many studies, the adsorption of heavy metals in wastewater by

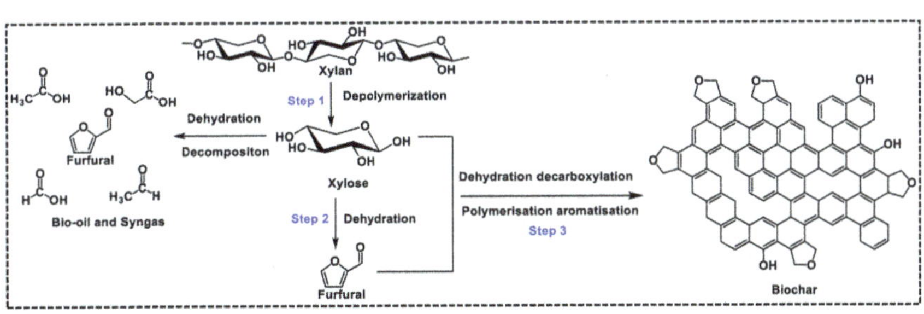

Figure 6.28. Mechanism of hemicellulose breakdown during pyrolysis and biochar formation. (Reproduced with permission from [14]. Copyright 2023 Elsevier.)

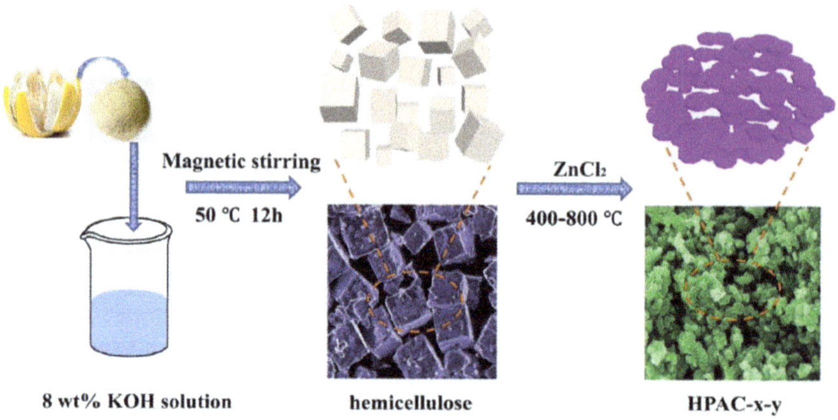

Figure 6.29. The preparation of hemicellulose porous carbon by one-step high-temperature carbonization. (Reproduced with permission from [137]. Copyright 2020 Elsevier.)

hemicellulose porous carbon follows the Langmuir adsorption model, that is, it is characterized by a monolayer adsorption. Heteroatom doping of hemicellulose porous carbon materials will effectively improve the adsorption capacity of ions or dyes, which due to the introduction of heteroatoms promotes electrostatic interaction, redox ability and metal ion chelating ability. As shown in figure 6.30, nitrogen-doped hydrothermal carbon material with Schiff base structures was successfully synthesized via ammonia-assisted hydrothermal treatment of hemicelluloses and used for the adsorption of toxic hexavalent chromium Cr(VI) [138]. Schiff base structures and amino groups were simultaneously introduced into carbon materials to improve the adsorption capacity.

6.6 Hemicellulose-based emulsifier

Hemicellulose exhibits significant potential for stabilizing emulsions due to its stability across a wide range of pH and ionic strength, as well as its resilience to thermal processing [139, 140]. The emulsion mechanism of hemicellulose was proposed and systematically studied by Mikkonen and co-workers [141–146]. In 2008, the potential use of GGM (obtained from the pulping process) as emulsifiers in food applications was first reported [144]. The turbidity of this GGM emulsion is higher than that of emulsions containing other mannans, such as locust bean gum galactomannans, konjac glucomannan, and corn AX, indicating its superior emulsifying properties. After the carboxymethylation modification, the surface charge of GGM increases, enhancing the emulsifying property [143]. The rheological analysis revealed that the emulsion stabilization mechanism of GGM is based on steric repulsion assisted by Pickering-type stabilization [145]. At high GGM concentrations, stabilization occurs via viscosity increments. These insights provide an understanding of and control over the processing conditions and application of GGM-stabilized emulsions. The hydrophilicity and hydrophobicity of different crystal planes in cellulose endow it with amphiphilic properties, whereas pure

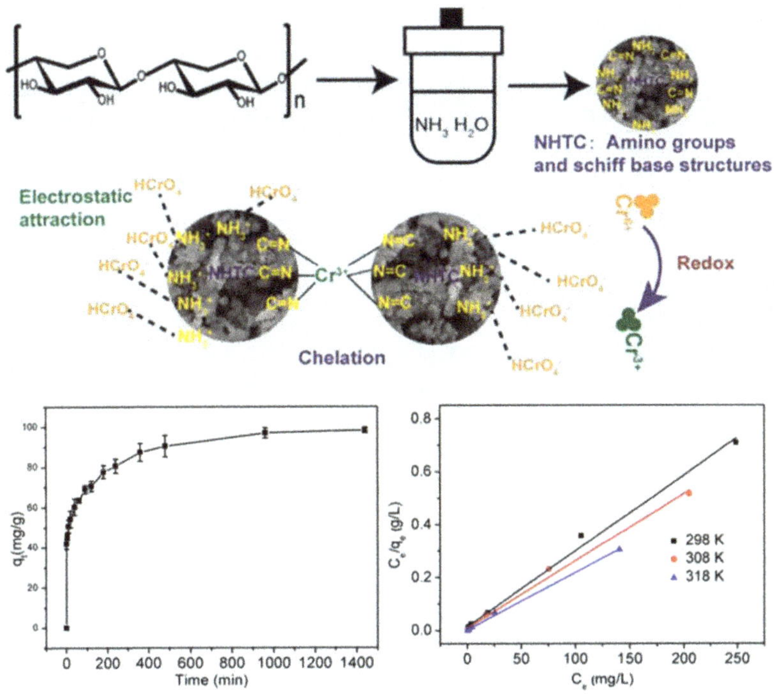

Figure 6.30. Preparation of nitrogen-doped hemicellulose porous carbon materials and the mechanism of adsorption of chromium ions. (Reproduced with permission from [138]. Copyright 2021 Elsevier.)

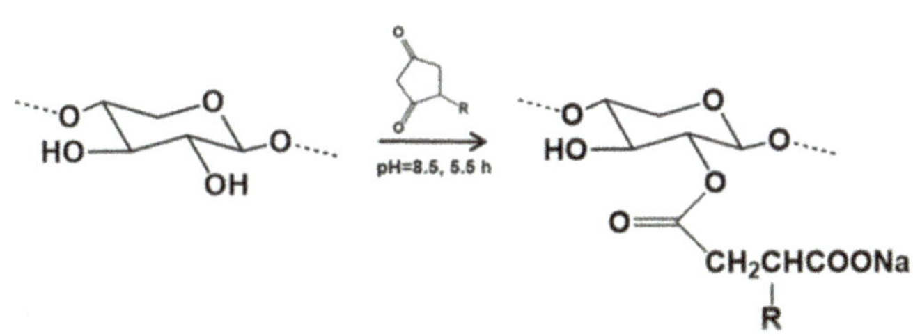

Figure 6.31. Succinic anhydride of glucuronic acid xylan. (Reproduced from [148]. CC BY 4.0.)

hemicellulose is not highly amphiphilic, which greatly limits its emulsifying properties [147]. Hence, improving the amphiphilicity of hemicellulose is crucial for its application as emulsifier. To date, two strategies have been used for improving the amphiphilicity of hemicellulose, including introduction of lignin by regulating the isolation method and molecular modification.

The presence of a large number of side groups on the hemicellulose molecular chain leads to its strong hydrophilicity, thus its lack of amphiphilicity. To improve the amphipathic character of glucuronoxylans and develop high quality natural-based/green emulsifiers, glucuronoxylans were chemically modified by alkenyl succinic anhydrides to produce a novel emulsifier, as shown in figure 6.31 [148].

The effects of DS and alkenyl chain length on the emulsifying properties of long-chain succinic anhydride modified glucuronoxylans were analyzed. Compared with glucuronoxylans emulsion, the dodecenyl succinic anhydride–glucuronoxylans emulsion showed significantly smaller droplet sizes, higher zeta potential in magnitude, higher emulsifying activity, and improved emulsion stability. When DS increased from 0.014 to 0.09, the emulsification performance of the DDSA was improved. When the DS increased to 0.09 or more, the emulsification performance was not changed significantly. Lahtinen and co-workers investigated the role of naturally occurring phenolic residues originating from lignin in GGM in stabilizing oil-in-water emulsions [141, 142, 149]. The phenolic residues attached to the GGM isolated by pressurized hot water extraction (PHWE), i.e. the PHWE GGM chain, improve the amphiphilicity of pure GGM, enabling the anchoring of PHWE GGM to the oil–water interface, forming a thick and dense interfacial layer [142].

In recent works, there has been a focus on the development and study of Pickering emulsions stabilized by xylan nanocrystals. First, xylan nanocrystals (XNCs) with dimensions of 25–60 nm were successfully prepared through oxalic acid hydrolysis of high-crystalline xylan as the raw material via a top-down approach (figure 6.32) [150]. To improve the hydrophobicity of the XNCs, 2-octen-1-ylsuccinic anhydride as an acylation agent was used to modulate the surface hydrophobicity of the XNCs. The XRD results indicated that succinylation had no significant effect on the crystalline structure of the XNCs. Compared with XNCs, succinylated XNCs showed a more remarkable emulsifying property over 7 days of storage at room temperature. Additionally, succinylated XNCs with a DS of 0.43 showed an outstanding emulsifying property for different oil phases with good stability over 7 days.

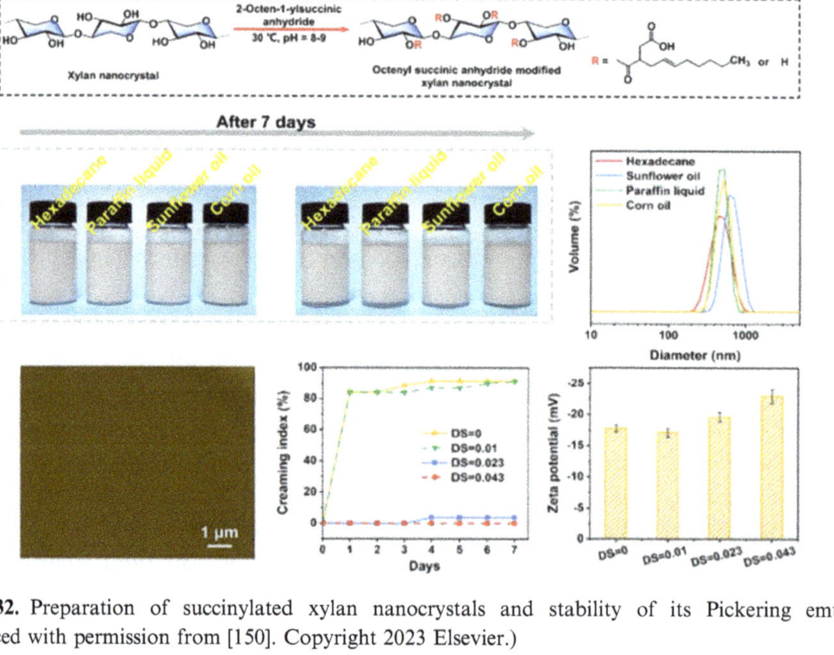

Figure 6.32. Preparation of succinylated xylan nanocrystals and stability of its Pickering emulsion. (Reproduced with permission from [150]. Copyright 2023 Elsevier.)

6.7 Hemicellulose-based room temperature phosphorescent materials

Organic phosphorescence is defined as a radiative transition between the different spin multiplicities of an organic molecule after excitation; here, we refer to the photoexcitation. Unlike fluorescence, it shows a long emission lifetime ($\sim\mu$s), large Stokes shift, and rich excited state properties, attracting considerable attention in organic electronics during recent years. Ultralong organic phosphorescence (UOP), a type of persistent luminescence in organic phosphors, normally shows an emission lifetime of over 100 ms according to the resolution limit of the naked eye [151]. However, artificial materials with UOP, especially with tunable multicolor afterglow, are hampered by complex synthesis and purification, poor processability, and sustainability-related issues. To overcome these drawbacks, room temperature phosphorescent materials based on natural polymers, such as cellulose, sodium alginate, and starch, have attracted significant attention. In a recent study, Lü *et al* [152] first reported a series of excitation- and time-dependent color-tuneable afterglow organic phosphorescence materials derived from hemicellulose, as shown in figure 6.33. In figure 6.33(b), the linear xylan with high crystallinity has the highest phosphorescence lifetime of 539 ms, which is due to the highly rigid conformation that significantly disturbs the intramolecular motions. Additionally, a water-soluble DAX with a phosphorescence lifetime of up to $\sim$700 ms was obtained by oxidation; the resultant foams or flexible and transparent films, inherit or even surpass the phosphorescence performance of pristine xylan, owing to the additional clusters of carbonyl groups. This work demonstrates a new application of hemicellulose and provides an ideal platform for developing color-tuneable, processable, sustainable, and large-scale organic phosphorescence materials.

6.8 Conclusion and outlook

Hemicelluloses, accounting for approximately 15%–35% of the total cell wall composition, are very promising natural polysaccharides that can be utilized as a feedstock for functional materials. This chapter summarized the preparation of hemicellulose-derived functional polymeric materials, including hydrogels and aerogels, films, carbon quantum dots, porous carbons, emulsifiers, and room temperature phosphorescent materials. Owing to its excellent barrier properties, biocompatibility, and biodegradability, hemicellulose-based films and coatings are promising alternatives to petroleum-based synthetic polymers in food and medical applications. However, their mechanical properties and water resistance still need to be improved further. The hemicellulose-based hydrogels can be divided into chemical hydrogels and physical hydrogels. However, it is still a challenge to construct multifunctional hemicellulose-based hydrogels with excellent anti-freezing, adhesion, conducting, and mechanical properties. In this case, when the optical, conductive and magnetic properties are incorporated, the obtained functional materials can find wider applications, such as in sensors, electrodes, and artificial muscle. Additionally, the preparation of hemicellulose-based carbon quantum dots with high quantum yield is still a challenge. The development of hemicellulose-derived excitation- and time-dependent color-tuneable afterglow organic phosphorescence materials provides a new

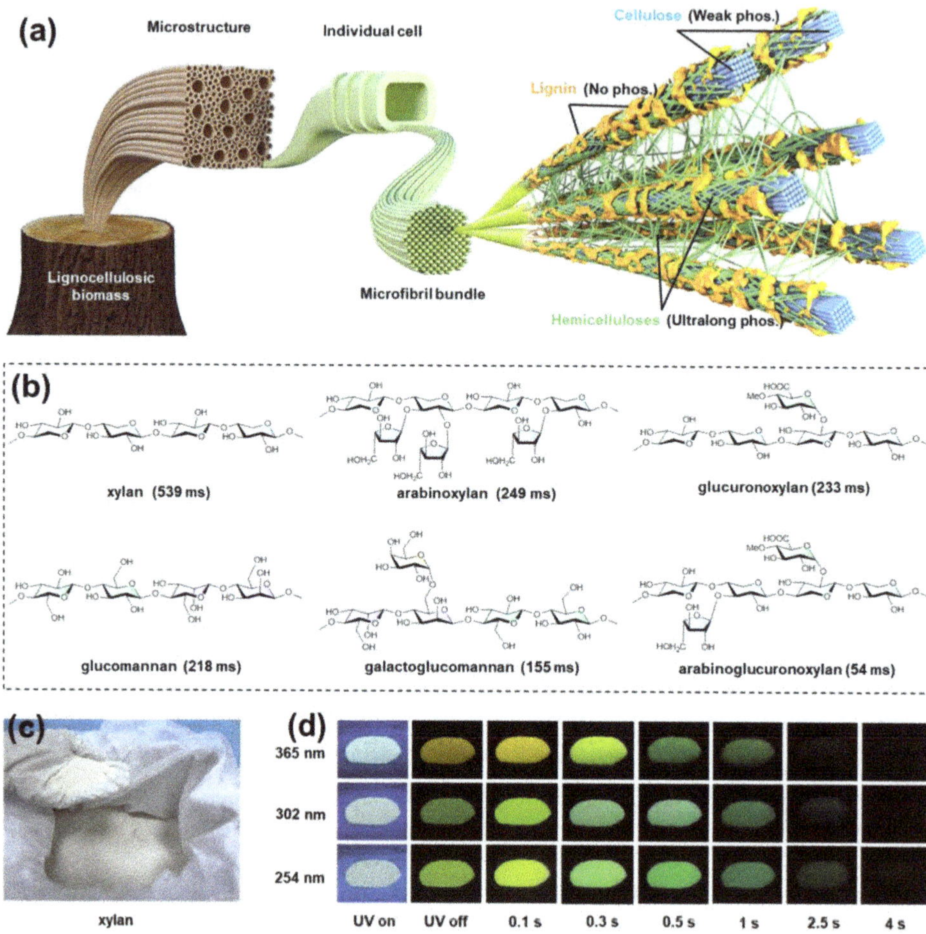

Figure 6.33. Hemicellulose-based durable, color-tuneable, processable, and sustainable phosphorescence material. (Reproduced with permission from [152]. Copyright 2022 Elsevier.)

application of hemicellulose. Therefore, there is significant potential for the efficient conversion and utilization of hemicellulose. Developing new chemical modification strategies for the preparation of hemicellulose derivatives with tailored functions is a crucial research direction. These advancements will be key to enhancing the conversion and utilization of hemicellulose.

Acknowledgments

The authors acknowledge funding from the Fundamental Research Funds for the Central Universities (QNTD202302), National Science Fund for Distinguished Young Scholars of China (32225034), National Natural Science Foundation of China (22308028), Postdoctoral Fellowship Program of China Postdoctoral Science Foundation (GZB20230062), and Ministry of Education, China-111 Project (BP0820033).

Bibliography

[1] Naik S N, Goud V V, Rout P K and Dalai A K 2010 Production of first and second generation biofuels: a comprehensive review *Renew. Sustain. Energy Rev.* **14** 578–97

[2] Li Z Q and Pan X J 2018 Strategies to modify physicochemical properties of hemicelluloses from biorefinery and paper industry for packaging material *Rev. Environ. Sci. Bio/Technol.* **17** 47–69

[3] Gírio F M, Fonseca C, Carvalheiro F, Duarte L C, Marques S and Bogel-Łukasik R 2010 Hemicelluloses for fuel ethanol: a review *Bioresour. Technol.* **101** 4775–800

[4] Scheller H V and Ulvskov P 2010 Hemicelluloses *Annu. Rev. Plant Biol.* **61** 263–89

[5] Sun R C, Sun X F and Tomkinson J 2003 Hemicelluloses and their derivatives *Hemicelluloses: Science and Technology* (Washington, DC: American Chemical Society) pp 2–22

[6] Ebringerová A 2005 Structural diversity and application potential of hemicelluloses *Macromol. Symp.* **232** 1–12

[7] Bouveng H O, Garegg P and Lindberg B 1960 Position of O-acetyl groups in birch xylan *Acta Chem. Scand.* **14** 742–8

[8] Rosell K G and Svensson S 1975 Studies of the distribution of the 4-*O*-methyl-D-glucuronic acid residues in birch xylan *Carbohydr. Res.* **42** 297–304

[9] Teleman A, Tenkanen M, Jacobs A and Dahlman O 2002 Characterization of *O*-acetyl-(4-*O*-methylglucurono) xylan isolated from birch and beech *Carbohydr. Res.* **337** 373–7

[10] Xu C L, Willfor S, Sundberg K, Pettersson C and Holmbom B 2007 Physico-chemical characterization of spruce galactoglucomannan solutions: stability, surface activity and rheology *Cellul. Chem. Technol.* **41** 51

[11] Izydorczyk M S and Biliaderis C G 1992 Influence of structure on the physicochemical properties of wheat arabinoxylan *Carbohydr. Polym.* **17** 237–47

[12] Izydorczyk M S and Biliaderis C G 1995 Cereal arabinoxylans: advances in structure and physicochemical properties *Carbohydr. Polym.* **28** 33–48

[13] Solhi L *et al* 2023 Understanding nanocellulose–water interactions: turning a detriment into an asset *Chem. Rev.* **23** 1925–2015

[14] Rao J, Lv Z W, Chen G G and Peng F 2023 Hemicellulose: structure, chemical modification, and application *Prog. Polym. Sci.* **140** 101675

[15] Liu X X, Lin Q X, Yan Y H, Peng F, Sun R C and Ren J L 2019 Hemicellulose from plant biomass in medical and pharmaceutical application: a critical review *Curr. Med. Chem.* **26** 2430–55

[16] Hansen N M L and Plackett D 2008 Sustainable films and coatings from hemicelluloses: a review *Biomacromolecules* **9** 1493–505

[17] Wichterle O and LÍM D 1960 Hydrophilic gels for biological use *Nature* **185** 117–8

[18] Peng X W, Zhong L X, Ren J L and Sun R C 2012 Highly effective adsorption of heavy metal ions from aqueous solutions by macroporous xylan-rich hemicelluloses-based hydrogel *J. Agric. Food Chem.* **60** 3909–16

[19] Dax D, Chávez M S, Xu C L, Willför S, Mendonça R T and Sánchez J 2014 Cationic hemicellulose-based hydrogels for arsenic and chromium removal from aqueous solutions *Carbohydr. Polym.* **111** 797–805

[20] Zhao W, Nugroho R W N, Odelius K, Edlund U, Zhao C and Albertsson A C 2015 *In situ* cross-linking of stimuli-responsive hemicellulose microgels during spray drying *ACS Appl. Mater. Interfaces* **7** 4202–15

[21] Sun X F, Wang H H, Jing Z X and Mohanathas R 2013 Hemicellulose-based pH-sensitive and biodegradable hydrogel for controlled drug delivery *Carbohydr. Polym.* **92** 1357–66

[22] Han T T, Song T, Pranovich A and Rojas O J 2022 Engineering a semi-interpenetrating constructed xylan-based hydrogel with superior compressive strength, resilience, and creep recovery abilities *Carbohydr. Polym.* **294** 119772

[23] Ali A, Hasan A and Negi Y S 2022 Effect of cellulose nanocrystals on xylan/chitosan/nano β-TCP composite matrix for bone tissue engineering *Cellulose* **29** 5689–709

[24] Ling Z, Ma J M, Zhang S, Shao L P, Wang C and Ma J F 2022 Stretchable and fatigue resistant hydrogels constructed by natural galactomannan for flexible sensing application *Int. J. Biol. Macromol.* **216** 193–202

[25] Gong X Q, Fu C L, Alam N, Ni Y H, Chen L H, Huang L L and Hu H C 2022 Tannic acid modified hemicellulose nanoparticle reinforced ionic hydrogels with multi-functions for human motion strain sensor applications *Ind. Crops Prod.* **176** 114412

[26] Kong W Q, Dai Q Q, Gao C D, Ren J L, Liu C and Sun R C 2018 Hemicellulose-based hydrogels and their potential application *Gels Horizons: From Science to Smart Materials* (Berlin: Springer) pp 87–127

[27] Talantikite M, Beury N, Moreau C and Cathala B 2019 Arabinoxylan/cellulose nanocrystal hydrogels with tunable mechanical properties *Langmuir* **35** 13427–34

[28] Schnell C N, Galván M V, Zanuttini M A and Mocchiutti P 2020 Hydrogels from xylan/chitosan complexes for the controlled release of diclofenac sodium *Cellulose* **27** 1465–81

[29] Gabrielii I, Gatenholm P, Glasser W G, Jain R K and Kenne L 2000 Separation, characterization and hydrogel-formation of hemicellulose from aspen wood *Carbohydr. Polym.* **43** 367–74

[30] Guan Y, Zhang B, Bian J, Peng F and Sun R C 2014 Nanoreinforced hemicellulose-based hydrogels prepared by freeze–thaw treatment *Cellulose* **21** 1709–21

[31] Guan Y, Bian J, Peng F, Zhang X M and Sun R C 2014 High strength of hemicelluloses based hydrogels by freeze/thaw technique *Carbohydr. Polym.* **101** 272–80

[32] Zhao W, Glavas L, Odelius K, Edlund U and Albertsson A C 2014 Facile and green approach towards electrically conductive hemicellulose hydrogels with tunable conductivity and swelling behavior *Chem. Mater.* **26** 4265–73

[33] Chang M M, Liu X X, Wang X H, Peng F and Ren J L 2021 Mussel-inspired adhesive hydrogels based on biomass-derived xylan and tannic acid cross-linked with acrylic acid with antioxidant and antibacterial properties *J. Mater. Sci.* **56** 14729–40

[34] Markstedt K, Xu W, Liu J, Xu C and Gatenholm P 2017 Synthesis of tunable hydrogels based on *O*-acetyl-galactoglucomannans from spruce *Carbohydr. Polym.* **157** 1349–57

[35] Farhat W, Venditti R, Becquart F, Ayoub A, Majesté J C, Taha M and Mignard N 2019 Synthesis and characterization of thermoresponsive xylan networks by Diels–Alder reaction *ACS Appl. Polym. Mater.* **1** 856–66

[36] Hu Y J, Li N, Yue P P, Chen G G, Hao X, Bian J and Peng F 2022 Highly antibacterial hydrogels prepared from amino cellulose, dialdehyde xylan, and Ag nanoparticles by a green reduction method *Cellulose* **29** 1055–67

[37] Wen J Y, Yang J Y, Wang W Y, Li M F, Peng F, Bian J and Sun R C 2020 Synthesis of hemicellulose hydrogels with tunable conductivity and swelling behavior through facile one-pot reaction *Int. J. Biol. Macromol.* **154** 1528–36

[38] Zhao W F, Odelius K, Edlund U, Zhao C S and Albertsson A C 2015 *In situ* synthesis of magnetic field-responsive hemicellulose hydrogels for drug delivery *Biomacromolecules* **16** 2522–8

[39] Seera S D K, Kundu D, Gami P, Naik P K and Banerjee T 2021 Synthesis and characterization of xylan-gelatin cross-linked reusable hydrogel for the adsorption of methylene blue *Carbohydr. Polym.* **256** 117520

[40] Gami P, Kundu D, Seera S D K and Banerjee T 2020 Chemically crosslinked xylan–β-cyclodextrin hydrogel for the *in vitro* delivery of curcumin and 5-fluorouracil *Int. J. Biol. Macromol.* **158** 18–31

[41] Elkihel A, Christie C, Vernisse C, Ouk T S, Lucas R, Chaleix V and Sol V 2021 Xylan-based cross-linked hydrogel for photodynamic antimicrobial chemotherapy *ACS Appl. Bio Mater.* **4** 7204–12

[42] Maleki L, Edlund U and Albertsson A-C 2015 Thiolated hemicellulose as a versatile platform for one-pot click-type hydrogel synthesis *Biomacromolecules* **16** 667–74

[43] Fu G Q, Zhang S C, Chen G G, Hao X, Bian J and Peng F 2020 Xylan-based hydrogels for potential skin care application *Int. J. Biol. Macromol.* **158** 244–50

[44] Guan Y, Chen J H, Qi X M, Chen G G, Peng F and Sun R C 2015 Fabrication of biopolymer hydrogel containing Ag nanoparticles for antibacterial property *Ind. Eng. Chem. Res.* **54** 7393–400

[45] Han Y M, Yuan L, Li G Y, Huang L H, Qin T F, Chu F X and Tang C B 2016 Renewable polymers from lignin via copper-free thermal click chemistry *Polymer* **83** 92–100

[46] Meyer J P, Adumeau P, Lewis J S and Zeglis B M 2016 Click chemistry and radiochemistry: the first 10 years *Bioconjug. Chem.* **27** 2791–807

[47] Deng Y L, Shavandi A, Okoro O V and Nie L 2021 Alginate modification via click chemistry for biomedical applications *Carbohydr. Polym.* **270** 118360

[48] Pahimanolis N, Kilpeläinen P, Master E, Ilvesniemi H and Seppälä J 2015 Novel thiol-amine- and amino acid functional xylan derivatives synthesized by thiol–ene reaction *Carbohydr. Polym.* **131** 392–8

[49] Gong X Q, Fu C L, Alam N, Ni Y H, Chen L H, Huang L L and Hu H C 2022 Preparation of hemicellulose nanoparticle-containing ionic hydrogels with high strength, self-healing, and UV resistance and their applications as strain sensors and asymmetric pressure sensors *Biomacromolecules* **23** 2272–9

[50] Voepel J, Edlund U, Albertsson A C and Percec V 2011 Hemicellulose-based multifunctional macroinitiator for single-electron-transfer mediated living radical polymerization *Biomacromolecules* **12** 253–9

[51] Xiao X Q, Guo F J, Mi H Y, Chang X, Yin H, Ji C C and Qiu J S 2024 Controlling dendrite growth with xylan-enhanced hydroxyl-rich hydrogel electrolyte for efficient Zn-ion energy storage *ACS Sustain. Chem. Eng.* **12** 470–9

[52] Liu X X, Chang M M, He B, Meng L, Wang X H, Sun R C, Ren J L and Kong F G 2019 A one-pot strategy for preparation of high-strength carboxymethyl xylan-g-poly(acrylic acid) hydrogels with shape memory property *J. Colloid Interface Sci.* **538** 507–18

[53] Zhang T T, Wang W, Zhao Y L, Bai H Y, Wen T, Kang S C, Song G S, Song S X and Komarneni S 2021 Removal of heavy metals and dyes by clay-based adsorbents: from natural clays to 1D and 2D nano-composites *Chem. Eng. J.* **420** 127574

[54] Liu X X, Chang M M, Zhang H and Ren J L 2022 Fabrication of bentonite reinforced dopamine grafted carboxymethyl xylan cross-linked with polyacrylamide hydrogels with adhesion properties *Colloids Surf.* A **647** 129024

[55] Briffa J, Sinagra E and Blundell R 2020 Heavy metal pollution in the environment and their toxicological effects on humans *Heliyon* **6** e04691
[56] Zhao G X, Li J X, Ren X M, Chen C L and Wang X K 2011 Few-layered graphene oxide nanosheets as superior sorbents for heavy metal ion pollution management *Environ. Sci. Technol.* **45** 10454–62
[57] Sud D, Mahajan G and Kaur M P 2008 Agricultural waste material as potential adsorbent for sequestering heavy metal ions from aqueous solutions—a review *Bioresour. Technol.* **99** 6017–27
[58] Ihsanullah, Abbas A, Al-Amer A M, Laoui T, Al-Marri M J, Nasser M S, Khraisheh M and Atieh M A 2016 Heavy metal removal from aqueous solution by advanced carbon nanotubes: critical review of adsorption applications *Sep. Purif. Technol.* **157** 141–61
[59] Zheng Y and Wang A 2009 Evaluation of ammonium removal using a chitosan-g-poly (acrylic acid)/rectorite hydrogel composite *J. Hazard. Mater.* **171** 671–7
[60] Xiang Z Y, Tang N, Jin X C and Gao W H 2022 Fabrications and applications of hemicellulose-based bio-adsorbents *Carbohydr. Polym.* **278** 118945
[61] Peng X W, Ren J L, Zhong L X, Peng F and Sun R C 2011 Xylan-rich hemicelluloses-graft-acrylic acid ionic hydrogels with rapid responses to pH, salt, and organic solvents *J. Agric. Food Chem.* **59** 8208–15
[62] Zheng Q X, Shang M S, Li X J, Jiang L M, Chen L, Long J, Jiao A Q, Ji H Y, Jin Z Y and Qiu C 2024 Advances in intelligent response and nano-enhanced polysaccharide-based hydrogels: material properties, response types, action mechanisms, applications *Food Hydrocoll.* **146** 109190
[63] Wang H Z and Zhang L M 2024 Intelligent biobased hydrogels for diabetic wound healing: a review *Chem. Eng. J.* **484** 149493
[64] Chen T, Liu H T, Dong C H, An Y Z, Liu J, Li J, Li X X, Si C L and Zhang M Y 2020 Synthesis and characterization of temperature/pH dual sensitive hemicellulose-based hydrogels from eucalyptus APMP waste liquor *Carbohydr. Polym.* **247** 116717
[65] Zhang W, Wen J Y, Ma M G, Li M F, Peng F and Bian J 2021 Anti-freezing, water-retaining, conductive, and strain-sensitive hemicellulose/polypyrrole composite hydrogels for flexible sensors *J. Mater. Res. Technol.* **14** 555–66
[66] Gao C D, Ren J L, Zhao C, Kong W Q, Dai Q Q, Chen Q F, Liu C and Sun R C 2016 Xylan-based temperature/pH sensitive hydrogels for drug controlled release *Carbohydr. Polym.* **151** 189–97
[67] Pahimanolis N, Sorvari A, Luong N D and Seppälä J 2014 Thermoresponsive xylan hydrogels via copper-catalyzed azide-alkyne cycloaddition *Carbohydr. Polym.* **102** 637–44
[68] Dai Q Q, Ren J L, Peng F, Chen X F, Gao C D and Sun R C 2016 Synthesis of acylated xylan-based magnetic Fe_3O_4 hydrogels and their application for H_2O_2 detection *Materials* **9** 690
[69] Sun X F, Liu B C, Jing Z X and Wang H H 2015 Preparation and adsorption property of xylan/poly(acrylic acid) magnetic nanocomposite hydrogel adsorbent *Carbohydr. Polym.* **118** 16–23
[70] Zhao W F, Glavas L, Odelius K, Edlund U and Albertsson A C 2014 A robust pathway to electrically conductive hemicellulose hydrogels with high and controllable swelling behavior *Polymer* **55** 2967–76
[71] Rahmanian V, Pirzada T, Wang S and Khan S A 2021 Cellulose-based hybrid aerogels: strategies toward design and functionality *Adv. Mater.* **33** 2102892

[72] Long L Y, Weng Y X and Wang Y Z 2018 Cellulose aerogels: synthesis, applications, and prospects *Polymers* **10** 623
[73] Mikkonen K S, Parikka K, Ghafar A and Tenkanen M 2013 Prospects of polysaccharide aerogels as modern advanced food materials *Trends Food Sci. Technol.* **34** 124–36
[74] Jin H, Nishiyama Y, Wada M and Kuga S 2004 Nanofibrillar cellulose aerogels *Colloids Surf.* A **240** 63–7
[75] Pakowski Z 2007 Modern methods of drying nanomaterials *Transp. Porous Media* **66** 19–27
[76] García-González C A, Alnaief M and Smirnova I 2011 Polysaccharide-based aerogels—promising biodegradable carriers for drug delivery systems *Carbohydr. Polym.* **86** 1425–38
[77] Köhnke T, Lin A, Elder T, Theliander H and Ragauskas A J 2012 Nanoreinforced xylan–cellulose composite foams by freeze-casting *Green Chem.* **14** 1864–9
[78] Mikkonen K S, Parikka K, Suuronen J P, Ghafar A, Serimaa R and Tenkanen M 2014 Enzymatic oxidation as a potential new route to produce polysaccharide aerogels *RSC Adv.* **4** 11884–92
[79] Berglund L, Forsberg F, Jonoobi M and Oksman K 2018 Promoted hydrogel formation of lignin-containing arabinoxylan aerogel using cellulose nanofibers as a functional biomaterial *RSC Adv.* **8** 38219–28
[80] Yan X Q, Pan J, Lv Z W, Jia S Y, Wen X, Peng P, Rao J and Peng F 2024 Xylan-assisted construction of anisotropic aerogel for pressure sensor *Chem. Eng. J.* **490** 151688
[81] Carosio F, Medina L, Kochumalayil J and Berglund L A 2021 Green and fire resistant nanocellulose/hemicellulose/clay foams *Adv. Mater. Interfaces* **8** 2101111
[82] MacLeod M, Arp H P H, Tekman M B and Jahnke A 2021 The global threat from plastic pollution *Science* **373** 61–5
[83] Smart C L and Whistler R L 1949 Films from hemicellulose acetates *Science* **110** 713–4
[84] Ibn Yaich A, Edlund U and Albertsson A C 2017 Transfer of biomatrix/wood cell interactions to hemicellulose-based materials to control water interaction *Chem. Rev.* **117** 8177–207
[85] Farhat W, Venditti R A, Hubbe M, Taha M, Becquart F and Ayoub A 2017 A review of water-resistant hemicellulose-based materials: processing and applications *ChemSusChem.* **10** 305–23
[86] Peng X W, Du F and Zhong L X 2019 Synthesis, characterization, and applications of hemicelluloses based eco-friendly polymer composites *Sustainable Polymer Composites and Nanocomposites* (Berlin: Springer) pp 1267–322
[87] Jin X C, Hu Z H, Wu S F, Song T, Yue F X and Xiang Z Y 2019 Promoting the material properties of xylan-type hemicelluloses from the extraction step *Carbohydr. Polym.* **215** 235–45
[88] Peresin M S, Kammiovirta K, Setälä H and Tammelin T 2012 Structural features and water interactions of etherified xylan thin films *J. Polym. Environ.* **20** 895–904
[89] Peng X W, Ren J L, Zhong L X and Sun R C 2011 Nanocomposite films based on xylan-rich hemicelluloses and cellulose nanofibers with enhanced mechanical properties *Biomacromolecules* **12** 3321–9
[90] Escalante A, Gonçalves A, Bodin A, Stepan A, Sandström C, Toriz G and Gatenholm P 2012 Flexible oxygen barrier films from spruce xylan *Carbohydr. Polym.* **87** 2381–7
[91] Stoklosa R J, Latona R J, Bonnaillie L M and Yadav M P 2019 Evaluation of arabinoxylan isolated from sorghum bran, biomass, and bagasse for film formation *Carbohydr. Polym.* **213** 382–92

[92] Wang S Y, Ren J L, Li W Y, Sun R C and Liu S J 2014 Properties of polyvinyl alcohol/xylan composite films with citric acid *Carbohydr. Polym.* **103** 94–9

[93] Gordobil O, Egüés I, Urruzola I and Labidi J 2014 Xylan–cellulose films: improvement of hydrophobicity, thermal and mechanical properties *Carbohydr. Polym.* **112** 56–62

[94] Huang C, Fang G G, Tao Y H, Meng X Z, Lin Y, Bhagia S, Wu X X, Yong Q and Ragauskas A J 2019 Nacre-inspired hemicelluloses paper with fire retardant and gas barrier properties by self-assembly with bentonite nanosheets *Carbohydr. Polym.* **225** 115219

[95] Chen G G, Hu Y J, Peng F, Bian J, Li M F, Yao C L and Sun R C 2018 Fabrication of strong nanocomposite films with renewable forestry waste/montmorillonite/reduction of graphene oxide for fire retardant *Chem. Eng. J.* **337** 436–45

[96] Mikkonen K S, Heikkinen S, Soovre A, Peura M, Serimaa R, Talja R A, Helén H, Hyvönen L and Tenkanen M 2009 Films from oat spelt arabinoxylan plasticized with glycerol and sorbitol *J. Appl. Polym. Sci.* **114** 457–66

[97] Zhang P and Whistler R L 2004 Mechanical properties and water vapor permeability of thin film from corn hull arabinoxylan *J. Appl. Polym. Sci.* **93** 2896–902

[98] Gröndahl M, Eriksson L and Gatenholm P 2004 Material properties of plasticized hardwood xylans for potential application as oxygen barrier films *Biomacromolecules* **5** 1528–35

[99] Saxena A, Elder T J, Kenvin J and Ragauskas A J 2010 High oxygen nanocomposite barrier films based on xylan and nanocrystalline cellulose *Nano-Micro Lett.* **2** 235–41

[100] Stevanic J S, Bergström E M, Gatenholm P, Berglund L and Salmén L 2012 Arabinoxylan/nanofibrillated cellulose composite films *J. Mater. Sci.* **47** 6724–32

[101] Xu J Y, Xia R R, Zheng L, Yuan T and Sun R C 2019 Plasticized hemicelluloses/chitosan-based edible films reinforced by cellulose nanofiber with enhanced mechanical properties *Carbohydr. Polym.* **224** 115164

[102] Zhang X Q, Wei Y, Chen M J, Xiao N Y, Zhang J and Liu C 2020 Development of functional chitosan-based composite films incorporated with hemicelluloses: effect on physicochemical properties *Carbohydr. Polym.* **246** 116489

[103] Liu X X, Chen X F, Ren J L, Chang M M, He B and Zhang C H 2019 Effects of nano-ZnO and nano-SiO_2 particles on properties of PVA/xylan composite films *Int. J. Biol. Macromol.* **132** 978–86

[104] Luo Y Q, Pan X Q, Ling Y Z, Wang X Y and Sun R C 2014 Facile fabrication of chitosan active film with xylan via direct immersion *Cellulose* **21** 1873–83

[105] Guan Y, Qi X M, Chen G G, Peng F and Sun R C 2016 Facile approach to prepare drug-loading film from hemicelluloses and chitosan *Carbohydr. Polym.* **153** 542–8

[106] Huang C, Fang G G, Deng Y J, Bhagia S, Meng X Z, Tao Y H, Yong Q and Ragauskas A J 2020 Robust galactomannan/graphene oxide film with ultra-flexible, gas barrier and self-clean properties *Composites* A **131** 105780

[107] Tao Y H, Huang C, Lai C H, Huang C X and Yong Q 2020 Biomimetic galactomannan/bentonite/graphene oxide film with superior mechanical and fire retardant properties by borate cross-linking *Carbohydr. Polym.* **245** 116508

[108] Guan Y, Zhang B, Tan X, Qi X M, Bian J, Peng F and Sun R C 2014 Organic–inorganic composite films based on modified hemicelluloses with clay nanoplatelets *ACS Sustain. Chem. Eng.* **2** 1811–8

[109] Kochumalayil J J, Bergenstråhle-Wohlert M, Utsel S, Wågberg L, Zhou Q and Berglund L A 2013 Bioinspired and highly oriented clay nanocomposites with a xyloglucan

biopolymer matrix: extending the range of mechanical and barrier properties *Biomacromolecules* **14** 84–91
[110] Chen G G, Qi X M, Guan Y, Peng F, Yao C L and Sun R C 2016 High strength hemicellulose-based nanocomposite film for food packaging applications *ACS Sustain. Chem. Eng.* **4** 1985–93
[111] Yaich A I, Edlund U and Albertsson A C 2014 Adapting wood hydrolysate barriers to high humidity conditions *Carbohydr. Polym.* **100** 135–42
[112] Wang Q, Yu Y and Liu J 2018 Preparations, characteristics and applications of the functional liquid metal materials *Adv. Eng. Mater.* **20** 1700781
[113] Zhang S L, Liu Y, Fan Q N, Zhang C, Zhou T F, Kalantar-Zadeh K and Guo Z P 2021 Liquid metal batteries for future energy storage *Energy Environ. Sci.* **14** 4177–202
[114] Dickey M D, Chiechi R C, Larsen R J, Weiss E A, Weitz D A and Whitesides G M 2008 Eutectic gallium-indium (EGain): a liquid metal alloy for the formation of stable structures in microchannels at room temperature *Adv. Funct. Mater.* **18** 1097–104
[115] Hu Y J, Hao X, Chen G G, Bian J, Li M F and Peng F 2022 Self-standing, photothermal-actuating, and motion-monitoring janus films one-pot synthesized by green carboxymethyl glucomannan/liquid metal nanoinks *ACS Appl. Mater. Interfaces* **14** 23717–25
[116] Šimkovic I, Gedeon O, Uhliariková I, Mendichi R and Kirschnerová S 2011 Xylan sulphate films *Carbohydr. Polym.* **86** 214–8
[117] Kisonen V, Xu C, Bollström R, Hartman J, Rautkoski H, Nurmi M, Hemming J, Eklund P and Willför S 2014 *O*-acetyl galactoglucomannan esters for barrier coatings *Cellulose* **21** 4497–509
[118] Simkovic I, Kelnar I, Uhliarikova I, Mendichi R, Mandalika A and Elder T 2014 Carboxymethylated-, hydroxypropylsulfonated- and quaternized xylan derivative films *Carbohydr. Polym.* **110** 464–71
[119] Mikkonen K S, Laine C, Kontro I, Talja R A, Serimaa R and Tenkanen M 2015 Combination of internal and external plasticization of hydroxypropylated birch xylan tailors the properties of sustainable barrier films *Eur. Polym. J.* **66** 307–18
[120] Kochumalayil J J, Zhou Q, Kasai W and Berglund L A 2013 Regioselective modification of a xyloglucan hemicellulose for high-performance biopolymer barrier films *Carbohydr. Polym.* **93** 466–72
[121] Börjesson M, Larsson A, Westman G and Ström A 2018 Periodate oxidation of xylan-based hemicelluloses and its effect on their thermal properties *Carbohydr. Polym.* **202** 280–7
[122] Zhang X Q, Chen M J, Liu c f, Zhang A P and Sun R C 2015 Homogeneous ring opening graft polymerization of ε-caprolactone onto xylan in dual polar aprotic solvents *Carbohydr. Polym.* **117** 701–9
[123] Liu R, Du J, Zhang Z P, Li H M, Lu J, Cheng Y, Lv Y N and Wang H S 2019 Preparation of polyacrylic acid-grafted-acryloyl/hemicellulose (PAA-g-AH) hybrid films with high oxygen barrier performance *Carbohydr. Polym.* **205** 83–8
[124] Fundador N G V, Enomoto-Rogers Y, Takemura A and Iwata T 2012 Syntheses and characterization of xylan esters *Polymer* **53** 3885–93
[125] Zhang X Q, Xiao N Y, Chen M J, Wei Y and Liu C 2020 Functional packaging films originating from hemicelluloses laurate by direct transesterification in ionic liquid *Carbohydr. Polym.* **229** 115336

[126] Yong Q W, Xu J Y, Wang L Y, Tirri T, Gao H J, Liao Y W, Toivakka M and Xu C L 2022 Synthesis of galactoglucomannan-based latex via emulsion polymerization *Carbohydr. Polym.* **291** 119565

[127] Jia S Y, Lv Z W, Rao J, Lü B Z, Chen G G, Bian J, Li M F and Peng F 2023 Xylan plastic *ACS Nano* **17** 13627–37

[128] Wareing T C, Gentile P and Phan A N 2021 Biomass-based carbon dots: current development and future perspectives *ACS Nano* **15** 15471–501

[129] Xu X Y, Ray R, Gu Y L, Ploehn H J, Gearheart L, Raker K and Scrivens W A 2004 Electrophoretic analysis and purification of fluorescent single-walled carbon nanotube fragments *J. Am. Chem. Soc.* **126** 12736–7

[130] Meng W X, Bai X, Wang B Y, Liu Z Y, Lu S Y and Yang B 2019 Biomass-derived carbon dots and their applications *Energy Environ. Mater.* **2** 172–92

[131] Liang Z C, Zeng L, Cao X D, Wang Q, Wang X H and Sun R C 2014 Sustainable carbon quantum dots from forestry and agricultural biomass with amplified photoluminescence by simple NH_4OH passivation *J. Mater. Chem.* C **2** 9760–6

[132] Han G D, Cai J H, Liu C, Ren J L, Wang X H, Yang J and Wang X Y 2021 Highly sensitive electrochemical sensor based on xylan-based Ag@CQDs-rGO nanocomposite for dopamine detection *Appl. Surf. Sci.* **541** 148566

[133] Feng X W, Han G D, Cai J H and Wang X Y 2022 Au@carbon quantum dots-MXene nanocomposite as an electrochemical sensor for sensitive detection of nitrite *J. Colloid Interface Sci.* **607** 1313–22

[134] Zhang Y Y, Cui X C, Wang X, Feng X Y, Deng Y K, Cheng W X, Xiong R H and Huang C B 2023 Xylan derived carbon dots composite Zif-8 and its immobilized carbon fibers membrane for fluorescence selective detection Cu^{2+} in real samples *Chem. Eng. J.* **474** 145804

[135] Tian W J, Zhang H Y, Duan X G, Sun H Q, Shao G S and Wang S B 2020 Porous carbons: structure-oriented design and versatile applications *Adv. Funct. Mater.* **30** 1909265

[136] Lee J, Kim J and Hyeon T 2006 Recent progress in the synthesis of porous carbon materials *Adv. Mater.* **18** 2073–94

[137] Lin H L, Liu Y P, Chang Z X, Yan S, Liu S C and Han S 2020 A new method of synthesizing hemicellulose-derived porous activated carbon for high-performance supercapacitors *Micropor. Mesopor. Mater.* **292** 109707

[138] Wei Y, Wang H H, Zhang X Q and Liu C F 2021 Ammonia-assisted hydrothermal carbon material with schiff base structures synthesized from factory waste hemicelluloses for Cr(VI) adsorption *J. Environ. Chem. Eng.* **9** 106187

[139] Mikkonen K S 2020 Strategies for structuring diverse emulsion systems by using wood lignocellulose-derived stabilizers *Green Chem.* **22** 1019–37

[140] Olorunsola E O, Akpabio E I, Adedokun M O and Ajibola D O 2018 Emulsifying properties of hemicelluloses *Science and Technology Behind Nanoemulsions* (London: IntechOpen) pp 639–59

[141] de Carvalho D M, Lahtinen M H, Bhattarai M, Lawoko M and Mikkonen K S 2021 Active role of lignin in anchoring wood-based stabilizers to the emulsion interface *Green Chem.* **23** 9084–98

[142] Lehtonen M, Merinen M, Kilpeläinen P O, Xu C, Willför S M and Mikkonen K S 2018 Phenolic residues in spruce galactoglucomannans improve stabilization of oil-in-water emulsions *J. Colloid Interface Sci.* **512** 536–47

[143] Mikkonen K S, Xu C, Berton-Carabin C and Schroën K 2016 Spruce galactoglucomannans in rapeseed oil-in-water emulsions: efficient stabilization performance and structural partitioning *Food Hydrocoll.* **52** 615–24

[144] Mikkonen K S, Tenkanen M, Cooke P, Xu C, Rita H, Willför S, Holmbom B, Hicks K B and Yadav M P 2009 Mannans as stabilizers of oil-in-water beverage emulsions *LWT-Food Sci. Technol.* **42** 849–55

[145] Mikkonen K S, Merger D, Kilpeläinen P, Murtomäki L, Schmidt U S and Wilhelm M 2016 Determination of physical emulsion stabilization mechanisms of wood hemicelluloses via rheological and interfacial characterization *Soft Matter* **12** 8690–700

[146] Bhattarai M, Pitkänen L, Kitunen V, Korpinen R, Ilvesniemi H, Kilpeläinen P O, Lehtonen M and Mikkonen K S 2019 Functionality of spruce galactoglucomannans in oil-in-water emulsions *Food Hydrocoll.* **86** 154–61

[147] Malaspina D C and Faraudo J 2019 Molecular insight into the wetting behavior and amphiphilic character of cellulose nanocrystals *Adv. Colloid Interface Sci.* **267** 15–25

[148] Hu Z H, Xiang Z Y and Lu F C 2019 Synthesis and emulsifying properties of long-chain succinic acid esters of glucuronoxylans *Cellulose* **26** 3713–24

[149] Lahtinen M H, Valoppi F, Juntti V, Heikkinen S, Kilpeläinen P O, Maina N H and Mikkonen K S 2019 Lignin-rich PHWE hemicellulose extracts responsible for extended emulsion stabilization *Front. Chem.* **7** 871

[150] Lv Z W, Rao J, Lü B Z, Chen G G, Hao X, Guan Y, Bian J and Peng F 2023 Microencapsulated phase change material via pickering emulsion based on xylan nanocrystal for thermoregulating application *Carbohydr. Polym.* **302** 120407

[151] Shi H F, Yao W, Ye W P, Ma H L, Huang W and An Z F 2022 Ultralong organic phosphorescence: from material design to applications *Acc. Chem. Res.* **55** 3445–59

[152] Lü B Z *et al* 2022 Natural ultralong hemicelluloses phosphorescence *Cell Rep. Phys. Sci.* **3** 101015

www.ingramcontent.com/pod-product-compliance
Ingram Content Group UK Ltd.
Pitfield, Milton Keynes, MK11 3LW, UK
UKHW051858100926
8556IPUK00005B/683

9 780750 361828